T0271264

ANALYSIS OF GRAVITATIONAL-WAVE DATA

Research in this field has grown considerably in recent years due to the commissioning of a world-wide network of large-scale detectors. This network collects a very large amount of data that is currently being analyzed and interpreted. This book introduces researchers entering the field, and researchers currently analyzing the data, to gravitational-wave data analysis.

An ideal starting point for studying the issues related to current gravitational-wave research, the book contains detailed derivations of the basic formulae related to the detectors' responses and maximum-likelihood detection. These derivations are much more complete and more pedagogical than those found in current research papers, and will enable readers to apply general statistical concepts to the analysis of gravitational-wave signals. It also discusses new ideas on devising the efficient algorithms needed to perform data analysis.

PIOTR JARANOWSKI is an Associate Professor in the Faculty of Physics at the University of Białystok, Poland. He has been a visiting scientist at the Max Planck Institute for Gravitational Physics and the Friedrich Schiller University of Jena, both in Germany, and the Institut des Hautes Études Scientifiques, France. He currently works in the field of gravitational-wave data analysis and general-relativistic problem of motion.

ANDRZEJ KRÓLAK is a Professor in the Institute of Mathematics at the Polish Academy of Sciences, Poland. He has twice been awarded the Second Prize by the Gravity Research Foundation (once with Bernard Schutz). He has been a visiting scientist at the Max Planck Institute for Gravitational Physics, Germany, and the Jet Propulsion Laboratory, USA. His field of research is gravitational-wave data analysis and general theory of relativity, and the phenomena predicted by this theory such as black holes and gravitational waves.

CAMBRIDGE MONOGRAPHS
ON PARTICLE PHYSICS,
NUCLEAR PHYSICS AND COSMOLOGY

General Editors: T. Ericson, P. V. Landshoff

ANALYSIS OF GRAVITATIONAL-WAVE DATA

PIOTR JARANOWSKI
University of Białystok, Poland

ANDRZEJ KRÓLAK
Polish Academy of Sciences, Poland

Shaftesbury Road, Cambridge CB2 8EA, United Kingdom

One Liberty Plaza, 20th Floor, New York, NY 10006, USA

477 Williamstown Road, Port Melbourne, VIC 3207, Australia

314–321, 3rd Floor, Plot 3, Splendor Forum, Jasola District Centre, New Delhi – 110025, India

103 Penang Road, #05–06/07, Visioncrest Commercial, Singapore 238467

Cambridge University Press is part of Cambridge University Press & Assessment, a department of the University of Cambridge.

We share the University's mission to contribute to society through the pursuit of education, learning and research at the highest international levels of excellence.

www.cambridge.org
Information on this title: www.cambridge.org/9780521864596

First published 2009

A catalogue record for this publication is available from the British Library

ISBN   978-0-521-86459-6   Hardback

Contents

Preface

Gravitational waves are predicted by Einstein's general theory of relativity. The only potentially detectable sources of gravitational waves are of astrophysical origin. So far the existence of gravitational waves has only been confirmed indirectly from radio observations of binary pulsars, notably the famous Hulse and Taylor pulsar PSR B1913+16 [1]. As gravitational waves are extremely weak, a very careful data analysis is required in order to detect them and extract useful astrophysical information. Any gravitational-wave signal present in the data will be buried in the noise of a detector. Thus the data from a gravitational-wave detector are realizations of a stochastic process. Consequently the problem of detecting gravitational-wave signals is a statistical one.

The purpose of this book is to introduce the reader to the field of gravitational-wave data analysis. This field has grown considerably in the past years as a result of commissioning a world-wide network of long arm interferometric detectors. This network together with an existing network of resonant detectors collects a very large amount of data that is currently being analyzed and interpreted. Plans exist to build more sensitive laser interferometric detectors and plans to build interferometric gravitational-wave detectors in space.

This book is meant both for researchers entering the field of gravitational-wave data analysis and the researchers currently analyzing the data. In our book we describe the basis of the theory of time series analysis, signal detection, and parameter estimation. We show how this theory applies to various cases of gravitational-wave signals. In our applications we usually assume that the noise in the detector is a Gaussian and stationary stochastic process. These assumptions will need to be verified in practice. In our presentation we focus on one very powerful method of detecting a signal in noise called the *maximum-likelihood* method. This method is optimal by several criteria and in the case of Gaussian

noise it consists of correlating the data with the template that is matched
to the expected signal. Robust methods of detecting signals in non-
Gaussian noise in the context of gravitational-wave data analysis are discussed in [2, 3]. In our book we do not discuss alternative data analysis
techniques such as time-frequency methods [4, 5], wavelets [6, 7, 8, 9],
Hough transform [10, 11], and suboptimal methods like those proposed
in [12, 13].

Early gravitational-wave data analysis was concerned with the detection of bursts originating from supernova explosions [14] and it consisted
mainly of analysis of coincidences among the detectors [15]. With the
growing interest in laser interferometric gravitational-wave detectors that
are broadband it was realized that sources other than supernovae can
also be detectable [16] and that they can provide a wealth of astrophysical information [17, 18]. For example, the analytic form of the gravitational-wave signal from a binary system is known to a good approximation in terms of a few parameters. Consequently, one can detect such
a signal by correlating the data with the waveform of the signal and
maximizing the correlation with respect to the parameters of the waveform. Using this method one can pick up a weak signal from the noise
by building a large signal-to-noise ratio over a wide bandwidth of the
detector [16]. This observation has led to a rapid development of the theory of gravitational-wave data analysis. It became clear that detectability
of sources is determined by an optimal signal-to-noise ratio, which is the
power spectrum of the signal divided by power spectrum of the noise
integrated over the bandwidth of the detector.

An important landmark was a workshop entitled *Gravitational Wave
Data Analysis* held in Dyffryn House and Gardens, St. Nicholas near
Cardiff, in July 1987 [19]. The meeting acquainted physicists interested in
analyzing gravitational-wave data with the basics of the statistical theory
of signal detection and its application to the detection of gravitational-
wave sources. As a result of subsequent studies the Fisher information
matrix was introduced to the theory of the analysis of gravitational-
wave data [20, 21]. The diagonal elements of the Fisher matrix give lower
bounds on the variances of the estimators of the parameters of the signal and can be used to assess the quality of the astrophysical information that can be obtained from detections of gravitational-wave signals
[22, 23, 24, 25]. It was also realized that the application of matched-
filtering to some signals, notably to periodic signals originating from neutron stars, will require extraordinarily large computing resources. This
gave a further stimulus to the development of optimal and efficient algorithms and data analysis methods [26].

A very important development was the paper [27], where it was realized
that for the case of coalescing binaries matched-filtering was sensitive to

very small post-Newtonian effects of the waveform. Thus, these effects can be detected. This leads to a much better verification of Einstein's theory of relativity and provides a wealth of astrophysical information that would make a laser interferometric gravitational-wave detector a true astronomical observatory complementary to those utilizing the electromagnetic spectrum. Further theoretical methods were introduced: to calculate the quality of suboptimal filters [28], to calculate the number of templates to do a search using matched-filtering [29], to determine the accuracy of templates required [30], to calculate the false alarm probability and thresholds [31]. An important point is the reduction of the number of parameters that one needs to search for in order to detect a signal. Namely, estimators of a certain type of parameters, called *extrinsic parameters*, can be found in a closed analytic form and consequently eliminated from the search. Thus a computationally intensive search needs only be performed over a reduced set of *intrinsic parameters* [21, 31, 32, 33].

The book is organized as follows. Chapter 1 introduces linearized Einstein equations and demonstrates how the gravitational waves arise in this approximation. Chapter 2 reviews the most important astrophysical sources of gravitational waves and presents in more detail calculations of two independent polarizations of waves from a compact binary system and emitted by a rotating triaxial ellipsoid. Also the main features of a gravitational-wave burst from a supernova explosion and of stochastic gravitational-wave background are considered. Chapter 3 is an introduction to the statistical theory of the detection of signals in noise and estimation of the signal's parameters. We discuss the basic properties of not only stationary stochastic processes but also certain aspects of non-stationary processes with applications to gravitational-wave data analysis. Chapter 4 is an introduction to the time series analysis. Chapter 5 studies the responses of detectors to gravitational waves for both ground-based and space-borne detectors. Chapter 6 introduces the maximum-likelihood detection method with an application to detection and estimation of parameters of gravitational-wave signals buried in the noise of the detector. The noise in the detector is assumed to be Gaussian and stationary. We consider both deterministic and stochastic signals. We also consider detection by networks of detectors. Chapter 7 presents an in detail application of the methods presented in Chapter 6 to the case of the periodic gravitational-wave signal from a rotating neutron star. The methodology explained in Chapter 7 can be used in the analysis of many other gravitational-wave signals such as signals originating from the white-dwarf binaries.

We would like to thank Gerhard Schäfer for reading parts of the manuscript of the book and for his useful remarks and comments.

Notation and conventions

General relativity. We adopt notation and conventions of the textbook by Misner, Thorne, and Wheeler [34]. Greek indices $\alpha, \beta, \ldots$ run from 0 to 3 and Latin indices $i, j, \ldots$ run from 1 to 3. We employ the Einstein summation convention. We use the spacetime metric of signature $(-1, 1, 1, 1)$, so the line element of the Minkowski spacetime in Cartesian (inertial) coordinates $(x^0 = ct, \ x^1 = x, \ x^2 = y, \ x^3 = z)$ reads

$$ds^2 = \eta_{\mu\nu} \, dx^\mu \, dx^\nu = -c^2 \, dt^2 + dx^2 + dy^2 + dz^2. \tag{1}$$

Numbers. $\mathbb{N} = \{1, 2, \ldots\}$ is the set of natural numbers (i.e. the positive integers); $\mathbb{Z} = \{\ldots, -2, -1, 0, 1, 2, \ldots\}$ denotes the set of all integers; $\mathbb{R}$ is the set of real numbers. $\mathbb{C}$ denotes the set of complex numbers; for the complex number $z = a + ib$ $(a, b \in \mathbb{R}, \ i = \sqrt{-1})$ the complex conjugate of z is denoted by $z^* := a - ib$ and $|z| := \sqrt{a^2 + b^2}$ is the modulus of z; the real and imaginary parts of z are denoted by $\Re(z) := a$ and $\Im(z) := b$.

3-vectors. For any 3-vectors $\mathbf{a} = (a^1, a^2, a^3)$ and $\mathbf{b} = (b^1, b^2, b^3)$ we define their usual Euclidean scalar product $\mathbf{a} \cdot \mathbf{b}$, and $|\mathbf{a}|$ denotes Euclidean length of a 3-vector $\mathbf{a}$:

$$\mathbf{a} \cdot \mathbf{b} := \sum_{i=1}^{3} a^i b^i, \quad |\mathbf{a}| := \sqrt{\mathbf{a} \cdot \mathbf{a}} = \sqrt{\sum_{i=1}^{3} (a^i)^2}. \tag{2}$$

Matrices are written in a sans serif font, e.g. M, N, ... Matrix multiplication is denoted by a dot, e.g. $\mathsf{M} \cdot \mathsf{N}$, and the superscript T stands for the matrix transposition, e.g. M^T.

Fourier transform. Different conventions are used for defining Fourier transform. In the general case the one-dimensional Fourier transform

$\hat{s} = \hat{s}(\xi)$ of a function $s = s(t)$ is defined as

$$\tilde{s}(\xi) := \sqrt{\frac{|b|}{(2\pi)^{1-a}}} \int_{-\infty}^{+\infty} e^{ib\xi t} s(t) \, dt, \qquad (3)$$

where a and b $(b \neq 0)$ are some real numbers. Then the solution of the above equation with respect to the function s [i.e. the inverse Fourier transform] reads

$$s(t) = \sqrt{\frac{|b|}{(2\pi)^{1+a}}} \int_{-\infty}^{+\infty} e^{-ib\xi t} \tilde{s}(\xi) \, d\xi. \qquad (4)$$

Throughout the book we employ two conventions. With the Fourier variable being the frequency f (measured in Hz) we use $a = 0$ and $b = -2\pi$, i.e.

$$\tilde{s}(f) := \int_{-\infty}^{+\infty} e^{-2\pi i f t} s(t) \, dt, \quad s(t) = \int_{-\infty}^{+\infty} e^{2\pi i f t} \tilde{s}(f) \, df. \qquad (5)$$

When angular frequency $\omega := 2\pi f$ is the Fourier variable we take $a = -1$ and $b = -1$, i.e.

$$\tilde{s}(\omega) := \frac{1}{2\pi} \int_{-\infty}^{+\infty} e^{-i\omega t} s(t) \, dt, \quad s(t) = \int_{-\infty}^{+\infty} e^{i\omega t} \tilde{s}(\omega) \, d\omega. \qquad (6)$$

Spectral densities. In Section 3.2 we introduce the *two-sided* spectral density $\bar{S}$ of a stationary stochastic process as a function of angular frequency ω, $\bar{S} = \bar{S}(\omega)$. The function $\bar{S}$ is the Fourier transform of the autocorrelation function R of the stochastic process, i.e.

$$\bar{S}(\omega) = \frac{1}{2\pi} \int_{-\infty}^{+\infty} e^{-i\omega t} R(t) \, dt, \quad -\infty < \omega < +\infty. \qquad (7)$$

In Section 3.4 the two-sided spectral density S as a function of frequency f is used, $S = S(f)$. The function S is also the Fourier transform of the function R,

$$S(f) = \int_{-\infty}^{+\infty} e^{-2\pi i f t} R(t) \, dt, \quad -\infty < f < +\infty. \qquad (8)$$

Comparing Eqs. (7) and (8) we see that the relation between the functions $\bar{S}$ and S is the following:

$$\bar{S}(\omega) = \frac{1}{2\pi} S\left(\frac{\omega}{2\pi}\right), \quad S(f) = 2\pi \bar{S}(2\pi f). \qquad (9)$$

The relation between *one-sided* and two-sided spectral densities is explained in Eqs. (3.93)–(3.97).

Dirac delta function, δ. We employ integral representation of δ function:

$$\delta(t) = \int_{-\infty}^{+\infty} e^{2\pi i f t}\, \mathrm{d}f = \frac{1}{2\pi} \int_{-\infty}^{+\infty} e^{i\omega t}\, \mathrm{d}\omega, \qquad (10)$$

as well as its periodic representation of the form:

$$\sum_{n=-\infty}^{\infty} \delta(x - 2n\pi) = \frac{1}{2\pi} \sum_{n=-\infty}^{\infty} e^{inx}. \qquad (11)$$

1

Overview of the theory of
gravitational radiation

In this chapter we very briefly review the theory of gravitational radiation. A detailed exposition of the theory can be found in many textbooks on general relativity, e.g. in Chapters 35–37 of [34], Chapter 9 of [35], or Chapter 7 of [36]. A detailed exposition of the theory of gravitational waves is contained in the recent monograph [37]. Reference [38] is an introductory review of the theory of gravitational radiation and Ref. [16] is an accessible review of different aspects of gravitational-wave research. Some parts of the present chapter closely follow Sections 9.2 and 9.3 of the review article [16].

The chapter begins (in Section 1.1) with a discussion of general relativity theory in the limit of weak gravitational fields. In this limit spacetime geometry is a small perturbation of the flat geometry of Minkowski spacetime. We restrict our considerations to coordinate systems in which the spacetime metric is the sum of the Minkowski metric and a small perturbation. We linearize Einstein field equations with respect to this perturbation and then we study two classes of coordinate transformations that preserve splitting the metric into the sum of Minkowski metric and its small perturbation: global Poincaré transformations and gauge transformations. Finally we discuss the harmonic gauge, which allows one to write the linearized Einstein field equations in the form of inhomogeneous wave equations for the metric perturbation.

In Sections 1.2–1.4 we introduce gravitational waves as time-dependent vacuum solutions of the linearized Einstein equations. In Section 1.2 we study the simplest such solution, namely a monochromatic plane gravitational wave. In Section 1.3 we introduce the TT coordinate system in which description of gravitational waves is especially simple. We first develop in some detail the TT coordinate system for monochromatic plane

1

gravitational waves and then we remark that the TT coordinates can be introduced for any kind of gravitational waves. In Section 1.4 we describe in the proper reference frame of an observer the effect of gravitational waves on freely falling particles.

Section 1.5 is devoted to the problem of defining gravitational waves in curved backgrounds. In Section 1.6 we present the energy–momentum tensor for gravitational waves. Section 1.7 discusses the generation of gravitational waves within the quadrupole formalism. It also contains approximate formulae for the rates of emission of energy and angular momentum carried away from the source by gravitational waves.

1.1 Linearized general relativity

If the gravitational field is weak, then it is possible to find a coordinate system (x^α) such that the components $g_{\mu\nu}$ of the metric tensor in this system will be small perturbations $h_{\mu\nu}$ of the flat Minkowski metric components $\eta_{\mu\nu}$:

$$g_{\mu\nu} = \eta_{\mu\nu} + h_{\mu\nu}, \quad |h_{\mu\nu}| \ll 1. \tag{1.1}$$

The coordinate system satisfying the condition (1.1) is sometimes called an *almost Lorentzian* coordinate system. For the almost Lorentzian coordinates (x^α) we will also use another notation:

$$x^0 \equiv c\,t, \quad x^1 \equiv x, \quad x^2 \equiv y, \quad x^3 \equiv z. \tag{1.2}$$

In the rest of Section 1.1 we assume that the indices of the $h_{\mu\nu}$ will be raised and lowered by means of the $\eta^{\mu\nu}$ and $\eta_{\mu\nu}$ (and *not* by $g^{\mu\nu}$ and $g_{\mu\nu}$). Therefore we have, e.g.,

$$h_\alpha{}^\beta = \eta^{\beta\mu} h_{\alpha\mu}, \quad h^{\alpha\beta} = \eta^{\alpha\mu} h_\mu{}^\beta = \eta^{\alpha\mu}\eta^{\beta\nu} h_{\mu\nu}. \tag{1.3}$$

1.1.1 Linearized Einstein equations

In the heart of general relativity there are the *Einstein field equations*,

$$G_{\mu\nu} = \frac{8\pi G}{c^4} T_{\mu\nu}, \tag{1.4}$$

which relate spacetime geometry expressed in terms of the *Einstein tensor* $G_{\mu\nu}$ with sources of a gravitational field represented by an *energy–momentum tensor* $T_{\mu\nu}$. The components of the Einstein tensor $G_{\mu\nu}$ are constructed from the components of the *Riemann curvature tensor* describing spacetime geometry in the following way. We first need to define the *Christoffel symbols* $\Gamma^\mu{}_{\alpha\beta}$, which depend on the spacetime metric components and their first derivatives:

$$\Gamma^\mu{}_{\alpha\beta} = \frac{1}{2}g^{\mu\nu}\left(\partial_\alpha g_{\beta\nu} + \partial_\beta g_{\nu\alpha} - \partial_\nu g_{\alpha\beta}\right). \tag{1.5}$$

The Riemann curvature tensor has components

$$R_{\mu\nu\rho\sigma} = g_{\rho\lambda}\left(\partial_\mu\Gamma^\lambda{}_{\nu\sigma} - \partial_\nu\Gamma^\lambda{}_{\mu\sigma} + \Gamma^\lambda{}_{\mu\eta}\Gamma^\eta{}_{\nu\sigma} - \Gamma^\lambda{}_{\nu\eta}\Gamma^\eta{}_{\mu\sigma}\right). \tag{1.6}$$

Next we introduce the symmetric *Ricci tensor*,

$$R_{\mu\nu} = g^{\rho\sigma}R_{\rho\mu\sigma\nu}, \tag{1.7}$$

and its trace known as the *Ricci scalar*,

$$R = g^{\mu\nu}R_{\mu\nu}. \tag{1.8}$$

Finally, the Einstein tensor is defined as

$$G_{\mu\nu} := R_{\mu\nu} - \frac{1}{2}g_{\mu\nu}R. \tag{1.9}$$

If condition (1.1) is satisfied, one can linearize the Einstein field equations (1.4) with respect to the small perturbation $h_{\mu\nu}$. To do this we start from linearizing the Christoffel symbols $\Gamma^\mu{}_{\alpha\beta}$. For the metric (1.1) they take the form

$$\Gamma^\mu{}_{\alpha\beta} = \frac{1}{2}\eta^{\mu\nu}\left(\partial_\alpha h_{\beta\nu} + \partial_\beta h_{\nu\alpha} - \partial_\nu h_{\alpha\beta}\right) + \mathcal{O}(h^2)$$

$$= \frac{1}{2}\left(\partial_\alpha h^\mu{}_\beta + \partial_\beta h^\mu{}_\alpha - \eta^{\mu\nu}\partial_\nu h_{\alpha\beta}\right) + \mathcal{O}(h^2). \tag{1.10}$$

Because Christoffel symbols are first-order quantities, the only contribution to the linearized Riemann tensor will come from the derivatives of the Christoffel symbols. Making use of Eqs. (1.6) and (1.10) we get

$$R_{\mu\nu\rho\sigma} = \frac{1}{2}\left(\partial_\rho\partial_\nu h_{\mu\sigma} + \partial_\sigma\partial_\mu h_{\nu\rho} - \partial_\sigma\partial_\nu h_{\mu\rho} - \partial_\rho\partial_\mu h_{\nu\sigma}\right) + \mathcal{O}(h^2). \tag{1.11}$$

Next we linearize the Ricci tensor. Making use of (1.11) we obtain

$$R_{\mu\nu} = \frac{1}{2}\left(\partial_\alpha\partial_\mu h^\alpha{}_\nu + \partial_\alpha\partial_\nu h^\alpha{}_\mu - \partial_\mu\partial_\nu h - \Box h_{\mu\nu}\right) + \mathcal{O}(h^2), \tag{1.12}$$

where h is the trace of the metric perturbation $h_{\mu\nu}$,

$$h := \eta^{\alpha\beta}h_{\alpha\beta}, \tag{1.13}$$

and where we have introduced the d'Alembertian operator $\Box$ in the flat Minkowski spacetime:

$$\Box := \eta^{\mu\nu}\partial_\mu\partial_\nu = -\frac{1}{c^2}\partial_t^2 + \partial_x^2 + \partial_y^2 + \partial_z^2. \tag{1.14}$$

Finally we linearize the Ricci scalar,

$$R = \eta^{\mu\nu} R_{\mu\nu} + \mathcal{O}(h^2) = \partial_\mu \partial_\nu h^{\mu\nu} - \Box h + \mathcal{O}(h^2). \qquad (1.15)$$

We are now ready to linearize the Einstein tensor. We get

$$G_{\mu\nu} = \frac{1}{2}\left(\partial_\mu\partial_\alpha h^\alpha{}_\nu + \partial_\nu\partial_\alpha h^\alpha{}_\mu - \partial_\mu\partial_\nu h - \Box h_{\mu\nu} + \eta_{\mu\nu}\left(\Box h - \partial_\alpha\partial_\beta h^{\alpha\beta}\right)\right)$$
$$+ \mathcal{O}(h^2). \qquad (1.16)$$

It is possible to simplify one bit the right-hand side of Eq. (1.16) by introducing the quantity

$$\bar{h}_{\mu\nu} := h_{\mu\nu} - \frac{1}{2}\eta_{\mu\nu} h. \qquad (1.17)$$

From the definition (1.17) follows that

$$h_{\mu\nu} = \bar{h}_{\mu\nu} - \frac{1}{2}\eta_{\mu\nu}\bar{h}, \qquad (1.18)$$

where $\bar{h} := \eta^{\alpha\beta}\bar{h}_{\alpha\beta}$ (let us also observe that $\bar{h} = -h$). Substituting the relation (1.18) into (1.16), one obtains

$$G_{\mu\nu} = \frac{1}{2}\left(\partial_\mu\partial_\alpha\bar{h}^\alpha{}_\nu + \partial_\nu\partial_\alpha\bar{h}^\alpha{}_\mu - \Box\bar{h}_{\mu\nu} - \eta_{\mu\nu}\partial_\alpha\partial_\beta\bar{h}^{\alpha\beta}\right) + \mathcal{O}(h^2). \quad (1.19)$$

Making use of Eq. (1.16) or Eq. (1.19) one can write down the linearized form of the Einstein field equations.

If spacetime (or its part) admits one almost Lorentzian coordinate system, then there exists in spacetime (or in its part) infinitely many almost Lorentzian coordinates. Below we describe two kinds of coordinate transformations leading from one almost Lorentzian coordinate system to another such system: global Poincaré transformations and gauge transformations.

1.1.2 Global Poincaré transformations

Let us consider the *global Poincaré* transformation leading from the "old" coordinates (x^α) to some "new" ones (x'^α). The new coordinates are linear inhomogeneous functions of the old coordinates,

$$x'^\alpha(x^\beta) = \Lambda^\alpha{}_\beta\, x^\beta + a^\alpha. \qquad (1.20)$$

Here the constant (i.e. independent of the spacetime coordinates x^μ) numbers $\Lambda^\alpha{}_\beta$ are the components of the matrix representing the special-relativistic Lorentz transformation, and the constant quantities a^α represent some translation in spacetime. The matrix $(\Lambda^\alpha{}_\beta)$ represents some

Lorentz transformation provided it fulfills the condition

$$\Lambda^{\alpha}{}_{\mu} \Lambda^{\beta}{}_{\nu} \eta_{\alpha\beta} = \eta_{\mu\nu}. \tag{1.21}$$

This condition means that the Lorentz transformation does not change the Minkowski metric. Let us recall that if the new coordinates are related to an observer that moves with respect to another observer related to the old coordinates along its x axis with velocity v, then the matrix built up from the coefficients $\Lambda^{\alpha}{}_{\beta}$ is the following

$$(\Lambda^{\alpha}{}_{\beta}) = \begin{pmatrix} \gamma & -v\gamma/c & 0 & 0 \\ -v\gamma/c & \gamma & 0 & 0 \\ 0 & 0 & 1 & 0 \\ 0 & 0 & 0 & 1 \end{pmatrix}, \quad \gamma := \left(1 - \frac{v^2}{c^2}\right)^{-1/2}. \tag{1.22}$$

The transformation inverse to that given in Eq. (1.20) leads from new (x'^{α}) to old (x^{α}) coordinates and is given by the relations

$$x^{\alpha}(x'^{\beta}) = (\Lambda^{-1})^{\alpha}{}_{\beta}(x'^{\beta} - a^{\beta}), \tag{1.23}$$

where the numbers $(\Lambda^{-1})^{\alpha}{}_{\beta}$ form the matrix inverse to that constructed from $\Lambda^{\alpha}{}_{\beta}$:

$$(\Lambda^{-1})^{\alpha}{}_{\beta} \Lambda^{\beta}{}_{\gamma} = \delta^{\alpha}_{\gamma}, \quad \Lambda^{\alpha}{}_{\beta} (\Lambda^{-1})^{\beta}{}_{\gamma} = \delta^{\alpha}_{\gamma}. \tag{1.24}$$

Let us also note that the matrix $((\Lambda^{-1})^{\alpha}{}_{\beta})$ also fulfills the requirement (1.21),

$$(\Lambda^{-1})^{\alpha}{}_{\mu} (\Lambda^{-1})^{\beta}{}_{\nu} \eta_{\alpha\beta} = \eta_{\mu\nu}. \tag{1.25}$$

Let us now assume that the old coordinates (x^{α}) are almost Lorentzian, so the decomposition (1.1) of the metric holds in these coordinates. Making use of the rule relating the components of the metric tensor in two coordinate systems,

$$g'_{\alpha\beta}(x') = \frac{\partial x^{\mu}}{\partial x'^{\alpha}} \frac{\partial x^{\nu}}{\partial x'^{\beta}} g_{\mu\nu}(x), \tag{1.26}$$

by virtue of Eqs. (1.1), (1.23), and (1.21) one easily obtains

$$g'_{\alpha\beta} = \eta_{\alpha\beta} + h'_{\alpha\beta}, \tag{1.27}$$

where we have defined

$$h'_{\alpha\beta} := (\Lambda^{-1})^{\mu}{}_{\alpha} (\Lambda^{-1})^{\nu}{}_{\beta} h_{\mu\nu}. \tag{1.28}$$

This last equation means that the metric perturbations $h_{\mu\nu}$ transform under the Poincaré transformation as the components of a $(0, 2)$ rank tensor. The result of Eqs. (1.27)–(1.28) also means that the new coordinate system (x'^{α}) will be almost Lorentzian, provided the numerical values of the matrix elements $(\Lambda^{-1})^\alpha{}_\beta$ are not too large, because then the condition $|h_{\mu\nu}| \ll 1$ also implies that $|h'_{\alpha\beta}| \ll 1$.

1.1.3 Gauge transformations

Another family of coordinate transformations, which lead from one almost Lorentzian coordinates to another such coordinates, consists of infinitesimal coordinate transformations known as *gauge transformations*. They are of the form

$$x'^\alpha = x^\alpha + \xi^\alpha(x^\beta), \qquad (1.29)$$

where the functions ξ^α are small in this sense, that

$$|\partial_\beta \xi^\alpha| \ll 1. \qquad (1.30)$$

Equations (1.29)–(1.30) imply that

$$\frac{\partial x'^\alpha}{\partial x^\beta} = \delta^\alpha_\beta + \partial_\beta \xi^\alpha, \qquad (1.31a)$$

$$\frac{\partial x^\alpha}{\partial x'^\beta} = \delta^\alpha_\beta - \partial_\beta \xi^\alpha + \mathcal{O}((\partial \xi)^2). \qquad (1.31b)$$

Let us now assume that the coordinates (x^α) are almost Lorentzian. Making use of Eqs. (1.1), (1.26), and (1.31), we compute the components of the metric in the (x'^α) coordinates:

$$g'_{\alpha\beta} = \eta_{\alpha\beta} + h_{\alpha\beta} - \partial_\alpha \xi_\beta - \partial_\beta \xi_\alpha + \mathcal{O}(h\,\partial\xi, (\partial\xi)^2), \qquad (1.32)$$

where we have introduced

$$\xi_\alpha := \eta_{\alpha\beta} \xi^\beta. \qquad (1.33)$$

The metric components $g'_{\alpha\beta}$ can thus be written in the form

$$g'_{\alpha\beta} = \eta_{\alpha\beta} + h'_{\alpha\beta} + \mathcal{O}(h\,\partial\xi, (\partial\xi)^2), \qquad (1.34)$$

where we have defined

$$h'_{\alpha\beta} := h_{\alpha\beta} - \partial_\alpha \xi_\beta - \partial_\beta \xi_\alpha. \qquad (1.35)$$

Because condition (1.30) is fulfilled, the new metric perturbation $h'_{\alpha\beta}$ is small, $|h'_{\alpha\beta}| \ll 1$, and the coordinates (x'^α) are almost Lorentzian. From Eq. (1.35), making use of the definition (1.17), one obtains the rule for how the metric perturbation $\bar{h}_{\alpha\beta}$ changes under the gauge transformation,

$$\bar{h}'_{\alpha\beta} = \bar{h}_{\alpha\beta} - \partial_\alpha \xi_\beta - \partial_\beta \xi_\alpha + \eta_{\alpha\beta}\,\partial_\mu \xi^\mu. \qquad (1.36)$$

1.1.4 Harmonic coordinates

Among almost Lorentzian coordinates one can choose coordinates for which the following additional *harmonic* gauge conditions are fulfilled:

$$\partial_\beta \bar{h}^{\beta\alpha} = 0. \tag{1.37}$$

Let us note that the conditions (1.37) can equivalently be written as

$$\eta^{\beta\gamma} \partial_\beta \bar{h}_{\gamma\alpha} = 0. \tag{1.38}$$

If these conditions are satisfied, the linearized Einstein tensor $G_{\mu\nu}$ from Eq. (1.19) reduces to

$$G_{\mu\nu} = -\frac{1}{2}\Box\bar{h}_{\mu\nu} + \mathcal{O}(h^2), \tag{1.39}$$

and the linearized Einstein field equations take the simple form of the wave equations in the flat Minkowski spacetime:

$$\Box\bar{h}_{\mu\nu} + \mathcal{O}(h^2) = -\frac{16\pi G}{c^4}T_{\mu\nu}. \tag{1.40}$$

Harmonic coordinates are not uniquely defined. The gauge conditions (1.37) [or (1.38)] are preserved by the Poincaré transformations (1.20), they are also preserved by the infinitesimal gauge transformations of the form (1.29), provided all the functions ξ^α satisfy homogeneous wave equations:

$$\Box\xi^\alpha = 0. \tag{1.41}$$

1.2 Plane monochromatic gravitational waves

The simplest way of introducing gravitational waves relies on studying vacuum solutions of linearized Einstein field equations in harmonic coordinates. In vacuum the energy–momentum tensor vanishes, $T_{\mu\nu} = 0$, and the linearized field equations in harmonic coordinates, Eqs. (1.40), reduce to homogeneous wave equations for all the components of the metric perturbation $\bar{h}_{\mu\nu}$:

$$\Box\bar{h}_{\mu\nu} = 0. \tag{1.42}$$

Time-dependent solutions of these equations can be interpreted as *weak* gravitational waves propagating through a region of spacetime where Eqs. (1.42) are valid, i.e. where the spacetime metric is almost Minkowskian.

The simplest solution of Eqs. (1.42) is a *monochromatic plane wave*, which is of the form

$$\bar{h}_{\mu\nu}(x^\alpha) = A_{\mu\nu} \cos\left(k_\alpha x^\alpha - \alpha_{(\mu)(\nu)}\right). \tag{1.43}$$

Here $A_{\mu\nu}$ and $\alpha_{(\mu)(\nu)}$ is the constant *amplitude* and the constant *initial phase*, respectively, of the $\mu\nu$ component of the wave, and k_α are another four real constants. We have encircled the indices μ and ν of the initial phases by parentheses to indicate that there is no summation over these indices on the right-hand side of Eq. (1.43). By substituting (1.43) into (1.42) one checks that the functions (1.43) are solutions of Eqs. (1.42) if and only if

$$\eta^{\alpha\beta} k_\alpha k_\beta = 0, \tag{1.44}$$

which means that if we define $k^\alpha := \eta^{\alpha\beta} k_\beta$, then k^α are the components of a null (with respect to the flat Minkowski metric) 4-vector.

Let us write the argument of the cosine function in (1.43) in a more explicit way. The contraction $k_\alpha x^\alpha$ can be written as

$$k_\alpha x^\alpha = k_0 \, x^0 + \sum_{i=1}^{3} k_i \, x^i = -k^0 \, x^0 + \sum_{i=1}^{3} k^i \, x^i = -c\, k^0 \, t + \mathbf{k} \cdot \mathbf{x}. \tag{1.45}$$

Here we have introduced the two 3-vectors: $\mathbf{k}$ with components (k^1, k^2, k^3) and $\mathbf{x}$ with components (x^1, x^2, x^3). If we additionally introduce the quantity $\omega := c\, k^0$, then the plane-wave solution (1.43) becomes

$$\bar{h}_{\mu\nu}(t, \mathbf{x}) = A_{\mu\nu} \cos\left(\omega t - \mathbf{k} \cdot \mathbf{x} + \alpha_{(\mu)(\nu)}\right). \tag{1.46}$$

We can assume, without loss of generality, that $\omega \geq 0$. Then ω is *angular frequency* of the wave; it is measured in radians per second. We will also use *frequency* f of the wave, measured in hertz (i.e. cycles per second). These two quantities are related to each other by the equation

$$\omega = 2\pi f. \tag{1.47}$$

The 3-vector $\mathbf{k}$ is known as a *wave vector*, it points to the direction in which the wave is propagating and its Euclidean length is related to the *wavelength* λ,

$$\lambda |\mathbf{k}| = 2\pi. \tag{1.48}$$

Equation (1.44) written in terms of ω and $\mathbf{k}$ takes the form

$$\omega = c|\mathbf{k}|. \tag{1.49}$$

This is the *dispersion relation* for gravitational waves. It implies that both the phase and the group velocity of the waves are equal to c.

Summing up: the solution (1.46) represents a plane gravitational wave with frequency $f = \omega/(2\pi)$ and wavelength $\lambda = 2\pi/|\mathbf{k}|$, which propagates through the 3-space in the direction of the 3-vector $\mathbf{k}$ with the speed of light.

Einstein field equations take the simple form (1.42) only if the harmonic gauge conditions (1.37) are satisfied. Therefore we must additionally require that the functions $\bar{h}_{\mu\nu}(x^\alpha)$ from Eq. (1.46) fulfill (1.37). Let us rewrite these functions in the form

$$\bar{h}_{\mu\nu}(t,\mathbf{x}) = C_{\mu\nu}\cos(\omega t - \mathbf{k}\cdot\mathbf{x}) + S_{\mu\nu}\sin(\omega t - \mathbf{k}\cdot\mathbf{x}), \qquad (1.50)$$

where we have introduced new quantities

$$C_{\mu\nu} := A_{\mu\nu}\cos\alpha_{(\mu)(\nu)}, \quad S_{\mu\nu} := -A_{\mu\nu}\sin\alpha_{(\mu)(\nu)}. \qquad (1.51)$$

Then the requirements (1.37) imply that

$$C_{\mu\nu}k^\nu = 0, \quad S_{\mu\nu}k^\nu = 0. \qquad (1.52)$$

Equations (1.52) provide constraints on the wave amplitudes $C_{\mu\nu}$ and $S_{\mu\nu}$: they must be orthogonal to the 4-vector k^μ. As a consequence the whole plane-wave solution $\bar{h}_{\mu\nu}$, Eq. (1.50), is orthogonal to k^μ:

$$\bar{h}_{\mu\nu}k^\nu = 0. \qquad (1.53)$$

The Poincaré transformations preserve the form of the plane-wave solution (1.43). Making use of the transformation rule (1.28) one can show that coordinate transformation (1.20) transforms the metric perturbations (1.43) into the new perturbations $\bar{h}'_{\mu\nu}$,

$$\bar{h}'_{\mu\nu}(x') = A'_{\mu\nu}\cos\left(k'_\alpha x'^\alpha - \alpha'_{(\mu)(\nu)}\right), \qquad (1.54)$$

where the new constants k'_α are related to the old ones k_α by

$$k'_\alpha = (\Lambda^{-1})^\beta{}_\alpha k_\beta, \qquad (1.55)$$

so they transform as components of a $(0,1)$ rank tensor, and the new amplitudes and initial phases are defined through the relations

$$A'_{\mu\nu}\cos\alpha'_{(\mu)(\nu)} = (\Lambda^{-1})^\alpha{}_\mu(\Lambda^{-1})^\beta{}_\nu A_{\alpha\beta}\cos(\alpha_{(\alpha)(\beta)} + k'_\sigma a^\sigma), \quad (1.56)$$

$$A'_{\mu\nu}\sin\alpha'_{(\mu)(\nu)} = (\Lambda^{-1})^\alpha{}_\mu(\Lambda^{-1})^\beta{}_\nu A_{\alpha\beta}\sin(\alpha_{(\alpha)(\beta)} + k'_\sigma a^\sigma). \quad (1.57)$$

Because the Poincaré transformations preserve the harmonicity conditions (1.37), as a consequence the orthogonality relations (1.53) are also preserved.

1.3 Description in the TT coordinate system

Equations (1.53) restrict the number of independent components of the gravitational plane wave from 10 to 6. These equations are a consequence of the harmonic gauge conditions (1.37), which are preserved by gauge transformations (1.29), provided each function ξ^α is a solution of homogeneous wave equation [see Eq. (1.41)]. Because we have at our disposal four functions ξ^α (for $\alpha = 0, 1, 2, 3$), we can use them to further restrict the number of independent components of the plane wave from 6 to 2. The choice of the functions ξ^α is equivalent to the choice of a coordinate system. We describe now the choice of ξ^α leading to the so called *transverse* and *traceless* (TT in short) coordinate system.

At each event in the spacetime region covered by some almost Lorentzian and harmonic coordinates (x^α) let us choose the timelike unit vector U^μ, $g_{\mu\nu}U^\mu U^\nu = -1$. Let us consider a gauge transformation generated by the functions ξ^α of the form

$$\xi^\alpha(t, \mathbf{x}) = B^\alpha \cos\left(\omega t - \mathbf{k} \cdot \mathbf{x} + \beta_{(\alpha)}\right), \qquad (1.58)$$

with $\omega = c|\mathbf{k}|$ and $\mathbf{k}$ the same as in the plane-wave solution (1.46). It is possible to choose the quantities B^α and $\beta_{(\alpha)}$ in such a way, that in the new almost Lorentzian and harmonic coordinates $x'^\alpha = x^\alpha + \xi^\alpha$ the following conditions are fulfilled

$$\bar{h}'_{\mu\nu}U'^\nu = 0, \qquad (1.59a)$$

$$\eta^{\mu\nu}\bar{h}'_{\mu\nu} = 0. \qquad (1.59b)$$

Furthermore, the gauge transformation based on the functions (1.58) preserves the condition (1.53):

$$\bar{h}'_{\mu\nu}k'^\nu = 0. \qquad (1.59c)$$

Let us also note that, as a consequence of Eq. (1.59b),

$$\bar{h}'_{\mu\nu} = h'_{\mu\nu}. \qquad (1.60)$$

Equations (1.59) define the TT coordinate system related to the 4-vector field, U'^μ. These equations comprise eight independent constraints on the components of the plane-wave solution $\bar{h}_{\mu\nu}$. This means that any plane monochromatic gravitational wave possesses two independent degrees of freedom, often also called the wave's *polarizations*.

The simplest way to describe the two gravitational-wave polarizations is by making more coordinate changes. We have used all the freedom related to the gauge transformations, but we are still able to perform global Lorentz transformations, which preserve equations (1.59) defining the TT gauge. One can first move to coordinates in which the vector U^μ

(from now we omit the primes in the coordinate names) has components $U^\mu = (1, 0, 0, 0)$. Then Eqs. (1.59a) imply

$$\bar{h}_{\mu 0} = 0. \tag{1.61}$$

Further, one can orient spatial coordinate axes such that the wave propagates in, say, the $+z$ direction. Then $\mathbf{k} = (0, 0, \omega/c)$, $k^\mu = (\omega/c, 0, 0, \omega/c)$, and Eqs. (1.59c) together with (1.61) give

$$\bar{h}_{\mu 3} = 0. \tag{1.62}$$

The last constraint on the plane-wave components $\bar{h}_{\mu\nu}$ provides Eq. (1.59b) supplemented by (1.61) and (1.62). It reads

$$\bar{h}_{11} + \bar{h}_{22} = 0. \tag{1.63}$$

It is common to use the following notation:

$$h_+ := \bar{h}_{11} = -\bar{h}_{22}, \quad h_\times := \bar{h}_{12} = \bar{h}_{21}. \tag{1.64}$$

The functions h_+ and $h_\times$ are called *plus* and *cross* polarization of the wave, respectively.

We will label all quantities computed in the TT coordinate system by super- or subscript "TT." Equations (1.61)–(1.63) allow us to rewrite the plane-wave solution (1.46) in TT coordinates in the following matrix form:

$$h_{\mu\nu}^{\mathrm{TT}}(t, \mathbf{x}) = \begin{pmatrix} 0 & 0 & 0 & 0 \\ 0 & h_+(t, \mathbf{x}) & h_\times(t, \mathbf{x}) & 0 \\ 0 & h_\times(t, \mathbf{x}) & -h_+(t, \mathbf{x}) & 0 \\ 0 & 0 & 0 & 0 \end{pmatrix}, \tag{1.65}$$

where the plus h_+ and the cross $h_\times$ polarizations of the plane wave with angular frequency ω traveling in the $+z$ direction are given by

$$h_+(t, \mathbf{x}) = A_+ \cos\left(\omega\left(t - \frac{z}{c}\right) + \alpha_+\right),$$

$$h_\times(t, \mathbf{x}) = A_\times \cos\left(\omega\left(t - \frac{z}{c}\right) + \alpha_\times\right). \tag{1.66}$$

Any gravitational wave can be represented as a superposition of plane monochromatic waves. Because the equations describing TT gauge,

$$\partial_\nu \bar{h}^{\mu\nu} = 0, \quad \bar{h}_{\mu\nu} U^\nu = 0, \quad \eta^{\mu\nu} \bar{h}_{\mu\nu} = 0, \tag{1.67}$$

are all linear in $\bar{h}_{\mu\nu}$, it is possible to find the TT gauge for *any* gravitational wave. If we restrict ourselves to plane (but not necessarily monochromatic) waves and orient the spatial axes of a coordinate system such that the wave propagates in the $+z$ direction, then Eqs. (1.61)–(1.63) are still valid. Moreover, because all the monochromatic components of the wave depend on spacetime coordinates only through the combination $t - z/c$, the same dependence will be valid also for the general wave. Therefore any weak plane gravitational wave propagating in the $+z$ direction is described in the TT gauge by the metric perturbation of the following matrix form:

$$h_{\mu\nu}^{\mathrm{TT}}(t, \mathbf{x}) = \begin{pmatrix} 0 & 0 & 0 & 0 \\ 0 & h_+(t - z/c) & h_\times(t - z/c) & 0 \\ 0 & h_\times(t - z/c) & -h_+(t - z/c) & 0 \\ 0 & 0 & 0 & 0 \end{pmatrix}. \qquad (1.68)$$

If we introduce the polarization tensors e^+ and $e^\times$ by means of equations

$$e_{xx}^+ = -e_{yy}^+ = 1, \quad e_{xy}^\times = e_{yx}^\times = 1, \quad \text{all other components zero,} \qquad (1.69)$$

then one can reconstruct the full gravitational-wave field from its plus and cross polarizations as

$$h_{\mu\nu}^{\mathrm{TT}}(t, \mathbf{x}) = h_+(t, \mathbf{x})\, e_{\mu\nu}^+ + h_\times(t, \mathbf{x})\, e_{\mu\nu}^\times. \qquad (1.70)$$

It turns out that $h_{\mu\nu}^{\mathrm{TT}}$ is a scalar under boosts and behaves like a spin-two field under rotations. This means that in two different reference frames related by a boost in some arbitrary direction (with the spatial axes of the frames unrotated relative to each other) the gravitational-wave fields $h_{\mu\nu}^{\mathrm{TT}}$ are the same; whereas if one rotates the x and y axes in the transverse plane of the wave through an angle ψ, the gravity-wave polarizations are changed to

$$\begin{pmatrix} h_+^{\mathrm{new}} \\ h_\times^{\mathrm{new}} \end{pmatrix} = \begin{pmatrix} \cos 2\psi & \sin 2\psi \\ -\sin 2\psi & \cos 2\psi \end{pmatrix} \begin{pmatrix} h_+^{\mathrm{old}} \\ h_\times^{\mathrm{old}} \end{pmatrix}. \qquad (1.71)$$

For further details see Section 2.3.2 of [38].

It is useful to write explicitly all the non-zero components of the gravitational-wave Riemann tensor in the TT coordinates. Components of the Riemann tensor linearized with respect to metric perturbations $h_{\mu\nu}$ are given by Eq. (1.11). Making use of Eqs. (1.60) and (1.68) one then computes

$$R_{x0x0}^{\mathrm{TT}} = -R_{y0y0}^{\mathrm{TT}} = -R_{x0xz}^{\mathrm{TT}} = R_{y0yz}^{\mathrm{TT}} = R_{xzxz}^{\mathrm{TT}} = -R_{yzyz}^{\mathrm{TT}} = -\frac{1}{2}\partial_0^2 h_+,$$

$$\qquad (1.72a)$$

$$R_{x0y0}^{\mathrm{TT}} = -R_{x0yz}^{\mathrm{TT}} = -R_{y0xz}^{\mathrm{TT}} = R_{xzyz}^{\mathrm{TT}} = -\frac{1}{2}\partial_0^2 h_\times. \qquad (1.72b)$$

All other non-zero components can be obtained by means of symmetries of the Riemann tensor.

1.4 Description in the observer's proper reference frame

Let us consider two nearby observers freely falling in the field of a weak and plane gravitational wave. The wave produces tiny variations in the proper distance between the observers. We describe now these variations from the point of view of one of the observers, which we call the *basic* observer. We endow this observer with his *proper reference frame*, which consists of a small Cartesian latticework of measuring rods and synchronized clocks. The time coordinate $\hat{t}$ of that frame measures proper time along the world line of the observer whereas the spatial coordinate $\hat{x}^i$ ($i = 1, 2, 3$)[1] measures proper distance along his ith axis. Spacetime coordinates $(\hat{x}^0 := c\,\hat{t}, \hat{x}^1, \hat{x}^2, \hat{x}^3)$ are *locally Lorentzian* along the whole geodesic of the basic observer, i.e. the line element of the metric in these coordinates has the form

$$ds^2 = -c^2\,d\hat{t}^2 + \delta_{ij}\,d\hat{x}^i\,d\hat{x}^j + \mathcal{O}\big((\hat{x}^i)^2\big)\,d\hat{x}^\alpha\,d\hat{x}^\beta, \tag{1.73}$$

so this line element deviates from the line element of the flat Minkowski spacetime by terms that are at least quadratic in the values of the spatial coordinates $\hat{x}^i$.

Let the basic observer $\mathcal{A}$ be located at the origin of his proper reference frame, so his coordinates are $\hat{x}^\alpha_{\mathcal{A}}(\hat{t}) = (c\,\hat{t}, 0, 0, 0)$. A neighboring observer $\mathcal{B}$ moves along the nearby geodesic and it possesses the coordinates $\hat{x}^\alpha_{\mathcal{B}}(\hat{t}) = (c\,\hat{t}, \hat{x}^1(\hat{t}), \hat{x}^2(\hat{t}), \hat{x}^3(\hat{t}))$. We define the *deviation vector* $\hat{\xi}^\alpha$ describing the instantaneous relative position of the observer $\mathcal{B}$ with respect to the observer $\mathcal{A}$:

$$\hat{\xi}^\alpha(\hat{t}) := \hat{x}^\alpha_{\mathcal{B}}(\hat{t}) - \hat{x}^\alpha_{\mathcal{A}}(\hat{t}) = \big(0, \hat{x}^1(\hat{t}), \hat{x}^2(\hat{t}), \hat{x}^3(\hat{t})\big). \tag{1.74}$$

The relative acceleration $D^2\hat{\xi}^\alpha/d\hat{t}^2$ of the observers is related to the spacetime curvature through the *equation of geodesic deviation*,

$$\frac{D^2\hat{\xi}^\alpha}{d\hat{t}^2} = -c^2\,\hat{R}^\alpha{}_{\beta\gamma\delta}\,\hat{u}^\beta\,\hat{\xi}^\gamma\,\hat{u}^\delta, \tag{1.75}$$

where $\hat{u}^\beta := d\hat{x}^\alpha/(c\,d\hat{t})$ is the 4-velocity of the basic observer. Let us note that all quantities in Eq. (1.75) are evaluated on the basic geodesic, so they are functions of the time coordinate $\hat{t}$ only. Equation (1.73) implies that Christoffel symbols $\hat{\Gamma}^\alpha{}_{\beta\gamma}$ vanish along the basic geodesic. Therefore their time derivative, $d\hat{\Gamma}^\alpha{}_{\beta\gamma}/d\hat{t}$, also vanishes there, and $D^2\hat{\xi}^\alpha/d\hat{t}^2 = d^2\hat{\xi}^\alpha/d\hat{t}^2$ along that geodesic. Also taking into account the fact that $\hat{u}^\alpha = (1, 0, 0, 0)$

---

[1] We will also use another notation for the spatial coordinates: $\hat{x}^1 \equiv \hat{x}$, $\hat{x}^2 \equiv \hat{y}$, $\hat{x}^3 \equiv \hat{z}$.

and making use of Eq. (1.74), from (1.75) one finally obtains

$$\frac{d^2 \hat{x}^i}{d\hat{t}^2} = -c^2 \, \hat{R}^i{}_{0j0} \hat{x}^j = -c^2 \, \hat{R}_{i0j0} \hat{x}^j + \mathcal{O}((\hat{x}^i)^3).$$
(1.76)

If one chooses the TT coordinates in such a way that the 4-velocity field needed to define TT coordinates coincides with the 4-velocity of our basic observer, then the TT coordinates (t, x^i) and the proper-reference-frame coordinates $(\hat{t}, \hat{x}^i)$ differ from each other in the vicinity of the basic observer's geodesic by quantities *linear* in h. It means that, up to the terms quadratic in h, the components of the Riemann tensor in both coordinate systems coincide,

$$\hat{R}_{i0j0} = R_{i0j0}^{TT} + \mathcal{O}(h^2).$$
(1.77)

Now we make use of the fact that in the TT coordinates $\bar{h}_{0\mu}^{TT} = 0$ [see Eq. (1.61)], then from the general formula (1.11) for the linearized Riemann tensor we get [remembering that in the TT coordinate system $\bar{h}_{\mu\nu}^{TT} = h_{\mu\nu}^{TT}$, Eq. (1.60)]

$$R_{i0j0}^{TT} = -\frac{1}{2c^2} \frac{\partial^2 h_{ij}^{TT}}{\partial t^2} + \mathcal{O}(h^2) = -\frac{1}{2c^2} \frac{\partial^2 h_{ij}^{TT}}{\partial \hat{t}^2} + \mathcal{O}(h^2).$$
(1.78)

In the derivation of the above relation we have not needed to assume that the wave is propagating in the $+z$ direction of the TT coordinate system, thus this relation is valid for the wave propagating in any direction. Collecting Eqs. (1.76)–(1.78) together, after neglecting terms $\mathcal{O}(h^2)$, one obtains

$$\frac{d^2 \hat{x}^i}{d\hat{t}^2} = \frac{1}{2} \frac{\partial^2 h_{ij}^{TT}}{\partial \hat{t}^2} \hat{x}^j,$$
(1.79)

where the second time derivative $\partial^2 h_{ij}^{TT}/\partial \hat{t}^2$ is to be evaluated along the basic geodesic $\hat{x} = \hat{y} = \hat{z} = 0$.

Let us now imagine that for times $\hat{t} \leq 0$ there were no waves $(h_{ij}^{TT} = 0)$ in the vicinity of the two observers and that the observers were at rest with respect to each other before the wave has come, so

$$\hat{x}^i(\hat{t}) = \hat{x}_0^i = \text{const.}, \qquad \frac{d\hat{x}^i}{d\hat{t}}(\hat{t}) = 0, \qquad \text{for } \hat{t} \leq 0.$$
(1.80)

At $\hat{t} = 0$ some wave arrives. We expect that $\hat{x}^i(\hat{t}) = \hat{x}_0^i + \mathcal{O}(h)$ for $\hat{t} > 0$, therefore, because we neglect terms $\mathcal{O}(h^2)$, Eq. (1.79) can be replaced by

$$\frac{d^2 \hat{x}^i}{d\hat{t}^2} = \frac{1}{2} \frac{\partial^2 h_{ij}^{TT}}{\partial \hat{t}^2} \hat{x}_0^i.$$
(1.81)

Making use of the initial conditions (1.80) we integrate (1.81). The result is

$$\hat{x}^i\left(\hat{t}\right) = \left(\delta_{ij} + \frac{1}{2}h_{ij}^{\mathrm{TT}}\left(\hat{t}\right)\right)\hat{x}_0^i, \quad \hat{t} > 0. \tag{1.82}$$

Let us orient the spatial axes of the proper reference frame such that the wave is propagating in the $+\hat{z}$ direction. Then we can employ Eq. (1.68) (where we put $z = 0$ and replace t by $\hat{t}$) to write Eqs. (1.82) in the more explicit form

$$\hat{x}\left(\hat{t}\right) = \hat{x}_0 + \frac{1}{2}\left(h_+\left(\hat{t}\right)\hat{x}_0 + h_\times\left(\hat{t}\right)\hat{y}_0\right), \tag{1.83a}$$

$$\hat{y}\left(\hat{t}\right) = \hat{y}_0 + \frac{1}{2}\left(h_\times\left(\hat{t}\right)\hat{x}_0 - h_+\left(\hat{t}\right)\hat{y}_0\right), \tag{1.83b}$$

$$\hat{z}\left(\hat{t}\right) = \hat{z}_0. \tag{1.83c}$$

Equations (1.83) show that the gravitational wave is *transverse*: it produces relative displacements of the test particles only in the plane perpendicular to the direction of the wave propagation.

We will study in more detail the effect of the gravitational wave on the cloud of freely falling particles. To do this let us imagine that the basic observer is checking the presence of the wave by observing some neighboring particles that form, before the wave arrives, a perfect ring in the $(\hat{x}, \hat{y})$ plane. Let the radius of the ring be r_0 and the center of the ring coincides with the origin of the observer's proper reference frame. Then the coordinates of any particle in the ring can be parametrized by some angle $\phi \in [0, 2\pi]$ such that they are equal

$$\hat{x}_0 = r_0 \cos\phi, \quad \hat{y}_0 = r_0 \sin\phi, \quad \hat{z}_0 = 0. \tag{1.84}$$

Equations (1.83) and (1.84) imply that in the field of a gravitational wave the $\hat{z}$ coordinates of all the ring's particles remain equal to zero: $\hat{z}\left(\hat{t}\right) = 0$, so only $\hat{x}$ and $\hat{y}$ coordinates of the particles should be analyzed.

1.4.1 Plus polarization

The gravitational wave is in the plus mode when $h_\times = 0$. Then, making use of Eqs. (1.83) and (1.84), one gets

$$\hat{x}\left(\hat{t}\right) = r_0 \cos\phi\left(1 + \frac{1}{2}h_+\left(\hat{t}\right)\right), \quad \hat{y}\left(\hat{t}\right) = r_0 \sin\phi\left(1 - \frac{1}{2}h_+\left(\hat{t}\right)\right). \tag{1.85}$$

Initially, before the wave arrives, the ring of the particles is perfectly circular. Does this shape change after the wave has arrived? One can

Fig. 1.1.   The effect of a plane monochromatic gravitational wave with $+$ polarization on a circle of test particles placed in a plane perpendicular to the direction
of the wave propagation. The plots show deformation of the circle measured in
the proper reference frame of the central particle at the instants of time equal
to $nT/4$ $(n = 0, 1, 2, \ldots)$, where T is the period of the gravitational wave.

treat Eqs. (1.85) as parametric equations of a certain curve, with ϕ being
the parameter. It is easy to combine the two equations (1.85) such that
ϕ is eliminated. The resulting equation reads

$$\frac{\hat{x}^2}{\left(a_+(\hat{t})\right)^2} + \frac{\hat{y}^2}{\left(b_+(\hat{t})\right)^2} = 1, \tag{1.86}$$

where

$$a_+(\hat{t}) := r_0\left(1 + \frac{1}{2}h_+(\hat{t})\right), \quad b_+(\hat{t}) := r_0\left(1 - \frac{1}{2}h_+(\hat{t})\right). \tag{1.87}$$

Equations (1.86)–(1.87) describe an ellipse with its center at the origin of
the coordinate system. The ellipse has semi-axes of the lengths $a_+(\hat{t})$ and
$b_+(\hat{t})$, which are parallel to the $\hat{x}$ or $\hat{y}$ axis, respectively. If $h_+(\hat{t})$ is the
oscillatory function, which changes its sign in time, then the deformation
of the initial circle into the ellipse is the following: in time intervals when
$h_+(\hat{t}) > 0$, the circle is stretched in the $\hat{x}$ direction and squeezed in the $\hat{y}$
direction, whereas when $h_+(\hat{t}) < 0$, the stretching is along the $\hat{y}$ axis and
the squeezing is along the $\hat{x}$ axis. This is illustrated in Fig. 1.1.

Let us now fix a single particle in the ring. The motion of this particle with respect to the origin of the proper reference frame is given by
Eqs. (1.85), for a fixed value of ϕ. What is the shape of the particle's
trajectory? It is again easy to combine Eqs. (1.85) in such a way that the
function $h_+(\hat{t})$ is eliminated. The result is

$$\frac{\hat{x}}{r_0\cos\phi} + \frac{\hat{y}}{r_0\sin\phi} - 2 = 0. \tag{1.88}$$

Equation (1.88) means that any single particle in the ring is moving
around its initial position along some straight line.

1.4.2 Cross polarization

The gravitational wave is in the crosss mode when $h_+ = 0$. Then from Eqs. (1.83) and (1.84) one gets

$$\hat{x}(\hat{t}) = r_0\left(\cos\phi + \frac{1}{2}\sin\phi\, h_\times(\hat{t})\right), \quad \hat{y}(\hat{t}) = r_0\left(\sin\phi + \frac{1}{2}\cos\phi\, h_\times(\hat{t})\right).$$

$$(1.89)$$

Let us now introduce in the $(\hat{x}, \hat{y})$ plane some new coordinates $(\hat{x}', \hat{y}')$. The new coordinates one gets from the old ones by rotation around the $\hat{z}$ axis by the angle of $\alpha = 45°$. Both coordinate systems are related to each other by the rotation matrix,

$$\begin{pmatrix} \hat{x}' \\ \hat{y}' \end{pmatrix} = \begin{pmatrix} \cos\alpha & \sin\alpha \\ -\sin\alpha & \cos\alpha \end{pmatrix} \begin{pmatrix} \hat{x} \\ \hat{y} \end{pmatrix} = \frac{\sqrt{2}}{2}\begin{pmatrix} 1 & 1 \\ -1 & 1 \end{pmatrix}\begin{pmatrix} \hat{x} \\ \hat{y} \end{pmatrix}. \qquad (1.90)$$

It is easy to rewrite Eqs. (1.89) in terms of the coordinates $(\hat{x}', \hat{y}')$. The result is

$$\hat{x}'(\hat{t}) = \frac{\sqrt{2}}{2}r_0(\sin\phi + \cos\phi)\left(1 + h_\times(\hat{t})\right), \qquad (1.91a)$$

$$\hat{y}'(\hat{t}) = \frac{\sqrt{2}}{2}r_0(\sin\phi - \cos\phi)\left(1 - h_\times(\hat{t})\right). \qquad (1.91b)$$

After eliminating from Eqs. (1.91) the parameter ϕ, one gets

$$\frac{\hat{x}'^2}{\left(a_\times(\hat{t})\right)^2} + \frac{\hat{y}'^2}{\left(b_\times(\hat{t})\right)^2} = 1, \qquad (1.92)$$

where

$$a_\times(\hat{t}) := r_0\left(1 + \frac{1}{2}h_\times(\hat{t})\right), \quad b_\times(\hat{t}) := r_0\left(1 - \frac{1}{2}h_\times(\hat{t})\right). \qquad (1.93)$$

Equations (1.92)–(1.93) have exactly the form of Eqs. (1.86)–(1.87). This means that the initial circle of particles is deformed into an ellipse with its center at the origin of the coordinate system. The ellipse has semi-axes of the lengths $a_\times(\hat{t})$ and $b_\times(\hat{t})$, which are inclined by an angle of $45°$ to the $\hat{x}$ or $\hat{y}$ axis, respectively. This is shown in Fig. 1.2.

1.5 Gravitational waves in the curved background

So far we have considered gravitational waves in such regions of spacetime, where the waves are the only non-negligible source of spacetime curvature. If there exist other sources it is not possible in a fully precise manner, because of the non-linear nature of relativistic gravity, to separate the

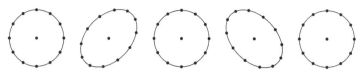

Fig. 1.2.   The effect of a plane monochromatic gravitational wave with × polarization on a circle of test particles placed in a plane perpendicular to the direction of the wave propagation. The plots show deformation of the circle measured in the proper reference frame of the central particle at the instants of time equal to $nT/4$ ($n = 0, 1, 2, \ldots$), where T is the period of the gravitational wave.

contribution of a gravitational wave to the spacetime curvature from the contributions of the other sources. Such separation can only be made approximately. We describe now a method of defining a gravitational wave that is a special case of a standard technique in mathematical physics called (among other names) *shortwave approximation*.

Let us consider a gravitational wave with a wavelength λ. This wave creates spacetime curvature that varies on the scale of the order of the reduced wavelength $\lambdabar$ of the wave, where

$$\lambdabar := \frac{\lambda}{2\pi}. \tag{1.94}$$

In many realistic astrophysical situations the lengthscale $\lambdabar$ is very short compared to lengthscales $\mathcal{L}$ on which all other non-gravitational-wave curvatures vary:

$$\lambdabar \ll \mathcal{L}. \tag{1.95}$$

This inequality allows one to split the full Riemann curvature tensor $R_{\alpha\beta\gamma\delta}$ into a background curvature $R^{\mathrm{b}}_{\alpha\beta\gamma\delta}$ and a gravitational-wave produced part $R^{\mathrm{gw}}_{\alpha\beta\gamma\delta}$. The background curvature $R^{\mathrm{b}}_{\alpha\beta\gamma\delta}$ is the average of the full Riemann tensor $R_{\alpha\beta\gamma\delta}$ over several gravitational-wave wavelengths

$$R^{\mathrm{b}}_{\alpha\beta\gamma\delta} := \langle R_{\alpha\beta\gamma\delta} \rangle, \tag{1.96}$$

whereas the gravitational-wave curvature $R^{\mathrm{gw}}_{\alpha\beta\gamma\delta}$ is the rapidly varying difference:

$$R^{\mathrm{gw}}_{\alpha\beta\gamma\delta} := R_{\alpha\beta\gamma\delta} - R^{\mathrm{b}}_{\alpha\beta\gamma\delta}. \tag{1.97}$$

It is possible to introduce a TT coordinate system for the gravitational wave propagating in the curved background. In this coordinate system the spacetime metric is nearly Minkowskian and can be written in the form

$$g_{\alpha\beta} = \eta_{\alpha\beta} + h^{\mathrm{b}}_{\alpha\beta} + h^{\mathrm{TT}}_{\alpha\beta}, \tag{1.98}$$

where $h^{\rm b}_{\alpha\beta}$ ($|h^{\rm b}_{\alpha\beta}| \ll 1$) is the background metric perturbation that varies on a long lengthscale $\mathcal{L}$, and $h^{\rm TT}_{\alpha\beta}$ ($|h^{\rm TT}_{\alpha\beta}| \ll 1$) is the gravity-wave metric perturbation that varies on a short lengthscale λ. The time–time $h^{\rm TT}_{00}$ and the space–time $h^{\rm TT}_{0i} = h^{\rm TT}_{i0}$ components of the gravity-wave perturbation vanish in the TT coordinate system, and if the wave propagates in the $+z$ direction, then the metric perturbation $h^{\rm TT}_{\alpha\beta}$ may be written in the form given in Eq. (1.68).

The extent of the TT coordinates in both time and space must be far smaller than the radius of background curvature. In typical astrophysical situations one can stretch TT coordinates over any small region compared to the distance at which curvature of the cosmological background of our universe becomes important ("the Hubble distance"), cutting out holes in the vicinities of black holes and neutron stars.

1.6 Energy–momentum tensor for gravitational waves

A fully satisfactory mathematical description of the energy carried by a gravitational wave was devised by Isaacson [39, 40], who introduced an energy–momentum tensor for gravitational waves. This tensor is obtained by averaging the squared gradient of the wave field over several wavelengths. In the TT gauge it has components

$$T^{\rm gw}_{\alpha\beta} = \frac{c^4}{32\pi G} \eta^{\mu\nu}\eta^{\rho\sigma} \left\langle \partial_\alpha h^{\rm TT}_{\mu\rho} \, \partial_\beta h^{\rm TT}_{\nu\sigma} \right\rangle. \tag{1.99}$$

The gravitational-wave energy–momentum tensor $T^{\rm gw}_{\alpha\beta}$, like the background curvature, is smooth on the lengthscale λ. If one additionally assumes that $h^{\rm TT}_{0\mu} = 0$, then Eq. (1.99) reduces to

$$T^{\rm gw}_{\alpha\beta} = \frac{c^4}{32\pi G} \sum_{i=1}^{3}\sum_{j=1}^{3} \left\langle \partial_\alpha h^{\rm TT}_{ij} \, \partial_\beta h^{\rm TT}_{ij} \right\rangle. \tag{1.100}$$

For the plane gravitational wave propagating in the $+z$ direction, the tensor $T^{\rm gw}_{\alpha\beta}$ takes the standard form for a bundle of zero-rest-mass particles moving at the speed of light in the $+z$ direction, which can be immediately demonstrated by means of Eqs. (1.68) and (1.100):

$$T^{\rm gw}_{00} = -T^{\rm gw}_{0z} = -T^{\rm gw}_{z0} = T^{\rm gw}_{zz} = \frac{c^2}{16\pi G} \left\langle \left(\frac{\partial h_+}{\partial t}\right)^2 + \left(\frac{\partial h_\times}{\partial t}\right)^2 \right\rangle \tag{1.101}$$

(all other components are equal to zero).

The energy–momentum tensor for gravitational waves defined in Eq. (1.99) has the same properties and plays the same role as the energy–momentum tensor for any other field in the background spacetime. It generates background curvature through the Einstein field equations (averaged over several wavelengths of the waves); it has vanishing divergence in regions where there is no wave generation, absorption, and scattering.

Let us compute the components of the energy–momentum tensor for the monochromatic plane wave with angular frequency ω. The gravitational-wave polarization functions h_+ and $h_\times$ for such a wave are given in Eqs. (1.66). Making use of them, from Eqs. (1.101) one gets

$$T_{00}^{\text{gw}} = -T_{0z}^{\text{gw}} = -T_{z0}^{\text{gw}} = T_{zz}^{\text{gw}} = \frac{c^2 \omega^2}{16\pi G}$$

$$\times \left(A_+^2 \left\langle \sin^2\left(\omega\left(t - \frac{z}{c}\right) + \alpha_+\right)\right\rangle + A_\times^2 \left\langle \sin^2\left(\omega\left(t - \frac{z}{c}\right) + \alpha_\times\right)\right\rangle \right).$$

$$(1.102)$$

Averaging the sine squared terms over one wavelength or one wave period gives $1/2$. After substituting this into (1.102), and replacing ω by the frequency $f = \omega/(2\pi)$ measured in hertz, one obtains

$$T_{00}^{\text{gw}} = -T_{0z}^{\text{gw}} = -T_{z0}^{\text{gw}} = T_{zz}^{\text{gw}} = \frac{\pi c^2 f^2}{8G}\left(A_+^2 + A_\times^2\right). \qquad (1.103)$$

Typical gravitational waves that we might expect to observe at Earth have frequencies between 10^{-4} and 10^4 Hz, and amplitudes of the order of $A_+ \sim A_\times \sim 10^{-22}$. The energy flux in the $+z$ direction for such waves can thus be estimated as

$$-T_{tz}^{\text{gw}} = -c\,T_{0z}^{\text{gw}} = 1.6 \times 10^{-6} \left(\frac{f}{1\,\text{Hz}}\right)^2 \frac{A_+^2 + A_\times^2}{(10^{-22})^2} \frac{\text{erg}}{\text{cm}^2\,\text{s}}.$$

1.7 Generation of gravitational waves and radiation reaction

Quadrupole formalism. The simplest technique for computing the gravitational-wave field $h_{\mu\nu}^{\text{TT}}$ is delivered by the famous *quadrupole formalism*. This formalism is especially important because it is highly accurate for many astrophysical sources of gravitational waves. It does not require, for high accuracy, any constraint on the strength of the source's internal gravity, but requires that internal motions inside the source are slow. This requirement implies that *the source's size L must be small compared to the reduced wavelength λ̄ of the gravitational waves it emits.*

Let us introduce a coordinate system (t, x_{s}^i) *centered on a gravitational-wave source* and let an observer *at rest* with respect to the source measure the gravitational-wave field $h_{\mu\nu}^{\mathrm{TT}}$ generated by that source. Let us further assume that the observer is situated within the *local wave zone* of the source, where the background curvature both of the source and of the external universe can be neglected. It implies (among other things) that the distance from the observer to the source is very large compared to the source's size. Then the quadrupole formalism allows one to write the gravitational-wave field in the following form:

$$h_{0\mu}^{\mathrm{TT}}(t, x_{\mathrm{s}}^i) = 0, \quad h_{ij}^{\mathrm{TT}}(t, x_{\mathrm{s}}^i) = \frac{2G}{c^4} \frac{1}{R} \frac{\mathrm{d}^2 \mathcal{J}_{ij}^{\mathrm{TT}}}{\mathrm{d}t^2} \left(t - \frac{R}{c}\right), \quad (1.104)$$

where $R := \sqrt{\delta_{ij} x_{\mathrm{s}}^i x_{\mathrm{s}}^j}$ is the distance from the point (x_{s}^i) where the gravitational-wave field is observed to the source's center, t is proper time measured by the observer, and $t - R/c$ is retarded time. The quantity $\mathcal{J}_{ij}$ is the source's *reduced mass quadrupole moment* (which we define below), and the superscript TT at $\mathcal{J}_{ij}$ means *algebraically project out and keep only the part that is transverse to the direction in which wave propagates and is traceless*. Quite obviously Eqs. (1.104) describe a *spherical* gravitational wave generated by the source located at the origin of the spatial coordinates.

Let $n^i := x_{\mathrm{s}}^i/R$ be the unit vector in the direction of wave propagation and let us define the projection operator $\mathcal{P}_j^i$ that projects 3-vectors to a 2-plane orthogonal to n^i,

$$\mathcal{P}_j^i := \delta_j^i - n^i n_j. \quad (1.105)$$

Then the TT part of the reduced mass quadrupole moment can be computed as (see Box 35.1 of [34])

$$\mathcal{J}_{ij}^{\mathrm{TT}} = \mathcal{P}_i^k \, \mathcal{P}_j^l \, \mathcal{J}_{kl} - \frac{1}{2} \mathcal{P}_{ij} \left(\mathcal{P}^{kl} \mathcal{J}_{kl}\right). \quad (1.106)$$

For the wave propagating in the $+z$ direction the unit vector n^i has components

$$n^x = n^y = 0, \quad n^z = 1. \quad (1.107)$$

Making use of Eqs. (1.105)–(1.106) one can then easily compute the TT projection of the reduced mass quadrupole moment. The result is

$$\mathcal{J}_{xx}^{\mathrm{TT}} = -\mathcal{J}_{yy}^{\mathrm{TT}} = \frac{1}{2}(\mathcal{J}_{xx} - \mathcal{J}_{yy}), \quad (1.108a)$$

$$\mathcal{J}_{xy}^{\mathrm{TT}} = \mathcal{J}_{yx}^{\mathrm{TT}} = \mathcal{J}_{xy}, \quad (1.108b)$$

$$\mathcal{J}_{zi}^{\mathrm{TT}} = \mathcal{J}_{iz}^{\mathrm{TT}} = 0 \quad \text{for} \quad i = x, y, z. \quad (1.108c)$$

Making use of these equations and the notation introduced in Eq. (1.64) one can write the following formulae for the plus and the cross polarizations of the wave progagating in the $+z$ direction of the coordinate system:

$$h_+(t, x_s^i) = \frac{G}{c^4 R}\left(\frac{\mathrm{d}^2 \mathcal{J}_{xx}}{\mathrm{d}t^2}\left(t - \frac{R}{c}\right) - \frac{\mathrm{d}^2 \mathcal{J}_{yy}}{\mathrm{d}t^2}\left(t - \frac{R}{c}\right)\right), \qquad (1.109\text{a})$$

$$h_\times(t, x_s^i) = \frac{2G}{c^4 R}\frac{\mathrm{d}^2 \mathcal{J}_{xy}}{\mathrm{d}t^2}\left(t - \frac{R}{c}\right). \qquad (1.109\text{b})$$

Polarization waveforms in the SSB reference frame. Let us now introduce another coordinate system (t, x^i) about which we assume that some solar-system-related observer measures the gravitational-wave field $h_{\mu\nu}^{\text{TT}}$ in the neighborhood of its spatial origin. In the following chapters of the book, where we discuss the detection of gravitational waves by solar-system-based detectors, the origin of these coordinates will be located at the *solar system barycenter* (SSB). Let us denote by $\mathbf{x}^*$ the constant 3-vector joining the origin of our new coordinates (x^i) (i.e. the SSB) with the origin of the (x_s^i) coordinates (i.e. the "center" of the source). We also assume that the spatial axes in both coordinate systems are parallel to each other, i.e. the coordinates x^i and x_s^i just differ by constant shifts determined by the components of the 3-vector $\mathbf{x}^*$,

$$x^i = x^{*i} + x_s^i. \qquad (1.110)$$

It means that in both coordinate systems the components $h_{\mu\nu}^{\text{TT}}$ of the gravitational-wave field are numerically the same.

Equations (1.108) and (1.109) are valid only when the z axis of the TT coordinate system is parallel to the 3-vector $\mathbf{x}^* - \mathbf{x}$ joining the observer located at $\mathbf{x}$ ($\mathbf{x}$ is the 3-vector joining the SSB with the observer's location) and the gravitational-wave source at the position $\mathbf{x}^*$. If one changes the location $\mathbf{x}$ of the observer, one has to rotate the spatial axes to ensure that Eqs. (1.108) and (1.109) are still valid. Let us now *fix*, in the whole region of interest, the direction of the $+z$ axis of both the coordinate systems considered here, by choosing it to be antiparallel to the 3-vector $\mathbf{x}^*$ (so for the observer located at the SSB the gravitational wave propagates along the $+z$ direction). To a very good accuracy one can assume that the size of the region where the observer can be located is very small compared to the distance from the SSB to the gravitational-wave source, which is equal to $r^* := |\mathbf{x}^*|$. Our assumption thus means that $r \ll r^*$, where $r := |\mathbf{x}|$. Then $\mathbf{x}^* = (0, 0, -r^*)$ and the Taylor expansion of $R = |\mathbf{x} - \mathbf{x}^*|$

and $\mathbf{n} := (\mathbf{x} - \mathbf{x}^*)/R$ around $\mathbf{x}^*$ reads

$$|\mathbf{x} - \mathbf{x}^*| = r^* \left(1 + \frac{z}{r^*} + \frac{x^2 + y^2}{2(r^*)^2} + \mathcal{O}((x^l/r^*)^3) \right), \tag{1.111a}$$

$$\mathbf{n} = \left(-\frac{x}{r^*} + \frac{xz}{2(r^*)^2}, -\frac{y}{r^*} + \frac{yz}{2(r^*)^2}, 1 - \frac{x^2 + y^2}{2(r^*)^2} \right) + \mathcal{O}((x^l/r^*)^3). \tag{1.111b}$$

Making use of Eqs. (1.105)–(1.106) and (1.111b) one can compute the TT projection of the reduced mass quadrupole moment at the point $\mathbf{x}$ in the direction of the unit vector $\mathbf{n}$ (which, in general, is not parallel to the $+z$ axis). One gets

$$\mathcal{I}_{xx}^{\mathrm{TT}} = -\mathcal{I}_{yy}^{\mathrm{TT}} = \frac{1}{2}(\mathcal{I}_{xx} - \mathcal{I}_{yy}) + \mathcal{O}(x^l/r^*), \tag{1.112a}$$

$$\mathcal{I}_{xy}^{\mathrm{TT}} = \mathcal{I}_{yx}^{\mathrm{TT}} = \mathcal{I}_{xy} + \mathcal{O}(x^l/r^*), \tag{1.112b}$$

$$\mathcal{I}_{zi}^{\mathrm{TT}} = \mathcal{I}_{iz}^{\mathrm{TT}} = \mathcal{O}(x^l/r^*) \quad \text{for} \quad i = x, y, z. \tag{1.112c}$$

To obtain the gravitational-wave field $h_{\mu\nu}^{\mathrm{TT}}$ in the coordinate system (t, x^i) one should plug Eqs. (1.112) into the formula (1.104). It is clear that if one neglects in Eqs. (1.112) the terms of the order of x^l/r^*, then in the whole region of interest covered by the *single* TT coordinate system, the gravitational-wave field $h_{\mu\nu}^{\mathrm{TT}}$ can be written in the form

$$h_{\mu\nu}^{\mathrm{TT}}(t, \mathbf{x}) = h_+(t, \mathbf{x}) \, e_{\mu\nu}^+ + h_\times(t, \mathbf{x}) \, e_{\mu\nu}^\times + \mathcal{O}(x^l/r^*), \tag{1.113}$$

where the polarization tensors e^+ and $e^\times$ are defined in Eq. (1.69), and the functions h_+ and $h_\times$ are of the form given in Eqs. (1.109).

Dependence of the $1/R$ factors in the *amplitudes* of the wave polarizations (1.109) on the observer's position $\mathbf{x}$ (with respect to the SSB) is usually negligible, so $1/R$ in the amplitudes can be replaced by $1/r^*$ [this is consistent with the neglection of the $\mathcal{O}(x^l/r^*)$ terms we have just made in Eqs. (1.112)]. This is not the case for the second time derivative of $\mathcal{I}_{ij}^{\mathrm{TT}}$ in (1.109), which determines the time evolution of the wave polarization *phases* and which is evaluated at the retarded time $t - R/c$. Here it is usually enough to take into account the first correction to r^* given by Eq. (1.111a). After taking all this into account the wave polarizations

(1.109) take the form

$$h_+(t, x^i) = \frac{G}{c^4\, r^*}\left(\frac{\mathrm{d}^2\mathcal{J}_{xx}}{\mathrm{d}t^2}\left(t - \frac{z + r^*}{c}\right) - \frac{\mathrm{d}^2\mathcal{J}_{yy}}{\mathrm{d}t^2}\left(t - \frac{z + r^*}{c}\right)\right), \quad (1.114a)$$

$$h_\times(t, x^i) = \frac{2G}{c^4\, r^*}\frac{\mathrm{d}^2\mathcal{J}_{xy}}{\mathrm{d}t^2}\left(t - \frac{z + r^*}{c}\right). \quad (1.114b)$$

The wave polarization functions (1.114) depend on the spacetime coordinates (t, x^i) only through the combination $t - z/c$, so they represent a *plane* gravitational wave propagating in the $+z$ direction.

Mass quadrupole moment of the source.   We shift now to the definition of the source's mass quadrupole moment $\mathcal{J}_{ij}$. We restrict only to situations when the source has weak internal gravity and small internal stresses, so Newtonian gravity is a good approximation to general relativity inside and near the source. Then $\mathcal{J}_{ij}$ is the symmetric and trace-free (STF) part of the second moment of the source's mass density ρ computed in a Cartesian coordinate system centered on the source:

$$\mathcal{J}_{ij}(t) := \left(\int \rho(x^k, t) x^i x^j \,\mathrm{d}^3 x\right)^{\mathrm{STF}} = \int \rho(x^k, t)\left(x^i x^j - \frac{1}{3}r^2\delta_{ij}\right)\mathrm{d}^3 x.$$

$$(1.115)$$

Equivalently, $\mathcal{J}_{ij}$ is the coefficient of the $1/r^3$ term in the multipolar expansion of the source's Newtonian gravitational potential Φ,

$$\Phi(t, x^k) = -\frac{GM}{r} - \frac{3G}{2}\frac{\mathcal{J}_{ij}(t) x^i x^j}{r^5} - \frac{5G}{2}\frac{\mathcal{J}_{ijk}(t) x^i x^j x^k}{r^7} + \cdots. \quad (1.116)$$

Gravitational-wave luminosities.   From the quadrupole formula (1.104) and Isaacson's formula (1.100) for the energy–momentum tensor of the gravitational waves, one can compute the fluxes of energy and angular momentum carried by the waves. After integrating these fluxes over a sphere surrounding the source in the local wave zone one obtains the rates $\mathcal{L}_E^{\mathrm{gw}}$ and $\mathcal{L}_{J_i}^{\mathrm{gw}}$ of emission respectively of energy and angular momentum:

$$\mathcal{L}_E^{\mathrm{gw}} = \frac{G}{5c^5}\sum_{i=1}^{3}\sum_{j=1}^{3}\left\langle\left(\frac{\mathrm{d}^3\mathcal{J}_{ij}}{\mathrm{d}t^3}\right)^2\right\rangle, \quad (1.117a)$$

$$\mathcal{L}_{J_i}^{\mathrm{gw}} = \frac{2G}{5c^5}\sum_{j=1}^{3}\sum_{k=1}^{3}\sum_{\ell=1}^{3}\epsilon_{ijk}\left\langle\frac{\mathrm{d}^2\mathcal{J}_{j\ell}}{\mathrm{d}t^2}\frac{\mathrm{d}^3\mathcal{J}_{k\ell}}{\mathrm{d}t^3}\right\rangle. \quad (1.117b)$$

Formula (1.117a) was first derived by Einstein [41, 42], whereas formula (1.117b) was discovered by Peters [43].

The laws of conservation of energy and angular momentum imply that *radiation reaction* should decrease the source's energy and angular momentum at rates just equal to minus rates given by Eqs. (1.117),

$$\frac{\mathrm{d}E^{\mathrm{source}}}{\mathrm{d}t} = -\mathcal{L}_E^{\mathrm{gw}}, \tag{1.118a}$$

$$\frac{\mathrm{d}J_i^{\mathrm{source}}}{\mathrm{d}t} = -\mathcal{L}_{J_i}^{\mathrm{gw}}. \tag{1.118b}$$

Justification of the validity of the *balance equations* [examples of which are Eqs. (1.118)] in general relativity is thorougly discussed e.g. in Section 6.15 of Ref. [44] (see also references therein).

2

Astrophysical sources of gravitational waves

It is convenient to split the expected astrophysical sources of gravitational waves into three main categories, according to the temporal behavior of the waveforms they produce: burst, periodic, and stochastic sources. In Sections 2.1–2.3 of the present chapter we enumerate some of the most typical examples of gravitational-wave sources from these categories (more detailed reviews can be found in [45], Section 9.4 of [16], and [46, 47]). Many sources of potentially detectable gravitational waves are related to compact astrophysical objects: white dwarfs, neutron stars, and black holes. The physics of compact objects is thoroughly studied in the monograph [48].

In the rest of the chapter we will perform more detailed studies of gravitational waves emitted by several important astrophysical sources. In Section 2.4 we derive gravitational-wave polarization functions h_+ and $h_\times$ for different types of waves emitted by binary systems. As an example of periodic waves we consider, in Section 2.5, gravitational waves coming from a triaxial ellipsoid rotating along a principal axis; we derive the functions h_+ and $h_\times$ for these waves. In Section 2.6 we relate the amplitude of gravitational waves emitted during a supernova explosion with the total energy released in gravitational waves and with the time duration and the frequency bandwidth of the gravitational-wave pulse. Finally in Section 2.7 we express the frequency dependence of the energy density of stationary, isotropic, and unpolarized stochastic background of gravitational waves in terms of their spectral density function.

2.1 Burst sources

2.1.1 Coalescing compact binaries

Binary systems consisting of any objects radiate gravitational waves and as a result of radiation reaction the distance between the components of the binary decreases. This results in a sinusoidal signal whose amplitude and frequency increases with time and which is called a *chirp*.

For the binary of circular orbits the characteristic dimensionless amplitude h_0 of the two gravitational-wave polarizations [see Eq. (2.40) and its derivation in Section 2.4 later in this chapter] is equal:

$$ h_0 = 2.6 \times 10^{-23} \left(\frac{\mathcal{M}}{M_\odot} \right)^{5/3} \left(\frac{f_{\mathrm{gw}}}{100 \text{ Hz}} \right)^{2/3} \left(\frac{R}{100 \text{ Mpc}} \right)^{-1}, \qquad (2.1) $$

where $\mathcal{M}$ is the *chirp mass* of the system [see Eq. (2.34) for the definition of $\mathcal{M}$ in terms of the individual masses of the binary components], f_{gw} is the frequency of gravitational waves (which is twice the orbital frequency), and R is the distance to the binary. The characteristic time $\tau_{\mathrm{gw}} := f_{\mathrm{gw}}/\dot{f}_{\mathrm{gw}}$ (where $\dot{f}_{\mathrm{gw}} := \mathrm{d}f_{\mathrm{gw}}/\mathrm{d}t$) for the increase of gravitational-wave frequency f_{gw} is given by [here Eq. (2.38) was employed]

$$ \tau_{\mathrm{gw}} = 8.0\,\mathrm{s} \left(\frac{\mathcal{M}}{M_\odot} \right)^{-5/3} \left(\frac{f_{\mathrm{gw}}}{100 \text{ Hz}} \right)^{-8/3}. \qquad (2.2) $$

Among all coalescing binaries the most interesting are those made of compact objects, neutron stars (NS), or black holes (BH), in the last few minutes of their inspiral. There are three kinds of such compact binaries: NS/NS, NS/BH, and BH/BH binaries. The final merger of the two NS in a NS/NS binary is a promising candidate for the trigger of some types of gamma-ray bursts. At the endpoint of a NS/BH inspiral, a neutron star can be tidally disrupted by its BH companion, and this disruption is another candidate for triggering gamma-ray bursts. For heavier BH/BH binaries, most of the detectable gravitational waves can come from the merger phase of the evolution as well as from the vibrational ringdown of the final BH.

The number densities per unit time of different compact binary coalescences are rather uncertain. Their estimates crucially depend on the event rate $\mathcal{R}_{\mathrm{Gal}}$ of binary mergers in our Galaxy. Recent studies (see Section 2.3 of [47] and references therein) give $10^{-6}\,\mathrm{yr}^{-1} \lesssim \mathcal{R}_{\mathrm{Gal}} \lesssim 5 \times 10^{-4}\,\mathrm{yr}^{-1}$ for NS/NS mergers, and $10^{-7}\,\mathrm{yr}^{-1} \lesssim \mathcal{R}_{\mathrm{Gal}} \lesssim 10^{-4}\,\mathrm{yr}^{-1}$ for NS/BH inspirals. For BH/BH binaries two distinct estimates can be made: one for the binaries not contained in globular and other types of dense star clusters ("in field" binaries), and the other for binaries from these clusters. The

BH/BH event rate estimates are: $10^{-7}\,\mathrm{yr}^{-1} \lesssim \mathcal{R}_{\mathrm{Gal}} \lesssim 10^{-5}\,\mathrm{yr}^{-1}$ for "in field" binaries, and $10^{-6}\,\mathrm{yr}^{-1} \lesssim \mathcal{R}_{\mathrm{Gal}} \lesssim 10^{-5}\,\mathrm{yr}^{-1}$ for binaries in clusters.

2.1.2 Supernovae

Neutron stars and black holes (of stellar masses) form in the gravitational collapse of a massive star, which leads to a supernova type II explosion ("core-collapse" supernova). Because of our incomplete knowledge of the process of collapse (we do not know how non-spherical the collapse might be in a typical supernova) and the diversity of emission mechanisms, we cannot predict the gravitational waveform from this event accurately. A gravitational-wave burst might be rather broad-band with frequency centered on 1 kHz, or it might contain a few cycles of radiation at a frequency anywhere between 100 Hz and 10 kHz, chirping up or down.

The dimensionless amplitude h_0 of the gravitational-wave pulse from supernova explosion can be estimated by [see Eq. (2.77) in Section 2.6]

$$h_0 \simeq 1.4 \times 10^{-21} \left(\frac{\Delta E_{\mathrm{gw}}}{10^{-2}\,M_\odot c^2} \right)^{1/2} \left(\frac{\tau}{1\,\mathrm{ms}} \right)^{-1/2} \left(\frac{\Delta f_{\mathrm{gw}}}{1\,\mathrm{kHz}} \right)^{-1} \left(\frac{R}{15\,\mathrm{Mpc}} \right)^{-1},$$

(2.3)

where ΔE_{gw} is the total energy carried away by gravitational waves during the explosion, τ is the duration of the pulse and Δf_{gw} is its frequency bandwidth, R is the distance to the source. The value of ΔE_{gw} is very uncertain, it can differ from the above quoted number (around $10^{-2}\,M_\odot c^2$) by orders of magnitude. We expect around ten such sources per year in the Virgo cluster of galaxies (15 Mpc is the distance to the center of the Virgo cluster).

For a recent review of the theory of core-collapse supernovae see [49], and recent studies of gravitational waves emitted during the core-collapse supernova can be found in [50, 51], see also the review article [52].

2.2 Periodic sources

The primary example of sources of periodic gravitational waves are spinning neutron stars. Because a rotating body, perfectly symmetric around its rotation axis, does not emit gravitational waves, the neutron star will emit waves only if it has some kind of asymmetry. Several mechanisms leading to such an asymmetry have been discovered and studied. These mechanisms include elastic deformations of the star's solid crust (or core) developed during the crystallization period of the crust and supported by anisotropic stresses in it. The strong magnetic field present in the star may not be aligned with the rotation axis, consequently the magnetic pressure distorts the entire star. Some mechanisms result in a triaxial neutron star rotating about a principal axis. Detailed computations of gravitational waves emitted by a triaxial ellipsoid rotating about a principal axis are presented in Section 2.5 later in the current chapter.

The dimensionless amplitude of the gravitational waves emitted by a rotating neutron star can be estimated by [this is Eq. (2.67) in which the physical constants are replaced by their numerical values]

$$h_0 = 4.2 \times 10^{-25} \frac{\varepsilon}{10^{-5}} \frac{\mathcal{I}_{\mathrm{s}}^{zz}}{10^{45}\,\mathrm{g\,cm^2}} \left(\frac{f_0}{100\,\mathrm{Hz}}\right)^2 \left(\frac{R}{10\,\mathrm{kpc}}\right)^{-1}, \qquad (2.4)$$

where ε is the star's ellipticity [defined in Eq. (2.63)], $\mathcal{I}_{\mathrm{s}}^{zz}$ is its moment of inertia around the rotation axis, f_0 is the rotational frequency of the star, and R is the distance to the star.

The LIGO Scientific Collaboration has recently imposed, using data from LIGO detectors, a non-trivial upper limit on h_0 for the Crab pulsar (PSR B0531+21 or PSR J0534+2200). The upper limit is $h_0^{95\%} = 3.4 \times 10^{-25}$ [53], where $h_0^{95\%}$ is the joint 95% upper limit on the gravitational-wave amplitude using uniform priors on all the parameters. This limit is substantially less than the spin-down upper limit $h_0^{\mathrm{sd}} = 1.4 \times 10^{-24}$ that can be inferred assuming that all the energy radiated by the Crab pulsar is due to gravitational-wave emission. This result assumes that the Crab's spin frequency $f_0 = 29.78\,\mathrm{Hz}$, spin-down rate $\dot{f}_0 = -3.7 \times 10^{-10}\,\mathrm{Hz\,s^{-1}}$, principal moment of inertia $\mathcal{I}_{\mathrm{s}}^{zz} = 10^{45}\,\mathrm{g\,cm^2}$, and distance $R = 2\,\mathrm{kpc}$. Moreover the analysis assumes that the gravitational-wave emission follows the observed radio timing.

2.3 Stochastic sources

A stochastic background of gravitational radiation arises from an extremely large number of weak, independent, and unresolved gravitational-wave sources. Such backgrounds may arise in the early universe from inflation, phase transitions, or cosmic strings. It may also arise from populations of astrophysical sources (e.g., radiation from many unresolved binary star systems). See Ref. [54] for a comprehensive review of stochastic gravitational-wave sources.

There is a broadband observational constraint on the stochastic background of gravitational waves that comes from a standard model of big-bang nucleosynthesis. This model provides remarkably accurate fits to the observed abundances of the light elements in the universe, tightly constraining a number of key cosmological parameters. One of the parameters constrained in this way is the expansion rate of the universe at the time of nucleosynthesis. This places a constraint on the energy density of the universe at that time, which in turn constrains the energy density in a cosmological background of gravitational radiation. This leads to the following bound [54], which is valid independently of the frequency f (and independently of the actual value of the Hubble expansion rate)

$$h_{100}^2\, \Omega_{\mathrm{gw}}(f) \lesssim 5 \times 10^{-6}, \tag{2.5}$$

where Ω_{gw} is the dimensionless ratio of the gravitational-wave energy density per logarithmic frequency interval to the closure density of the universe [see Eq. (2.88)] and h_{100} is related to the Hubble constant H_0 by

$$H_0 = h_{100} \times 100\, \frac{\mathrm{km\, s}^{-1}}{\mathrm{Mpc}}. \tag{2.6}$$

From observations it follows that h_{100} almost certainly lies in the range $0.6 \lesssim h_{100} \lesssim 0.8$ (see e.g. Ref. [55]). In terms of the nucleosynthesis bound (2.5) we have the following numerical expression for the characteristic dimensionless amplitude h_{c} of the gravitational-wave stochastic background [see Eq. (2.92)]:

$$h_{\mathrm{c}}(f) \simeq 2.8 \times 10^{-23} \left(\frac{h_{100}^2\, \Omega_{\mathrm{gw}}(f)}{5 \times 10^{-6}} \right)^{1/2} \left(\frac{f}{100\,\mathrm{Hz}} \right)^{-1}. \tag{2.7}$$

There are other, tighter constraints on $h_{100}^2\, \Omega_{\mathrm{gw}}(f)$ that come from observed isotropy of the cosmic microwave background and timing of the millisecond pulsars, but they are valid for very small frequencies, well below the bands of the existing and planned gravitational-wave detectors.

2.4 Case study: binary systems

The *relativistic two-body problem*, i.e. the problem of describing the dynamics and gravitational radiation of two extended bodies interacting gravitationally according to general relativity theory, is very difficult (see the review articles [44, 56]). Among all binary systems the most important sources of gravitational waves are binaries made of *compact* objects: white dwarfs, neutron stars, or black holes. Only compact objects can reach separations small enough and relative velocities large enough to enter the essentially relativistic regime in which using Einstein's equations is unavoidable. There are two approaches to studying the two-body problem in its relativistic regime. The first approach is to solve Einstein's equations numerically. Another is to employ an analytic approximation scheme. Among the many approximation schemes that were developed, the most effective approach turned out to be the *post-Newtonian* (PN) one, in which the gravitational field is assumed to be weak and relative velocities of the bodies generating the field are assumed to be small (so it is a weak-field and slow-motion approximation).

We focus here on binary systems made of black holes. The time evolution of a black-hole binary driven by gravitational-wave emission can be split into three stages [57]: adiabatic inspiral, merger or plunge, and ringdown. In the inspiral phase, the orbital period is much shorter than the time scale over which orbital parameters change. This stage can be accurately modeled by PN approximations. During the merger stage gravitational-wave emission is so strong that the evolution of the orbit is no longer adiabatic, and the black holes plunge together to form a single black hole. Full numerical simulations are needed to understand this phase. Finally, in the ringdown phase the gravitational waves emitted can be well modeled by *quasi-normal modes* of the final Kerr black hole.

Only recently there was a remarkable breakthrough in numerical simulations of binary black-hole mergers (see the review article [57]). Here we are more interested in explicit approximate analytical results concerning motion and gravitational radiation of compact binary systems. Such results were obtained by perturbative solving Einstein field equations within the PN approximation of general relativity.

Post-Newtonian approximate results. Post-Newtonian calculations provide equations of motion of binary systems and rates of emission of energy and angular momentum carried by gravitational waves emitted by the binary (gravitational-wave luminosities) in the form of the PN series, i.e. the power series of the ratio v/c, where v is the typical orbital velocity of the binary members. Let us mention that the most higher-order PN explicit results were obtained under the assumption that the binary members can be modeled as *point particles*. Different PN formalisms are presented in Refs. [58, 59, 60, 61, 62, 63, 64, 65, 66, 67].

The PN dynamics of binary systems can be split into a conservative part and a dissipative part connected with the radiation-reaction effects. The conservative dynamics can be derived from an autonomous Hamiltonian. Equations of motion of compact binary systems made of *non-rotating* bodies (which can be modeled as *spinless* point particles) were explicitly derived up to the 3.5PN order, i.e. up to the terms of the order $(v/c)^7$ beyond Newtonian equations of motion. The details of derivations of the most complicated 3PN and 3.5PN contributions to the equations of motion can be found in Refs. [68, 69, 70, 71, 72, 73, 74, 75, 76, 77, 78, 79, 80, 81]. The rate of emission of gravitational-wave energy by binary systems was computed up to the terms of the order $(v/c)^7$ beyond the leading-order quadrupole formula for binaries along *quasi-circular* orbits

[82, 83, 84, 85, 86, 87, 88] and up to the terms of the order $(v/c)^6$ beyond the leading-order contribution for general *quasi-elliptical* orbits [89].

Analytical results concerning compact binaries made of *rotating* bodies were also obtained in the form of a PN expansion. There are various spin-dependent contributions to the PN-expanded equations of motion and gravitational-wave luminosities of a binary system: terms exist that are linear, quadratic, cubic, etc., in the individual spins $\mathbf{S}_1$ and $\mathbf{S}_2$ of the bodies. Gravitational spin-orbit effects (i.e. effects that are linear in the spins) were analytically obtained to the next-to-leading order terms (i.e. one PN order beyond the leading-order contribution) both in the orbital and spin precessional equations of motion [90, 91, 92] and in the rate of emission of gravitational-wave energy [93, 94]. Higher-order in the spins effects (i.e. the effects of the order of $\mathbf{S}_1 \cdot \mathbf{S}_2$, $\mathbf{S}_1^2$, $\mathbf{S}_2^2$, etc.) were recently discussed in Refs. [95, 96, 97, 98, 99].

PN effects modify not only the orbital phase evolution of the binary, but also the amplitudes of the two independent wave polarizations h_+ and $h_\times$. In the case of binaries along quasi-circular orbits the PN amplitude corrections were explicitly computed up to the 3PN order beyond the leading-order formula given in Eqs. (2.26) (see [100] and references therein). For quasi-elliptical orbits the 2PN-accurate corrections to wave polarizations were computed in Ref. [101].

It is not an easy task to obtain the wave polarizations h_+ and $h_\times$ as explicit functions of time, taking into account all known higher-order PN corrections. To do this in the case of quasi-elliptical orbits one has to deal with three different time scales: orbital period, periastron precession, and radiation-reaction time scale. Explicit results concerning the "phasing" of the inspiraling binaries along quasi-elliptical orbits were obtained e.g. in Refs. [102, 103] (see also references therein).

Effective-one-body approach.   We should also mention here the effective-one-body (EOB) formalism, which provides an accurate quasi-analytical description of the motion and radiation of a coalescing black-hole binary at all stages of its evolution, from adiabatic inspiral to ringdown. The "quasi-analytical" means here that to compute a waveform within the EOB approach one needs to solve numerically some explicitly given ordinary differential equations (ODEs). Numerical solving ODEs is extremely fast, contrary to computationally very expensive $(3 + 1)$-dimensional numerical relativity simulations of merging black holes.

The core of the EOB formalism is mapping, through the use of invariant (i.e. gauge-independent) functions, the real two-body problem (two spinning masses orbiting around each other) onto an "effective one-body" problem: one spinless mass moving in some "effective" background metric, which is a deformation of the Kerr metric. The EOB approach was introduced at the 2PN level for non-rotating bodies in [104, 105]. The method was then extended to the 3PN level in [106], and spin effects were included in [107, 108]. See Ref. [109] for a comprehensive introduction to the EOB formalism.

2.4.1 Newtonian binary dynamics

In the rest of this section we consider a binary system made of two bodies with masses m_1 and m_2 (we will always assume $m_1 \geq m_2$). We introduce

$$M := m_1 + m_2, \quad \mu := \frac{m_1 m_2}{M}, \tag{2.8}$$

so M is the total mass of the system and μ is its reduced mass. It is also useful to introduce the dimensionless *symmetric mass ratio*

$$\eta := \frac{m_1 m_2}{M^2} = \frac{\mu}{M}. \tag{2.9}$$

The quantity η satisfies $0 \leq \eta \leq 1/4$, the case $\eta = 0$ corresponds to the test-mass limit and $\eta = 1/4$ describes equal-mass binary. We start from deriving the wave polarization functions h_+ and $h_\times$ for waves emitted by a binary system in the case when the dynamics of the binary can reasonably be described within the Newtonian theory of gravitation.

Let $\mathbf{r}_1$ and $\mathbf{r}_2$ denote the position vectors of the bodies, i.e. the 3-vectors connecting the origin of some reference frame with the bodies. We introduce the *relative* position vector,

$$\mathbf{r}_{12} := \mathbf{r}_1 - \mathbf{r}_2. \tag{2.10}$$

The *center-of-mass* reference frame is defined by the requirement that

$$m_1 \mathbf{r}_1 + m_2 \mathbf{r}_2 = \mathbf{0}. \tag{2.11}$$

Solving Eqs. (2.10) and (2.11) with respect to $\mathbf{r}_1$ and $\mathbf{r}_2$ one gets

$$\mathbf{r}_1 = \frac{m_2}{M} \mathbf{r}_{12}, \quad \mathbf{r}_2 = -\frac{m_1}{M} \mathbf{r}_{12}. \tag{2.12}$$

In the center-of-mass reference frame we introduce the spatial coordinates (x_c, y_c, z_c) such that the total orbital angular momentum vector $\mathbf{J}$ of the binary is directed along the $+z_c$ axis. Then the trajectories of both bodies lie in the (x_c, y_c) plane, so the position vector $\mathbf{r}_a$ of the ath body $(a = 1, 2)$ has components $\mathbf{r}_a = (x_{ca}, y_{ca}, 0)$, and the relative position vector components are $\mathbf{r}_{12} = (x_{c12}, y_{c12}, 0)$, where $x_{c12} := x_{c1} - x_{c2}$ and $y_{c12} := y_{c1} - y_{c2}$. It is convenient to introduce in the coordinate (x_{c12}, y_{c12}) plane the usual polar coordinates (r, ϕ):

$$x_{c12} = r \cos \phi, \quad y_{c12} = r \sin \phi. \tag{2.13}$$

Within Newtonian gravity the orbit of the relative motion is an ellipse (here we consider only gravitationally bound binaries). We place the focus of the ellipse at the origin of the (x_{c12}, y_{c12}) coordinates. In polar coordinates the ellipse is described by the equation

$$r(\phi) = \frac{a(1 - e^2)}{1 + e \cos(\phi - \phi_0)}, \tag{2.14}$$

where a is the semi-major axis, e is the eccentricity, and ϕ_0 is the azimuthal angle of the orbital periapsis. The time dependence of the relative motion is determined by Kepler's equation,

$$\dot{\phi} = \frac{2\pi}{P}(1 - e^2)^{-3/2}\left(1 + e \cos(\phi - \phi_0)\right)^2, \tag{2.15}$$

where P is the orbital period of the binary,

$$P = 2\pi\sqrt{\frac{a^3}{GM}}. \tag{2.16}$$

The binary's binding energy E and the modulus J of its total angular orbital momentum are related to the parameters of the relative orbit through the equations

$$E = -\frac{GM\mu}{2a}, \quad J = \mu\sqrt{GMa(1 - e^2)}. \tag{2.17}$$

Let us now introduce the TT "wave" coordinates (x_w, y_w, z_w) in which the gravitational wave is traveling in the $+z_w$ direction and with the origin at the SSB. The line along which the plane tangential to the celestial sphere at the location of the binary's center-of-mass [this plane is parallel to the (x_w, y_w) plane] intersects the orbital (x_c, y_c) plane is called the *line of nodes*. Let us adjust the center-of-mass and the wave coordinates in such a way that the x axes of both coordinate systems are parallel to each other and to the line of nodes. Then the relation between these coordinates is determined by the rotation matrix S,

$$\begin{pmatrix} x_w \\ y_w \\ z_w \end{pmatrix} = \begin{pmatrix} x_w^* \\ y_w^* \\ z_w^* \end{pmatrix} + \mathsf{S}\begin{pmatrix} x_c \\ y_c \\ z_c \end{pmatrix}, \quad \mathsf{S} := \begin{pmatrix} 1 & 0 & 0 \\ 0 & \cos\iota & \sin\iota \\ 0 & -\sin\iota & \cos\iota \end{pmatrix}, \tag{2.18}$$

where (x_w^*, y_w^*, z_w^*) are the components of the vector $\mathbf{x}^*$ joining the SSB and the binary's center-of-mass, and ι ($0 \leq \iota \leq \pi$) is the angle between the orbital angular momentum vector $\mathbf{J}$ of the binary and the line of sight (i.e. the $+z_w$ axis). We assume that the center-of-mass of the binary is at rest with respect to the SSB.

In the center-of-mass reference frame the binary's *moment of inertia tensor* has changing in time components that are equal

$$\mathcal{I}_{c}^{ij}(t) = m_1\, x_{c1}^{i}(t)\, x_{c1}^{j}(t) + m_2\, x_{c2}^{i}(t)\, x_{c2}^{j}(t). \tag{2.19}$$

Making use of Eqs. (2.12) one easily computes that the matrix $\mathcal{I}_c$ built from the $\mathcal{I}_c^{ij}$ components of the inertia tensor equals

$$\mathcal{I}_c(t) = \mu \begin{pmatrix} (x_{c12}(t))^2 & x_{c12}(t)\, y_{c12}(t) & 0 \\ x_{c12}(t)\, y_{c12}(t) & (y_{c12}(t))^2 & 0 \\ 0 & 0 & 0 \end{pmatrix}. \tag{2.20}$$

The inertia tensor components in wave coordinates are related to those in center-of-mass coordinates through the relation

$$\mathcal{I}_{w}^{ij}(t) = \sum_{k=1}^{3}\sum_{\ell=1}^{3} \frac{\partial x_{w}^{i}}{\partial x_{c}^{k}} \frac{\partial x_{w}^{j}}{\partial x_{c}^{\ell}} \mathcal{I}_{c}^{k\ell}(t), \tag{2.21}$$

which in matrix notation reads

$$\mathcal{I}_w(t) = \mathsf{S} \cdot \mathcal{I}_c(t) \cdot \mathsf{S}^{\mathsf{T}}. \tag{2.22}$$

To obtain the wave polarization functions h_+ and $h_\times$ we plug the components of the binary's inertia tensor [computed by means of Eqs. (2.18), (2.20), and (2.22)] into general equations (1.114) [note that in Eqs. (1.114) the components $\mathcal{J}_{ij}$ of the reduced quadrupole moment can be replaced by the components $\mathcal{I}_{ij}$ of the inertia tensor, compare the definitions (1.115) and (2.56)]. We get [here $R = |\mathbf{x}^*|$ is the distance to the binary's center-of-mass and $t_r = t - (z_w + R)/c$ is the retarded time]

$$h_+(t, \mathbf{x}) = \frac{G\mu}{c^4 R}\Bigg(\sin^2\iota\big(\dot{r}(t_r)^2 + r(t_r)\ddot{r}(t_r)\big)$$

$$+ (1 + \cos^2\iota)\big(\dot{r}(t_r)^2 + r(t_r)\ddot{r}(t_r) - 2r(t_r)^2\dot\phi(t_r)^2\big)\cos 2\phi(t_r)$$

$$- (1 + \cos^2\iota)\big(4r(t_r)\dot{r}(t_r)\dot\phi(t_r) + r(t_r)^2\ddot\phi(t_r)\big)\sin 2\phi(t_r)\Bigg), \tag{2.23a}$$

$$h_\times(t, \mathbf{x}) = \frac{2G\mu}{c^4 R}\cos\iota\Bigg(\big(4r(t_r)\dot{r}(t_r)\dot\phi(t_r) + r(t_r)^2\ddot\phi(t_r)\big)\cos 2\phi(t_r)$$

$$+ \big(\dot{r}(t_r)^2 + r(t_r)\ddot{r}(t_r) - 2r(t_r)^2\dot\phi(t_r)^2\big)\sin 2\phi(t_r)\Bigg). \tag{2.23b}$$

Polarization waveforms without radiation-reaction effects. Gravitational waves emitted by the binary diminish the binary's binding energy and total orbital angular momentum. This makes the orbital parameters changing in time. Let us first assume that these changes are so slow that they can be neglected during the time interval in which the observations are performed. Making use of Eqs. (2.15) and (2.14) one can then eliminate from the formulae (2.23) the first and the second time derivatives of r and ϕ, treating the parameters of the orbit, a and e, as constants. The result is the following [110]:

$$h_+(t, \mathbf{x}) = \frac{4G^2 M \mu}{c^4 Ra(1 - e^2)} \Big(A_0(t_{\mathrm{r}}) + A_1(t_{\mathrm{r}}) e + A_2(t_{\mathrm{r}}) e^2 \Big), \qquad (2.24\mathrm{a})$$

$$h_\times(t, \mathbf{x}) = \frac{4G^2 M \mu}{c^4 Ra(1 - e^2)} \cos \iota \Big(B_0(t_{\mathrm{r}}) + B_1(t_{\mathrm{r}}) e + B_2(t_{\mathrm{r}}) e^2 \Big), \qquad (2.24\mathrm{b})$$

where the functions A_i and B_i, $i = 0, 1, 2$, are defined as follows

$$A_0(t) = -\frac{1}{2}(1 + \cos^2 \iota) \cos 2\phi(t), \qquad (2.25\mathrm{a})$$

$$A_1(t) = \frac{1}{4} \sin^2 \iota \cos(\phi(t) - \phi_0)$$

$$\qquad - \frac{1}{8}(1 + \cos^2 \iota)\Big(5 \cos(\phi(t) + \phi_0) + \cos(3\phi(t) - \phi_0)\Big), \qquad (2.25\mathrm{b})$$

$$A_2(t) = \frac{1}{4} \sin^2 \iota - \frac{1}{4}(1 + \cos^2 \iota) \cos 2\phi_0, \qquad (2.25\mathrm{c})$$

$$B_0(t) = -\sin 2\phi(t), \qquad (2.25\mathrm{d})$$

$$B_1(t) = -\frac{1}{4}\Big(\sin(3\phi(t) - \phi_0) + 5 \sin(\phi(t) + \phi_0)\Big), \qquad (2.25\mathrm{e})$$

$$B_2(t) = -\frac{1}{2} \sin 2\phi_0. \qquad (2.25\mathrm{f})$$

In the case of circular orbits the eccentricity e vanishes and the wave polarization functions (2.24) simplify to

$$h_+(t, \mathbf{x}) = -\frac{2G^2 M \mu}{c^4 Ra}(1 + \cos^2 \iota) \cos 2\phi(t_{\mathrm{r}}), \qquad (2.26\mathrm{a})$$

$$h_\times(t, \mathbf{x}) = -\frac{4G^2 M \mu}{c^4 Ra} \cos \iota \sin 2\phi(t_{\mathrm{r}}), \qquad (2.26\mathrm{b})$$

where a is the radius of the relative circular orbit.

Polarization waveforms with radiation-reaction effects. Let us now consider short enough observational intervals that it is necessary to take into account the effects of radiation reaction changing the parameters of the bodies' orbits. In the leading order the rates of emission of energy and angular momentum carried by gravitational waves are given by the formulae (1.117). We plug into them the Newtonian trajectories of the bodies

and perform time averaging over one orbital period of the motion. We get
[43, 111]

$$\mathcal{L}_E^{\text{gw}} = \frac{32}{5} \frac{G^4 \mu^2 M^3}{c^5 a^5 (1 - e^2)^{7/2}} \left(1 + \frac{73}{24} e^2 + \frac{37}{96} e^4 \right), \tag{2.27a}$$

$$\mathcal{L}_J^{\text{gw}} = \frac{32}{5} \frac{G^{7/2} \mu^2 M^{5/2}}{c^5 a^{7/2} (1 - e^2)^2} \left(1 + \frac{7}{8} e^2 \right). \tag{2.27b}$$

It is easy to obtain equations describing the leading-order time evolution
of the relative orbit parameters a and e. We first differentiate both sides of
Eqs. (2.17) with respect to time treating a and e as functions of time. Then
the rates of changing the binary's binding energy and angular momentum,
dE/dt and dJ/dt, one replaces [according to Eqs. (1.118)] by *minus* the
rates of emission of respectively energy $\mathcal{L}_E^{\text{gw}}$ and angular momentum $\mathcal{L}_J^{\text{gw}}$
carried by gravitational waves, given in Eqs. (2.27). The two equations
obtained can be solved with respect to the derivatives da/dt and de/dt:

$$\frac{da}{dt} = -\frac{64}{5} \frac{G^3 \mu M^2}{c^5 a^3 (1 - e^2)^{7/2}} \left(1 + \frac{73}{24} e^2 + \frac{37}{96} e^4 \right), \tag{2.28a}$$

$$\frac{de}{dt} = -\frac{304}{15} \frac{G^3 \mu M^2}{c^5 a^4 (1 - e^2)^{5/2}} e \left(1 + \frac{121}{304} e^2 \right). \tag{2.28b}$$

Let us note two interesting features of the evolution described by the
above equations: (i) Eq. (2.28b) implies that initially circular orbits (for
which $e = 0$) remain circular during the evolution; (ii) the eccentricity e of
the relative orbit decreases with time much more rapidly than the semi-
major axis a, so gravitational-wave reaction induces the *circularization*
of the binary orbits (roughly, when the semi-major axis is halved, the
eccentricity goes down by a factor of three [43]).

Evolution of the orbital parameters caused by the radiation reaction
obviously influences gravitational waveforms. To see this influence we
restrict our considerations to binaries along *circular* orbits. The decay
of the radius a of the relative orbit is given by the equation

$$\frac{da}{dt} = -\frac{64}{5} \frac{G^3 \mu M^2}{c^5 a^3} \tag{2.29}$$

[this is Eq. (2.28a) with $e = 0$]. Integration of this equation leads to

$$a(t) = a_0 \left(\frac{t_c - t}{t_c - t_0} \right)^{1/4}, \tag{2.30}$$

where $a_0 := a(t = t_0)$ is the initial value of the orbital radius and $t_c - t_0$
is the formal "lifetime" or "coalescence time" of the binary, i.e. the time
after which the distance between the bodies is zero. It means that t_c is

the moment of the coalescence, $a(t_c) = 0$; it is equal to

$$t_c = t_0 + \frac{5}{256} \frac{c^5 a_0^4}{G^3 \mu M^2}. \tag{2.31}$$

If one neglects radiation-reaction effects, then along the Newtonian circular orbit the time derivative of the orbital phase, i.e. the orbital *angular frequency* ω, is constant and given by [see Eqs. (2.15) and (2.16)]

$$\omega := \dot{\phi} = \frac{2\pi}{P} = \sqrt{\frac{GM}{a^3}}. \tag{2.32}$$

We now assume that the inspiral of the binary system caused by the radiation reaction is *adiabatic*, i.e. it can be thought of as a sequence of circular orbits with radii $a(t)$ which slowly diminish in time according to Eq. (2.30). Adiabacity thus means that Eq. (2.32) is valid during the inspiral, but now angular frequency ω is a function of time, because of time dependence of the orbital radius a. Making use of Eqs. (2.30)–(2.32) we get the following equation for the time evolution of the instantaneous orbital angular frequency during the adiabatic inspiral:

$$\omega(t) = \omega_0 \left(1 - \frac{256(G\mathcal{M})^{5/3}}{5c^5} \omega_0^{8/3}(t - t_0) \right)^{-3/8}, \tag{2.33}$$

where $\omega_0 := \omega(t = t_0)$ is the initial value of the orbital angular frequency and where we have introduced the new parameter $\mathcal{M}$ (with dimension of a mass) called a *chirp mass* of the system:

$$\mathcal{M} := \mu^{3/5} M^{2/5}. \tag{2.34}$$

To obtain the polarization waveforms h_+ and $h_\times$, which describe gravitational radiation emitted during the adiabatic inspiral of the binary system, we use the general formulae (2.23). In these formulae we neglect all the terms proportional to $\dot{r} \equiv \dot{a}$ or $\ddot{r} \equiv \ddot{a}$ (note that the radial coordinate r is identical to the radius a of the relative circular orbit, $r \equiv a$). By virtue of Eq. (2.32) $\dot{\phi} \propto \dot{a}$, therefore we also neglect terms $\propto \ddot{\phi}$. All of these simplifications lead to the following formulae [where Eq. (2.32) was employed again to replace a by the orbital frequency ω]:

$$h_+(t, \mathbf{x}) = -\frac{4(G\mathcal{M})^{5/3}}{c^4 R} \frac{1 + \cos^2 \iota}{2} \omega(t_r)^{2/3} \cos 2\phi(t_r), \tag{2.35a}$$

$$h_\times(t, \mathbf{x}) = -\frac{4(G\mathcal{M})^{5/3}}{c^4 R} \cos \iota \, \omega(t_r)^{2/3} \sin 2\phi(t_r). \tag{2.35b}$$

Making use of Eq. (2.32) one can express the time dependence of the waveforms (2.35) in terms of the single function $a(t)$ [explicitly given in

Eq. (2.30)]:

$$h_+(t, \mathbf{x}) = -\frac{4G^2\mu M}{c^4 R} \frac{1 + \cos^2 \iota}{2} \frac{1}{a(t_r)} \cos 2 \left(\sqrt{GM} \int_{t_0}^{t_r} a(t')^{-3/2} \, \mathrm{d}t' + \phi_0 \right),$$
(2.36a)

$$h_\times(t, \mathbf{x}) = -\frac{4G^2\mu M}{c^4 R} \cos \iota \frac{1}{a(t_r)} \sin 2 \left(\sqrt{GM} \int_{t_0}^{t_r} a(t')^{-3/2} \, \mathrm{d}t' + \phi_0 \right),$$
(2.36b)

where $\phi_0 := \phi(t = t_0)$ is the initial value of the orbital phase of the binary.

It is sometimes useful to analyze the chirp waveforms h_+ and $h_\times$ in terms of the instantaneous frequency f_{gw} of the gravitational waves emitted by a coalescing binary system. The frequency f_{gw} is defined as

$$f_{\mathrm{gw}}(t) := 2 \times \frac{1}{2\pi} \frac{\mathrm{d}\phi(t)}{\mathrm{d}t},$$
(2.37)

where the extra factor 2 is due to the fact that [see Eqs. (2.35)] the instantaneous phase of the gravitational waveforms h_+ and $h_\times$ is twice the instantaneous phase $\phi(t)$ of the orbital motion (so the gravitational-wave frequency f_{gw} is twice the orbital frequency). Making use of Eqs. (2.30)–(2.32) one finds that for the binary with circular orbits the gravitational-wave frequency f_{gw} reads

$$f_{\mathrm{gw}}(t) = \frac{5^{3/8} c^{15/8}}{8\pi G^{5/8}} \frac{1}{\mathcal{M}^{5/8}} (t_c - t)^{-3/8}.$$
(2.38)

The chirp waveforms given in Eqs. (2.36) can be rewritten in terms of the gravitational-wave frequency f_{gw} as follows

$$h_+(t, \mathbf{x}) = -h_0(t_r) \frac{1 + \cos^2 \iota}{2} \cos \left(2\pi \int_{t_0}^{t_r} f_{\mathrm{gw}}(t') \, \mathrm{d}t' + 2\phi_0 \right),$$
(2.39a)

$$h_\times(t, \mathbf{x}) = -h_0(t_r) \cos \iota \sin \left(2\pi \int_{t_0}^{t_r} f_{\mathrm{gw}}(t') \, \mathrm{d}t' + 2\phi_0 \right),$$
(2.39b)

where $h_0(t)$ is the changing in time amplitude of the waveforms,

$$h_0(t) := \frac{4\pi^{2/3} G^{5/3}}{c^4} \frac{\mathcal{M}^{5/3}}{R} f_{\mathrm{gw}}(t)^{2/3}.$$
(2.40)

2.4.2 Post-Newtonian binary dynamics

Here we consider in more detail the 3.5PN-accurate time evolution of the orbital phase in the case of a binary consisting of *non-rotating* bodies that move along *quasi-circular* orbits. The numerical value of the binary's

Hamiltonian, i.e. the binding energy of the system, is computed up to the 3PN order. It is useful to express the binding energy in terms of a dimensionless quantity x, which is connected with the angular frequency $\omega := d\phi/dt$ along circular orbits (here ϕ is the orbital phase of the system):

$$x := \frac{(GM\omega)^{2/3}}{c^2}. \tag{2.41}$$

The energy E of the binary in the center-of-mass reference frame and for circular orbits reads:

$$E = -\frac{\mu c^2 x}{2}\left(1 + e_{1PN}\, x + e_{2PN}\, x^2 + e_{3PN}\, x^3 + \mathcal{O}\!\left(x^4\right)\right), \tag{2.42}$$

where $-\frac{1}{2}\mu c^2 x$ is the Newtonian binding energy of the system, and e_{nPN} (with $n = 1, 2, 3$) are the fractional PN corrections to the Newtonian value. These corrections depend solely on the symmetric mass ratio η and they are equal

$$e_{1PN} = -\frac{3}{4} - \frac{1}{12}\eta, \tag{2.43a}$$

$$e_{2PN} = -\frac{27}{8} + \frac{19}{8}\eta - \frac{1}{24}\eta^2, \tag{2.43b}$$

$$e_{3PN} = -\frac{675}{64} + \left(\frac{34\,445}{576} - \frac{205}{96}\pi^2\right)\eta - \frac{155}{96}\eta^2 - \frac{35}{5184}\eta^3. \tag{2.43c}$$

Gravitational-wave luminosity of the system can also be written in the form of the post-Newtonian expansion. It is known up to the 3.5PN order and, for circular orbits, it reads:

$$\mathcal{L} = \frac{32c^5}{5G}\eta^2 x^5\left(1 + \ell_{1PN}\, x + 4\pi\, x^{3/2} + \ell_{2PN}\, x^2 + \ell_{2.5PN}\, x^{5/2}\right.$$
$$\left. + \left(\ell_{3PN} - \frac{856}{105}\ln(16x)\right)x^3 + \ell_{3.5PN}\, x^{7/2} + \mathcal{O}\!\left(x^4\right)\right), \tag{2.44}$$

where $\frac{32c^5}{5G}\eta^2 x^5$ is the dominant value of the luminosity, the coefficients ℓ_{nPN} describe different PN fractional corrections to this dominant value, they again depend only on η and are equal

$$\ell_{1PN} = -\frac{1247}{336} - \frac{35}{12}\eta, \tag{2.45a}$$

$$\ell_{2PN} = -\frac{44\,711}{9072} + \frac{9271}{504}\eta + \frac{65}{18}\eta^2, \tag{2.45b}$$

$$\ell_{2.5PN} = \left(-\frac{8191}{672} - \frac{535}{24}\eta\right)\pi, \tag{2.45c}$$

$$\ell_{3PN} = \frac{6643\,739\,519}{69\,854\,400} + \frac{16}{3}\pi^2 - \frac{1712}{105}\gamma_E \qquad (2.45d)$$

$$+ \left(\frac{41}{48}\pi^2 - \frac{134\,543}{7776}\right)\eta - \frac{94\,403}{3024}\eta^2 - \frac{775}{324}\eta^3, \qquad (2.45e)$$

$$\ell_{3.5PN} = \left(-\frac{16\,285}{504} + \frac{214\,745}{1728}\eta + \frac{193\,385}{3024}\eta^2\right)\pi, \qquad (2.45f)$$

where γ_E is the Euler constant [see Eq. (A.2) in Appendix A for its definition and numerical value].

The time evolution of the angular frequency ω (or the dimensionless
parameter x) of the system moving along circular orbits is determined
assuming the validity of the balance equation

$$\frac{dE}{dt} = -\mathcal{L}, \qquad (2.46)$$

which states that the binding energy E of the system is diminishing in
time with the rate being equal minus the gravitational-wave luminosity
of the binary. After substitution of (2.42) and (2.44) into the balance
equation, it becomes the first-order ordinary differential equation for the
function $x = x(t)$. It is convenient to introduce the dimensionless time
variable

$$\hat{t} := \frac{c^3}{GM}t, \qquad (2.47)$$

then Eq. (2.46) is the first-order ordinary differential equation for the
function $x = x(\hat{t})$. Let $x(\hat{t}_a) = x_a$ for some initial time $\hat{t} = \hat{t}_a$, then the
solution to Eq. (2.46) can be written in the form

$$\hat{t} - \hat{t}_a = \mathcal{F}(x_a) - \mathcal{F}(x), \qquad (2.48)$$

where we have introduced the dimensionless indefinite integral

$$\mathcal{F}(x) := \frac{c^3}{GM}\int \frac{1}{\mathcal{L}}\frac{dE}{dx}\,dx. \qquad (2.49)$$

After expanding the integrand in power series of the variable x, the integral (2.49) can be explicitly computed. The result is given in Eq. (A.1)
of Appendix A. One can then perturbatively invert Eq. (2.48) to get x
as a function of $\hat{t}$. To display this inversion it is convenient to introduce
another dimensionless time variable τ,

$$\tau(\hat{t}) := \frac{1}{5}\eta\left(\hat{t}_a + \mathcal{F}(x_a) - \hat{t}\right). \qquad (2.50)$$

Then inversion of Eq. (2.48) can symbolically be written as

$$x(\hat{t}) = \mathcal{T}\left(\tau(\hat{t})\right). \qquad (2.51)$$

The explicit form of the function $\mathcal{T}$ is given in Eq. (A.4) of Appendix A.

Making use of Eqs. (2.41) and (2.46) one gets the differential equation for the phase ϕ as a function of the parameter x:

$$\frac{\mathrm{d}\phi}{\mathrm{d}x} = \frac{\mathrm{d}\phi/\mathrm{d}t}{\mathrm{d}x/\mathrm{d}t} = -\frac{c^3}{GM} \frac{x^{3/2}}{\mathcal{L}} \frac{\mathrm{d}E}{\mathrm{d}x}. \tag{2.52}$$

The solution to this equation reads

$$\phi(x) = \phi_a + \mathcal{G}(x) - \mathcal{G}(x_a), \tag{2.53}$$

where $\phi_a := \phi(x_a)$ is the initial value of the phase ϕ and the dimensionless function $\mathcal{G}$ is defined by means of the following indefinite integral:

$$\mathcal{G}(x) := -\frac{c^3}{GM} \int \frac{x^{3/2}}{\mathcal{L}} \frac{\mathrm{d}E}{\mathrm{d}x} \, \mathrm{d}x. \tag{2.54}$$

After expanding the integrand in Eq. (2.54) in power series with respect to x, the integration in (2.54) can be performed and the function $\mathcal{G}$ can be given in explicit form [which can be found in Eq. (A.5) of Appendix A]. Collecting Eqs. (2.51) and (2.53) together we obtain the formula for the orbital phase ϕ as a function of the time variable $\hat{t}$,

$$\phi(\hat{t}) = \phi_a + \mathcal{G}\big(\mathcal{T}\big(\tau(\hat{t})\big)\big) - \mathcal{G}(x_a). \tag{2.55}$$

The right-hand side of this equation can be expanded in a PN series. The explicit 3.5PN-accurate form of this expansion is given in Eq. (A.6) of Appendix A.

2.5 Case study: a rotating triaxial ellipsoid

As an example of a source emitting periodic gravitational waves we consider here a torque-free rigid body rotating around one of its principal axes (so the body does not precess). We use Newtonian mechanics and Newtonian theory of gravitation to describe the dynamics of the body, so we employ the weak-field, slow-motion, and small-stress approximation of general relativity. Gravitational waves from rotating and precessing Newtonian rigid bodies are studied, e.g. in [112, 113].

Let us thus consider a rigid triaxial ellipsoid rotating around a principal axis. In the reference frame corotating with the star we introduce the spatial coordinates (x_s, y_s, z_s), the origin of the coordinate system coincides with the center of the ellipsoid and the axes are parallel to the principal axes of the star. Let the star rotate around the $+z_s$ axis. In these coordinates the star's moment of inertia tensor has constant components $\mathcal{I}_s^{ij}$. If the gravitational field inside and near the star is weak enough to be accurately described by the Newtonian gravity, the components $\mathcal{I}_s^{ij}$ can be calculated as volume integrals [here $\rho(x_s^k)$ denotes the mass density of the star]

$$\mathcal{I}_s^{ij} := \int \rho(x_s^k)\, x_s^i\, x_s^j\, \mathrm{d}^3 x_s. \tag{2.56}$$

We arrange these components into a diagonal matrix $\mathcal{I}_s$,

$$\mathcal{I}_s := \begin{pmatrix} \mathcal{I}_s^{xx} & 0 & 0 \\ 0 & \mathcal{I}_s^{yy} & 0 \\ 0 & 0 & \mathcal{I}_s^{zz} \end{pmatrix}. \tag{2.57}$$

We next introduce some inertial reference frame with coordinates (x_i, y_i, z_i). Its origin also coincides with the center of the ellipsoid and the z_i axis is parallel to the angular momentum vector $\mathbf{J}$ of the star, so $z_i \equiv z_s$. The relation between (x_i, y_i, z_i) and (x_s, y_s, z_s) coordinates involves a time-dependent rotation matrix $\mathsf{R}(t)$,

$$\begin{pmatrix} x_i \\ y_i \\ z_i \end{pmatrix} = \mathsf{R}(t) \begin{pmatrix} x_s \\ y_s \\ z_s \end{pmatrix}, \quad \mathsf{R}(t) := \begin{pmatrix} \cos\phi(t) & -\sin\phi(t) & 0 \\ \sin\phi(t) & \cos\phi(t) & 0 \\ 0 & 0 & 1 \end{pmatrix}, \tag{2.58}$$

where the function $\phi(t)$ is the angle between, say, the axes x_i and x_s, so it determines the instantaneous rotational phase of the star.

Let $\mathbf{n}$ be a unit vector pointing from the center of the ellipsoid towards the solar system barycenter (SSB). We uniquely fix the position of the x_i axis in space by requiring that (i) it lies in the 2-plane spanned by the vectors $\mathbf{e}_{z_i}$ (where $\mathbf{e}_{z_i}$ is a vector of unit length along the $+z_i$ axis) and $\mathbf{n}$, and (ii) the scalar product of the vectors $\mathbf{e}_{x_i}$ and $\mathbf{n}$ is non-negative, $\mathbf{e}_{x_i} \cdot \mathbf{n} \geq 0$. The position of the y_i axis is then uniquely determined by the condition $\mathbf{e}_{x_i} \times \mathbf{e}_{y_i} = \mathbf{e}_{z_i}$.

Finally we introduce the TT "wave" coordinates (x_w, y_w, z_w) related to the traveling gravitational wave. The origin of these coordinates is at the SSB. The wave travels in the $+z_w$ direction (so $\mathbf{e}_{z_w} \equiv \mathbf{n}$) and we choose the y_w axis to be parallel to the y_i axis. The relation between (x_w, y_w, z_w) and (x_i, y_i, z_i) coordinates is then given by

$$\begin{pmatrix} x_w \\ y_w \\ z_w \end{pmatrix} = \begin{pmatrix} x_w^* \\ y_w^* \\ z_w^* \end{pmatrix} + \mathsf{S} \begin{pmatrix} x_i \\ y_i \\ z_i \end{pmatrix}, \quad \mathsf{S} := \begin{pmatrix} \cos\iota & 0 & -\sin\iota \\ 0 & 1 & 0 \\ \sin\iota & 0 & \cos\iota \end{pmatrix}, \tag{2.59}$$

where (x_w^*, y_w^*, z_w^*) are the components of the position vector $\mathbf{x}^*$ of the star's center with respect to the SSB, and ι $(0 \leq \iota \leq \pi)$ is the angle between the angular momentum vector $\mathbf{J}$ of the star and the line of sight (i.e. the 3-vector $\mathbf{n}$). We assume that the center of the star does not move with respect to the SSB.

To compute the two independent wave polarizations h_+ and $h_\times$ we need to calculate the components of the star's inertia tensor in (x_w, y_w, z_w)

coordinates. To do this we use the transformation rule of tensor components, which leads to the equation

$$\mathcal{I}_{\mathrm{w}}^{ij} = \sum_{k=1}^{3}\sum_{\ell=1}^{3} \frac{\partial x_{\mathrm{w}}^i}{\partial x_{\mathrm{s}}^k}\frac{\partial x_{\mathrm{w}}^j}{\partial x_{\mathrm{s}}^\ell}\mathcal{I}_{\mathrm{s}}^{k\ell} = \sum_{k=1}^{3}\sum_{\ell=1}^{3}\sum_{m=1}^{3}\sum_{n=1}^{3} \frac{\partial x_{\mathrm{w}}^i}{\partial x_{\mathrm{i}}^m}\frac{\partial x_{\mathrm{i}}^m}{\partial x_{\mathrm{s}}^k}\frac{\partial x_{\mathrm{w}}^j}{\partial x_{\mathrm{i}}^n}\frac{\partial x_{\mathrm{i}}^n}{\partial x_{\mathrm{s}}^\ell}\mathcal{I}_{\mathrm{s}}^{k\ell}.$$

(2.60)

Making use of Eqs. (2.58) and (2.59) we rewrite this relation in matrix form,

$$\mathcal{I}_{\mathrm{w}}(t) = \mathsf{S}\cdot\mathsf{R}(t)\cdot\mathcal{I}_{\mathrm{s}}\cdot\mathsf{R}(t)^{\mathrm{T}}\cdot\mathsf{S}^{\mathrm{T}}.$$

(2.61)

From Eq. (2.61) we compute the components $\mathcal{I}_{\mathrm{w}}^{ij}$ of the inertia tensor in wave coordinates, and plug them into Eqs. (1.114) from Chapter 1 [note that in Eqs. (1.114) the components $\mathcal{J}_{ij}$ of the reduced quadrupole moment can be replaced by the components $\mathcal{I}_{ij}$ of the inertia tensor, compare the definitions (1.115) and (2.56)]. We obtain

$$h_+(t,\mathbf{x}) = -\frac{4G}{c^4 r^*}\varepsilon\,\mathcal{I}_{\mathrm{s}}^{zz}\frac{1+\cos^2\iota}{2}\left((\dot\phi(t_{\mathrm{r}}))^2\cos 2\phi(t_{\mathrm{r}}) + \frac{1}{2}\ddot\phi(t_{\mathrm{r}})\sin 2\phi(t_{\mathrm{r}})\right),$$

(2.62a)

$$h_\times(t,\mathbf{x}) = -\frac{4G}{c^4 r^*}\varepsilon\,\mathcal{I}_{\mathrm{s}}^{zz}\cos\iota\left((\dot\phi(t_{\mathrm{r}}))^2\sin 2\phi(t_{\mathrm{r}}) - \frac{1}{2}\ddot\phi(t_{\mathrm{r}})\cos 2\phi(t_{\mathrm{r}})\right),$$

(2.62b)

where $r^* = |\mathbf{x}^*|$, $t_{\mathrm{r}} = t - (z_{\mathrm{w}} + r^*)/c$ is the retarded time, and we have also introduced here the *ellipticity* of the star defined by

$$\varepsilon := \frac{\mathcal{I}_{\mathrm{s}}^{xx} - \mathcal{I}_{\mathrm{s}}^{yy}}{\mathcal{I}_{\mathrm{s}}^{zz}}.$$

(2.63)

We expect that a neutron star rotates almost uniformly during a typical observation period. It means that it is reasonable to model the time evolution of the star's rotational phase $\phi(t)$ by the first few terms of its Taylor expansion around some initial time t_0,

$$\phi(t) = \phi_0 + 2\pi\left(f_0(t-t_0) + \frac{1}{2!}\dot f_0(t-t_0)^2 + \frac{1}{3!}\ddot f_0(t-t_0)^3 + \cdots\right).$$

(2.64)

Here $\phi_0 := \phi(t_0)$ is the initial value of the phase, f_0 is the instantaneous rotational frequency at $t = t_0$, and $\dot f_0$, $\ddot f_0$, ..., is the first, the second, ..., time derivative of the rotational frequency evaluated at $t = t_0$. It is natural to take $t_0 = -r^*/c$, then Eq. (2.64) leads to the relation

$$\phi(t_{\mathrm{r}}) = \phi\left(t - \frac{z_{\mathrm{w}} + r^*}{c}\right)$$

$$= \phi_0 + 2\pi\left(f_0\left(t - \frac{z_{\mathrm{w}}}{c}\right) + \frac{1}{2!}\dot f_0\left(t - \frac{z_{\mathrm{w}}}{c}\right)^2 + \cdots\right).$$

(2.65)

It is clear that now $f_0, \dot{f}_0, \ldots$, can be interpreted as respectively the rotational frequency, the first time derivative of $f_0, \ldots$, *observed* at $t = 0$ by the fictitious observer located at the SSB (where $z_w = 0$).

Usually the corrections to the amplitudes of the wave polarizations h_+ and $h_\times$ that are caused by the non-uniformity of the star's rotation are small and can be neglected. This means that in Eqs. (2.62) the terms $\propto \ddot{\phi}(t_r)$ can be neglected and the derivative $\dot{\phi}(t_r)$ can be replaced by $2\pi f_0$. After these simplifications the wave polarizations can be rewritten in the form

$$h_+(t, \mathbf{x}) = -h_0 \frac{1 + \cos^2 \iota}{2} \cos 2\phi(t_r), \qquad (2.66a)$$

$$h_\times(t, \mathbf{x}) = -h_0 \cos \iota \sin 2\phi(t_r), \qquad (2.66b)$$

where we have introduced the dimensionless amplitude

$$h_0 := \frac{16\pi^2 G}{c^4} \frac{\varepsilon \, \mathcal{I}_s^{zz} \, f_0^2}{r^*}. \qquad (2.67)$$

2.6 Case study: supernova explosion

In this section we estimate the dimensionless amplitude of gravitational waves emitted during a supernova explosion by relating it to the total energy released in gravitational waves during the explosion and also to the time duration and the frequency bandwidth of the gravitational-wave pulse.

Let us start from surrounding the supernova by a two-sphere of radius large enough that the sphere is located within the local wave zone of the source. Let us choose a point on the sphere. In the vicinity of the point the flux of energy (i.e. energy transported in unit time per unit area) carried away by gravitational waves is equal to

$$F_{gw} = \frac{dE_{gw}}{dt \, dS} = \sum_{i=1}^{3} T_{gw}^{ti} \, n_i = -c \sum_{i=1}^{3} T_{0i}^{gw} n^i, \qquad (2.68)$$

where $T_{gw}^{\alpha\beta}$ is the energy–momentum tensor of gravitational waves and $n_i = n^i$ is the unit 3-vector normal to the surface of the two-sphere. The value of contraction over the index i in Eq. (2.68) does not depend on the orientation of the spatial axes of the coordinate system we use, therefore let us orient the z axis such that it has the direction of the unit 3-vector n^i. In the small region (of area dS) around our fixed point the gravitational wave can be treated as planar, i.e. we can employ here

Eq. (1.101) describing components of the plane gravitational wave (traveling in the $+z$ direction). Taking all this into account we rewrite Eq. (2.68) as

$$F_{\text{gw}} = \frac{c^3}{16\pi G} \left\langle \left(\frac{\partial h_+}{\partial t}\right)^2 + \left(\frac{\partial h_\times}{\partial t}\right)^2 \right\rangle. \tag{2.69}$$

The total energy ΔE_{gw} radiated in gravitational waves we obtain integrating the flux (2.69) over the two-sphere $\mathbb{S}^2$ and time,

$$\Delta E_{\text{gw}} = \frac{c^3}{16\pi G} \oint_{\mathbb{S}^2} \mathrm{d}S \int_{-\infty}^{\infty} \mathrm{d}t \left(\left(\frac{\partial h_+}{\partial t}\right)^2 + \left(\frac{\partial h_\times}{\partial t}\right)^2\right). \tag{2.70}$$

Because we integrate over time here, we can omit the averaging denoted by $\langle \cdots \rangle$ in Eq. (2.69). Making use of Parseval's theorem we replace time integration in (2.70) by integration over frequency,

$$\Delta E_{\text{gw}} = \oint_{\mathbb{S}^2} \mathrm{d}S \int_{-\infty}^{\infty} \mathrm{d}f\, \mathcal{F}_{\text{gw}}(f), \tag{2.71}$$

where $\mathcal{F}_{\text{gw}}$ is gravitational-wave energy radiated per unit area and per unit frequency. Making use of Eq. (2.70) we get

$$\mathcal{F}_{\text{gw}}(f) = \frac{\pi c^3}{4G} f^2 \left(|\tilde{h}_+(f)|^2 + |\tilde{h}_\times(f)|^2\right), \tag{2.72}$$

where we have denoted by a tilde the Fourier transform and we have used that $\tilde{\dot{s}} = 2\pi i f \tilde{s}$ for any function s. Finally, performing in Eq. (2.71) integration over the sphere of radius R, we obtain the following formula for the total energy ΔE_{gw} radiated in gravitational waves:

$$\Delta E_{\text{gw}} = \frac{\pi^2 c^3}{G} R^2 \int_{-\infty}^{\infty} f^2 \left(|\tilde{h}_+(f)|^2 + |\tilde{h}_\times(f)|^2\right) \mathrm{d}f. \tag{2.73}$$

We expect the gravitational wave from a supernova explosion to be a pulse of time duration τ and frequency bandwidth Δf_{gw}. The frequency bandwidth we define through the relation

$$\Delta f_{\text{gw}}^2 = \frac{\int_{-\infty}^{\infty} f^2 \left(|\tilde{h}_+(f)|^2 + |\tilde{h}_\times(f)|^2\right) \mathrm{d}f}{\int_{-\infty}^{\infty} \left(|\tilde{h}_+(f)|^2 + |\tilde{h}_\times(f)|^2\right) \mathrm{d}f}. \tag{2.74}$$

Then, using Eqs. (2.73)–(2.74) and Parseval's theorem again, the energy ΔE_{gw} can be written as

$$\Delta E_{\text{gw}} = \frac{\pi^2 c^3}{G} R^2 \Delta f_{\text{gw}}^2 \int_{-\infty}^{\infty} \left(h_+^2(t) + h_\times^2(t)\right) \mathrm{d}t. \tag{2.75}$$

We also define the dimensionless "effective" amplitude h_0 of the pulse as

$$h_0 := \sqrt{\frac{1}{\tau} \int_{-\infty}^{\infty} \left(h_+^2(t) + h_\times^2(t)\right) \mathrm{d}t}, \qquad (2.76)$$

where τ is the duration of the pulse. Making use of Eqs. (2.75) and (2.76) we obtain the following expression for the amplitude h_0:

$$h_0 = \frac{1}{\pi c} \sqrt{\frac{G\Delta E_{gw}}{c\tau}} \frac{1}{\Delta f_{gw}} \frac{1}{R}. \qquad (2.77)$$

2.7 Case study: stochastic background

In this section (which follows Refs. [114, 115]) we will relate the distribution in frequency of the energy of stochastic gravitational waves to statistical properties of gravitational-wave background expressed by its *spectral density* function. We will also express the frequency dependence of the characteristic dimensionless amplitude of stochastic gravitational waves by the energy density of waves with different frequencies. We consider here a stochastic gravitational-wave background which is *stationary*, *isotropic*, and *unpolarized* (see Section IIB of Ref. [115] for a discussion of these assumptions).

The energy density ρ_{gw} of gravitational waves is the $(0,0)$ component of the energy–momentum tensor $T_{\alpha\beta}^{gw}$ related with propagating waves [see Eq. (1.100)],

$$\rho_{gw} = T_{00}^{gw} = \frac{c^2}{32\pi G} \sum_{i=1}^{3} \sum_{j=1}^{3} \left\langle \left(\frac{\partial h_{ij}^{TT}}{\partial t}\right)^2 \right\rangle. \qquad (2.78)$$

The TT metric perturbation h_{ij}^{TT} induced by stochastic gravitational waves at a given spacetime point $(t, \mathbf{x})$ can be expanded as follows

$$h_{ij}^{TT}(t, \mathbf{x}) = \sum_A \oint_{\mathbb{S}^2} \mathrm{d}\Omega \, h_A(t, \mathbf{x}; \theta, \phi) \, e_{ij}^A(\theta, \phi). \qquad (2.79)$$

Here the sum is over two polarizations $A = +, \times$ of the waves, the integration is over the Euclidean two-sphere $\mathbb{S}^2$ parametrized by the standard polar θ and azimuthal ϕ angles (thus $\mathrm{d}\Omega = \sin\theta \, \mathrm{d}\theta \, \mathrm{d}\phi$ is the surface element on the two-sphere), $h_A(t, \mathbf{x}; \theta, \phi)$ represents gravitational waves (where A is polarization) propagating in the direction defined by the angles θ and ϕ, and $e_{ij}^A(\theta, \phi)$ are the polarization tensors for the plus and cross polarization states of the wave. The quantity $h_A(t, \mathbf{x}; \theta, \phi)$ can further be decomposed into monochromatic components,

$$h_A(t, \mathbf{x}; \theta, \phi) = \int_{-\infty}^{\infty} \tilde{h}_A(f; \theta, \phi) \exp\left(2\pi i f\left(t - \frac{\mathbf{n} \cdot \mathbf{x}}{c}\right)\right) df, \quad (2.80)$$

where $\tilde{h}_A(f; \theta, \phi)$ are the *Fourier amplitudes* of the wave propagating in the direction of the unit vector $\mathbf{n}$ with components

$$\mathbf{n} = (\cos \phi \sin \theta, \sin \phi \sin \theta, \cos \theta). \quad (2.81)$$

The Fourier amplitudes $\tilde{h}_A(f; \theta, \phi)$ are arbitrary complex functions that satisfy, as a consequence of the reality of h_{ij}^{TT}, the relation $\tilde{h}_A(-f; \theta, \phi) = \tilde{h}_A^*(f; \theta, \phi)$, where the asterisk denotes complex conjugation. Equations (2.79) and (2.80) describe the plane-wave expansion of the metric perturbation h_{ij}^{TT}.

The polarization tensors e_{ij}^A were introduced in Eqs. (1.69), where their components in the coordinate system in which the wave is propagating in the $+z$ direction are given. If the wave propagates in the direction of the vector $\mathbf{n}$ arbitrarily oriented with respect to the coordinate axes, then the components of the tensors e_{ij}^A can be expressed in terms of the components of the two unit vectors $\mathbf{p}$ and $\mathbf{q}$ orthogonal to $\mathbf{n}$ and to each other,

$$e_{ij}^+ = p_i p_j - q_i q_j, \quad e_{ij}^\times = p_i q_j + q_i p_j. \quad (2.82)$$

The components of the vectors $\mathbf{p}$ and $\mathbf{q}$ can be chosen as follows

$$\mathbf{p} = (\sin \phi, -\cos \phi, 0), \quad \mathbf{q} = (\cos \phi \cos \theta, \sin \phi \cos \theta, -\sin \theta). \quad (2.83)$$

Making use of Eqs. (2.82)–(2.83) one shows that

$$\sum_A \sum_{i=1}^{3} \sum_{j=1}^{3} \oint_{\mathbb{S}^2} d\Omega \left(e_{ij}^A(\theta, \phi)\right)^2 = 16\pi. \quad (2.84)$$

We now assume that each Fourier amplitude $\tilde{h}_A(f; \theta, \phi)$ is a *Gaussian* random process with *zero mean*. Consequently the gravitational-wave metric perturbation h_{ij}^{TT} given by Eqs. (2.79) and (2.80), which depends linearly on the Fourier amplitudes, is also a Gaussian random process with zero mean. Thus the statistical properties of the gravitational-wave background are completely specified by the expectation value of the product of two different Fourier amplitudes, $\mathrm{E}\{\tilde{h}_A^*(f; \theta, \phi) \tilde{h}_{A'}(f'; \theta', \phi')\}$. Assuming in addition that the gravitational-wave background is *stationary*, *isotropic*, and *unpolarized*, this expectation value satisfies the following relation:

$$E\{\tilde{h}_A^*(f;\theta,\phi)\,\tilde{h}_{A'}(f';\theta',\phi')\} = \delta_{AA'}\,\frac{\delta^2(\theta,\phi;\theta',\phi')}{4\pi}\,\frac{1}{2}\delta(f-f')S_{\text{gw}}(f),$$

(2.85)

where $\delta^2(\theta,\phi;\theta',\phi') := \delta(\phi-\phi')\delta(\cos\theta - \cos\theta')$ is the covariant Dirac delta function on the two-sphere and S_{gw} is the *one-sided spectral density* of the stochastic background.

Making use of Eqs. (2.79) and (2.80), (2.84) and (2.85), and assuming that the averaging denoted by angle brackets in Eq. (2.78) can be replaced by taking the expectation value E, we compute

$$\sum_{i=1}^{3}\sum_{j=1}^{3}\left\langle\left(\frac{\partial h_{ij}^{\text{TT}}}{\partial t}\right)^2\right\rangle = 16\pi^2\int_0^\infty f^2\,S_{\text{gw}}(f)\,\mathrm{d}f.$$

(2.86)

If the stochastic background is anisotropic, the spectral density S_{gw} would, in addition, depend on the angles (θ,ϕ), and if the stochastic background is polarized, S_{gw} would be a function of the polarization states A as well. The total energy density ρ_{gw} of the gravitational-wave background takes, by virtue of Eqs. (2.78) and (2.86), the form

$$\rho_{\text{gw}} = \frac{\pi c^2}{2G}\int_0^\infty f^2 S_{\text{gw}}(f)\,\mathrm{d}f.$$

(2.87)

The strength of the stochastic gravitational waves is usually described by the dimensionless ratio Ω_{gw} of the gravitational-wave energy density per *logarithmic* frequency interval to the closure density ρ_{cl} of the universe,

$$\Omega_{\text{gw}}(f) := \frac{1}{\rho_{\text{cl}}}\frac{\mathrm{d}\bar{\rho}_{\text{gw}}(f)}{\mathrm{d}\ln f},$$

(2.88)

where $\mathrm{d}\bar{\rho}_{\text{gw}}(f)$ is the energy density in the waves with frequencies from f to $f+\mathrm{d}f$. The critical energy density required (today) to close the universe is given by (here H_0 is the today's value of the Hubble expansion rate)

$$\rho_{\text{cl}} = \frac{3c^2 H_0^2}{8\pi G}.$$

(2.89)

Consequently the spectral density of the gravitational-wave background can be expressed, using Eqs. (2.87)–(2.89), as

$$S_{\text{gw}}(f) = \frac{3H_0^2}{4\pi^2}\frac{\Omega_{\text{gw}}(f)}{f^3}.$$

(2.90)

It is sometimes convenient to express the strength of the gravitational-wave stochastic background by a dimensionless amplitude h_c defined as

$$h_c(f) := \sqrt{2fS_{gw}(f)}. \qquad (2.91)$$

By virtue of Eq. (2.90) the characteristic amplitude h_c can be expressed in terms of the dimensionless quantity Ω_{gw},

$$h_c(f) = \sqrt{\frac{3}{2}\frac{H_0}{\pi}}\frac{\sqrt{\Omega_{gw}(f)}}{f}. \qquad (2.92)$$

3

Statistical theory of signal detection

Data from a gravitational-wave detector are realizations of a stochastic (or random) process, thus in order to analyze them we need a statistical model. In this chapter we present a theory of the detection of signals in noise and an estimation of the signal's parameters from a statistical point of view. We begin in Section 3.1 with a brief introduction to random variables and in Section 3.2 we present the basic concepts of stochastic processes. A comprehensive introduction to mathematical statistics can be found, for example, in the texts [116, 117]. Our treatment follows the monograph [118]. Other expositions can be found in the texts [119, 120]. A general introduction to stochastic processes is given in [121]. Advanced treatment of the subject can be found in [122, 123].

In Section 3.3 we present the problem of hypothesis testing and in Section 3.4 we discuss its application to the detection of deterministic signals in Gaussian noise. Section 3.5 is devoted to the problem of estimation of stochastic signals. Hypothesis testing is discussed in detail in the monograph [124]. Classical expositions of the theory of signal detection in noise can be found in the monographs [125, 126, 127, 128, 129, 130, 131].

In Section 3.6 we introduce the subject of parameter estimation and present several statistical concepts relevant for this problem. Parameter estimation is discussed in detail in Ref. [132], and Ref. [133] contains a concise account. In Section 3.7 we discuss topics related to the non-stationary stochastic processes. They include the resampled stochastic processes because of their importance for the analysis of periodic gravitational-wave signals modulated by the motion of the detector with respect to the solar system barycenter [26, 31, 134, 135] and cyclostationary processes because they occur in the analysis of the gravitational-wave background from white-dwarf binaries [136].

3.1 Random variables

Probability theory is concerned with random experiments, each modeled by some *probability space*, which is a triplet $(\mathcal{X}, \mathcal{O}, P)$, where the set $\mathcal{X}$ is called an *observation set* (or a *sample space*), $\mathcal{O}$ is a family of subsets of $\mathcal{X}$ called *observation events* and P is a *probability distribution* (or *probability measure*) on $\mathcal{O}$.

The observation set is a set of values that our data can assume. Usually $\mathcal{X}$ is n-dimensional Euclidean space $\mathbb{R}^n$ or a discrete set $\Gamma = \{\gamma_1, \gamma_2, \ldots\}$. Observation events $\mathcal{O}$ are subsets of $\mathcal{X}$ to which we can assign consistent probabilities. We demand that $\mathcal{O}$ is a *σ-algebra* of sets, i.e. the collection of subsets of $\mathcal{X}$ such that (i) $\mathcal{X} \in \mathcal{O}$; (ii) if $A \in \mathcal{O}$, then its complement $A' := \mathcal{X} - \mathcal{O}$ in $\mathcal{X}$ is also in $\mathcal{O}$; (iii) for any countable family A_k ($k = 1, 2, \ldots$) of members of $\mathcal{O}$ their union belongs to $\mathcal{O}$. From the conditions (i)–(iii) it also follows that the intersection of a finite or countable family of members of $\mathcal{O}$ belongs to $\mathcal{O}$. In the case of the observation set $\mathcal{X} = \mathbb{R}^n$ we consider open intervals in $\mathbb{R}^n$ of the form

$$\{\mathbf{y} = (y_1, \ldots, y_n) \in \mathbb{R}^n : a_1 < y_1 < b_1, \ldots, a_n < y_n < b_n\}, \quad (3.1)$$

where a_i and b_i ($i = 1, \ldots, n$) are arbitrary real numbers and we usually take $\mathcal{O}$ as the smallest σ-algebra containing all intervals of the type (3.1). This σ-algebra is called *Borel σ-algebra* and it is denoted by B^n. For the case of a discrete observation set Γ, the σ-algebra $\mathcal{O}$ is the set of all subsets of Γ (it is called the *power set* of Γ and is denoted by 2^Γ).

The probability distribution P on σ-algebra $\mathcal{O}$ in the space $\mathcal{X}$ is a function assigning a number $P(A) \geq 0$ to any set $A \in \mathcal{O}$ such that $P(\mathcal{X}) = 1$ and such that for any countable family of disjoint sets $A_k \in \mathcal{O}$ ($k = 1, 2, \ldots$) we have

$$P\left(\bigcup_{k=1}^{\infty} A_k\right) = \sum_{k=1}^{\infty} P(A_k). \quad (3.2)$$

For probability spaces $(\mathbb{R}^n, B^n, P)$ usually there exists a *probability density function* $p : \mathbb{R}^n \to [0, \infty)$ such that the probability $P(A)$ of the event $A \in B^n$ is given by

$$P(A) = \int_A p(\mathbf{y}) \, d\mathbf{y}, \quad (3.3)$$

where $\mathbf{y}$ are some coordinates on $\mathbb{R}^n$.

Example 1 (Probability space for a die roll). *The sample space $\mathcal{X}$ is the set of all outcomes of a die roll, $\mathcal{X} = \{1, 2, 3, 4, 5, 6\}$. The observation space $\mathcal{O}$ is the set of all subsets A of $\mathcal{X}$ (there are 2^6 of them).*

The probability measure P can be defined as $P(A) := |A|/6$, where $|A|$ is the number of elements in the event A (called the cardinality *of the set A). For example the event of rolling an even number is $A = \{2, 4, 6\}$, its cardinality $|A| = 3$, and the probability $P(A)$ of this event is $1/2$.*

A *random variable* X is a real-valued function $X : \mathcal{X} \to \mathbb{R}$, whose domain is the observation set $\mathcal{X}$, such that

$$\{\xi \in \mathcal{X} : X(\xi) \leq x\} \in \mathcal{O} \quad \text{for all } x \in \mathbb{R}.$$

The *cumulative distribution function* (often also just called the *distribution function*) P_X of the random variable X is defined as

$$P_X(x) := P(X \leq x), \tag{3.4}$$

where $P(X \leq x)$ denotes the probability of the event $\{\xi \in \mathcal{X} : X(\xi) \leq x\}$. The cumulative distribution function P_X is a non-decreasing function on the real axis $\mathbb{R}$ with the properties that $\lim_{x \to -\infty} P_X(x) = 0$ and $\lim_{x \to +\infty} P_X(x) = 1$. It is also a *right-continuous* function.

A random variable X is *discrete*, if there exists (finite or countably infinite) a set $W = \{x_1, x_2, \ldots\} \subset \mathbb{R}$ such that it holds

$$p_i := P(\{\xi \in \mathcal{X} : X(\xi) = x_i\}) > 0 \text{ for } x_i \in W, \text{ and } \sum_i p_i = 1. \tag{3.5}$$

The cumulative distribution function of a discrete random variable is a piecewise constant function. A *continuous* random variable X is defined as one with a continuous cumulative distribution function P_X. We will only consider continuous random variables such that the derivative $\mathrm{d}P_X/\mathrm{d}x$ exists everywhere; this derivative is called the *probability density function* (pdf) p_X of the random variable X:

$$p_X(x) = \frac{\mathrm{d}P_X(x)}{\mathrm{d}x}. \tag{3.6}$$

As any P_X is non-decreasing, $p_X(x) \geq 0$ everywhere. The pdf p_X is also normalized,

$$\int_{-\infty}^{\infty} p_X(x)\,\mathrm{d}x = 1. \tag{3.7}$$

The most important characteristics of the random variable are its expectation value and variance (or standard deviation). The *expectation value* (or *mean*) of the discrete random variable X [defined through Eq. (3.5)] is defined as

$$\mathrm{E}\{X\} := \sum_{x_i \in W} x_i\, p_i. \tag{3.8}$$

If one defines a random variable $Y := g(X)$ (where g is a deterministic function), then the expectation value of Y can be computed as

$$E\{Y\} = E\{g(X)\} = \sum_{x_i \in W} g(x_i)\, p_i. \tag{3.9}$$

The expectation value of the continuous random variable X possessing the pdf p_X is defined as

$$E\{X\} := \int_{-\infty}^{\infty} x\, p_X(x)\, dx. \tag{3.10}$$

For a random variable $Y := g(X)$ its expectation value can be computed from the formula

$$E\{Y\} = E\{g(X)\} = \int_{-\infty}^{\infty} g(x)\, p_X(x)\, dx. \tag{3.11}$$

The *variance* of the random variable X is

$$\mathrm{Var}\{X\} \equiv \sigma_X^2 := E\left\{(X - E\{X\})^2\right\}. \tag{3.12}$$

Its square root σ_X is the *standard deviation*. The standard deviation measures the spread of the random variable around its mean value.

Let us consider two random variables X and Y defined on the same probability space $(\mathcal{X}, \mathcal{O}, P)$. The *covariance* between random variables X and Y is defined as

$$\mathrm{Cov}(X, Y) := E\{(X - E\{X\})(Y - E\{Y\})\}. \tag{3.13}$$

Random variables X, Y for which $E\{XY\} = E\{X\}E\{Y\}$ are called *uncorrelated*. If $E\{XY\} = 0$, then the random variables X, Y are called *orthogonal*.

Let the random variables X and Y be continuous and let p_X and p_Y be the pdf of respectively X and Y, and let $p_{X,Y}$ be the *joint* pdf of the variables X, Y. Then the *conditional* pdf $p_{X|Y}$ of X given that $Y = y$ is equal to

$$p_{X|Y}(x|Y = y) = \frac{p_{X,Y}(x, y)}{p_Y(y)}. \tag{3.14}$$

Random variables X, Y are said to be *independent* provided $p_{X,Y}(x, y) = p_X(x)\, p_Y(y)$. For independent random variables $p_{X|Y}(x|Y = y) = p_X(x)$.

A *complex* random variable Z can be defined as $Z := X + iY$, where X and Y are real random variables. The pdf p_Z of Z is defined as the joint pdf of the real random variables X and Y, i.e.

$$p_Z(z) := p_{X,Y}(x, y), \quad \text{where } z = x + iy, \tag{3.15}$$

so p_Z is a real-valued (and non-negative) function. The mean of a complex random variable Z is

$$\mathrm{E}\{Z\} := \int_{-\infty}^{\infty} \mathrm{d}x \int_{-\infty}^{\infty} \mathrm{d}y \, (x + iy) \, p_{X,Y}(x,y) = \mathrm{E}\{X\} + i\,\mathrm{E}\{Y\}, \quad (3.16)$$

and the variance is, by definition,

$$\mathrm{Var}\{Z\} \equiv \sigma_Z^2 := \mathrm{E}\{|Z - \mathrm{E}\{Z\}|^2\} = \mathrm{E}\{|Z|^2\} - |\mathrm{E}\{Z\}|^2. \quad (3.17)$$

For two complex random variables $W := U + iV$ and $Z := X + iY$ their joint pdf is

$$p_{W,Z}(w,z) := p_{U,V,X,Y}(u,v,x,y), \quad \text{where } w = u + iv, \ z = x + iy,$$
$$(3.18)$$

and the covariance between W and Z is defined as

$$\mathrm{Cov}(W,Z) := \mathrm{E}\{(W - \mathrm{E}\{W\})(Z - \mathrm{E}\{Z\})^*\}. \quad (3.19)$$

Probability distributions of the data collected by a detector and of the quantities that we derive from the data are, in general, unknown and may be very complex. However, it is frequently possible to assume that certain events have equal probabilities, that they are statistically independent, or that there is a relative order of certain infinitesimal probabilities. Here we give examples of the three probability distributions that are most frequently encountered in data analysis.

1. The *binomial* distribution $b(p,n)$ with

$$P(X = x) = \binom{n}{x} p^x (1-p)^{n-x}, \quad x \in \{0, \ldots, n\}, \quad 0 \le p \le 1.$$
$$(3.20)$$

This is the distribution of the total number of successes in n independent trials when the probability of success for each trial is p. The mean and variance of the binomial distribution are respectively equal to np and $np(1-p)$.

2. The *Poisson* distribution $P(\tau)$ with

$$P(X = x) = \frac{\tau^x}{x!} e^{-\tau}, \quad x \in \mathbb{Z}, \quad \tau > 0. \quad (3.21)$$

This is the distribution of the number of events occurring in a fixed interval of time (or space) if the probability of more than one occurrence in a very short interval is of a smaller order of magnitude than that of a single occurrence, and if the numbers of events in non-overlapping intervals are statistically independent. The mean and variance of the Poisson distribution are both equal to τ. Thus the parameter τ of the Poisson distribution is the expected number of events in an interval.

3. The *normal* (or *Gaussian*) distribution $N(\mu, \sigma^2)$ with probability density function

$$p(x) = \frac{1}{\sqrt{2\pi}\,\sigma} \exp\left(-\frac{(x-\mu)^2}{2\sigma^2}\right), \quad -\infty < x, \mu < \infty, \quad \sigma > 0,$$

(3.22)

where μ is the mean of the distribution and σ^2 is its variance. Under very general conditions, which are made precise by *central limit theorems*, this is the approximate distribution of the sum of a large number of independent random variables when the relative contribution of each term to the sum is small.

3.2 Stochastic processes

Let T be a subset of real numbers, $T \subset \mathbb{R}$. A *stochastic* (or *random*) *process* $x(t)$, $t \in T$, is a family of random variables $x(t)$, labeled by the numbers $t \in T$, all defined on the same probability space $(\mathcal{X}, \mathcal{O}, P)$. For each finite subset $\{t_1, \ldots, t_n\} \subset T$ the random variables $x(t_1), \ldots, x(t_n)$ have a joint n-dimensional cumulative distribution function F defined by

$$F_{t_1, \ldots, t_n}(x_1, \ldots, x_n) = P\big(x(t_1) \leq x_1, \ldots, x(t_n) \leq x_n\big). \quad (3.23)$$

When T is a set of discrete points, $T = \{t_1, t_2, \ldots\}$, the stochastic process is called a *random sequence* and its values $x(t_k)$ are denoted by x_k ($k = 1, 2, \ldots$). Very often the variable t is just time and then the random sequence is called the *time series*. For example data from a gravitational-wave detector (such as resonant bar or laser interferometer) are time series.

The stochastic process $x(t)$, $t \in T$, is *Gaussian* if the cumulative distribution function $F_{t_1, \ldots, t_n}$ is Gaussian for any n and any $t_1, \ldots, t_n \in T$. The stochastic process $x(t)$, $t \in T$, is called *stationary* (sometimes referred to as *strongly stationary*, *strictly stationary*, or *completely stationary*) if all the finite-dimensional cumulative distribution functions (3.23) defining $x(t)$ remain the same if the set of points $\{t_1, \ldots, t_n\} \subset T$ is shifted by an arbitrary constant amount τ, i.e. if

$$F_{t_1+\tau, \ldots, t_n+\tau}(x_1, \ldots, x_n) = F_{t_1, \ldots, t_n}(x_1, \ldots, x_n) \quad (3.24)$$

for any $n, t_1, \ldots, t_n$, and τ. In other words, the probabilistic structure of a stationary process is invariant under a shift of the parameter t.

To study properties of stochastic processes it is useful to consider their simplest numerical characteristics – the *moments* of probability distributions defining the process. In general the moments are defined by

$$\mu_{m_1,\ldots,m_n} := \mathrm{E}\left\{x(t_1)^{m_1}\cdots x(t_n)^{m_n}\right\}$$

$$= \int_{-\infty}^{\infty}\cdots\int_{-\infty}^{\infty} x_1^{m_1}\cdots x_n^{m_n}\,\mathrm{d}F_{t_1,\ldots,t_n}(x_1,\ldots,x_n). \qquad (3.25)$$

The integral on the right-hand side of (3.25) is an n-fold *Stieltjes integral*. The one-fold Stieltjes integral

$$\int_a^b f(x)\,\mathrm{d}F(x) \qquad (3.26)$$

is defined as the limit of the quantity

$$\sum_{k=1}^{n} f(x_k')[F(x_k) - F(x_{k-1})], \qquad (3.27)$$

as $\max_k |x_k - x_{k-1}| \to 0$, where

$$a = x_0 < x_1 < \ldots < x_{n-1} < x_n = b, \qquad x_{k-1} \le x_k' \le x_k. \qquad (3.28)$$

The improper integral

$$\int_{-\infty}^{\infty} f(x)\,\mathrm{d}F(x) \qquad (3.29)$$

is defined as the double limit

$$\lim_{\substack{a\to-\infty \\ b\to+\infty}} \int_a^b f(x)\,\mathrm{d}F(x). \qquad (3.30)$$

n-fold Stieltjes integrals are defined in an analogous way.

In the case when the cumulative distribution function $F_{t_1,\ldots,t_n}$ is differentiable, it has the joint probability density function

$$f_{t_1,\ldots,t_n}(x_1,\ldots,x_n) := \frac{\partial^n F_{t_1,\ldots,t_n}(x_1,\ldots,x_n)}{\partial x_1 \ldots \partial x_n}, \qquad (3.31)$$

then the integral (3.25) can be written as an ordinary Riemann integral

$$\mu_{m_1,\ldots,m_n} = \mathrm{E}\left\{x(t_1)^{m_1}\cdots x(t_n)^{m_n}\right\}$$

$$= \int_{-\infty}^{\infty}\cdots\int_{-\infty}^{\infty} x_1^{m_1}\cdots x_n^{m_n} f_{t_1,\ldots,t_n}(x_1,\ldots,x_n)\,\mathrm{d}x_1\ldots\mathrm{d}x_n.$$

$$(3.32)$$

The simplest moments are the first- and second-order moments μ_1 and $\mu_{1,1}$. The first-order moment μ_1 is called the *mean* value of the stochastic process $x(t)$ and is denoted by $m(t)$,

$$m(t) := \mathrm{E}\left\{x(t)\right\}. \tag{3.33}$$

The second-order moment $\mu_{1,1}$ is called the *autocorrelation function* and is denoted by $K(t, s)$,

$$K(t, s) := \mathrm{E}\left\{x(t)x(s)\right\}. \tag{3.34}$$

In addition to the autocorrelation function K it is useful to define the *autocovariance function* C of the stochastic process $x(t)$,

$$C(t, s) := \mathrm{E}\left\{\left(x(t) - m(t)\right)\left(x(s) - m(s)\right)\right\} = K(t, s) - m(t)\,m(s). \tag{3.35}$$

If the stochastic process $x(t)$ is stationary, its mean value is a constant m that does not depend on time t,

$$\mathrm{E}\left\{x(t)\right\} = m = \mathrm{const.}, \tag{3.36}$$

and the autocorrelation function $K(t, s)$ depends only on the difference $t - s$, i.e. there exists a function R of one variable such that

$$\mathrm{E}\left\{x(t)x(s)\right\} = R(t - s). \tag{3.37}$$

A stochastic process $x(t)$ is said to be *wide-sense stationary* (sometimes referred to as *weakly stationary, second-order stationary,* or *covariance stationary*) if it fulfills conditions (3.36) and (3.37). In the important case of the Gaussian stochastic processes the first two moments completely specify all the distributions (3.23) and thus a wide-sense stationary Gaussian process is also strictly stationary. The autocorrelation function R of the stationary stochastic process has the following properties that are simple consequences of its definition:

$$R(0) > 0, \tag{3.38a}$$

$$R(-\tau) = R(\tau), \tag{3.38b}$$

$$|R(\tau)| \leq R(0). \tag{3.38c}$$

One can also consider *complex* stationary stochastic processes for which the autocorrelation function is defined as

$$R(\tau) := \mathrm{E}\left\{x(t + \tau)\,x(t)^*\right\}. \tag{3.39}$$

The complex-valued autocorrelation function (3.39) satisfies the properties (3.38a) and (3.38c) and the property (3.38b) is replaced by

$$R(-\tau) = R(\tau)^*. \tag{3.40}$$

Let us now consider a complex stochastic process given by the formula

$$x(t) := x_0\, f(t), \quad -\infty < t < \infty, \tag{3.41}$$

where x_0 is a zero-mean complex random variable and f is a deterministic complex-valued function of t. The stochastic process x of the form given in Eq. (3.41) is stationary if and only if it has the form

$$x(t) = x_0 \, e^{i\omega t}, \quad -\infty < t < \infty, \tag{3.42}$$

where ω is a real constant [118]. The autocorrelation function of the process (3.42) equals

$$R(\tau) = E\{|x_0|^2\} \, e^{i\omega\tau} = b_0 \, e^{i\omega\tau}, \tag{3.43}$$

where $b_0 := E\{|x_0 - E\{x_0\}|^2\} = E\{|x_0|^2\}$ is the variance of the complex random variable x_0. We can obtain a more general stationary stochastic process by superposition of an arbitrary number of processes of the form (3.42):

$$x(t) = \sum_{k=1}^{\infty} x_k \, e^{i\omega_k t}, \quad -\infty < t < \infty, \tag{3.44}$$

where the complex random variables x_k fulfill the conditions $E\{x_k\} = 0$ and $E\{x_k x_l^*\} = 0$ for $k \neq l$. The autocorrelation function of the above process is given by

$$R(\tau) = \sum_{k=1}^{\infty} b_k \, e^{i\omega_k \tau}, \tag{3.45}$$

where $b_k := E\{|x_k - E\{x_k\}|^2\} = E\{|x_k|^2\}$ are variances of the random variables x_k. We assume that $\sum_{k=1}^{\infty} b_k < \infty$, so the series on the right-hand side of Eq. (3.45) is absolutely convergent. A stationary process of the form (3.44) is called a *process with a discrete spectrum*, and the set of numbers $\{\omega_1, \omega_2, \ldots\}$ is called the *spectrum* of the process.

In the special case when the complex random variables x_k are Gaussian, their amplitudes $|x_k|$ have a *Rayleigh distribution* with a pdf of the form

$$p(x) = 2\frac{x}{b_k} \, e^{-x^2/b_k}, \quad x \geq 0, \tag{3.46}$$

while their phases $\phi_k := \arg(x_k)$ have a *uniform distribution* over the interval $[0, 2\pi)$.

It can be shown that *any* complex stationary stochastic process can be obtained as the limit of a sequence of processes with discrete spectra. This result is known as a spectral representation of a stationary stochastic process which is also called the *Cramèr representation*. Its precise statement is as follows [120].

Theorem 1 (The Cramèr representation). *Let $x(t)$, $-\infty < t < \infty$, be a zero-mean stochastically continuous complex stationary process. Then* *there exists an orthogonal process $Z(\omega)$ such that for all t, $x(t)$ may be written in the form,*

$$x(t) = \int_{-\infty}^{\infty} e^{i\omega t} dZ(\omega), \qquad (3.47)$$

the integral being defined in the mean-square sense. The process $Z(\omega)$ has the following properties:

1. $E\{dZ(\omega)\} = 0$ *for all ω;*

2. $E\{|dZ(\omega)|^2\} = dF(\omega)$ *for all ω, where $F(\omega)$ is the (non-normalized) integrated spectrum of $x(t)$;*

3. *for any two distinct frequencies ω, ω' ($\omega \neq \omega'$),*

$$\text{Cov}\{dZ(\omega), dZ(\omega')\} = E\{dZ(\omega) dZ(\omega')^*\} = 0. \qquad (3.48)$$

Another very important characterization of stationary stochastic processes is the existence of a well-defined spectrum. This is the content of the Wiener–Khinchin theorem which can be stated as follows [120].

Theorem 2 (The Wiener–Khinchin theorem). *A necessary and sufficient condition for $\rho(\tau)$ to be the autocorrelation function of a stochastically continuous stationary process $x(t)$, is that there exists a function $F(\omega)$, having the properties of a cumulative distribution function on $(-\infty, \infty)$ [i.e. $F(-\infty) = 0$, $F(\infty) = 1$, and F is non-decreasing], such that, for all τ, $\rho(\tau)$ may be expressed in the form,*

$$\rho(\tau) = \int_{-\infty}^{\infty} e^{i\omega \tau} dF(\omega). \qquad (3.49)$$

The necessary part of the above theorem follows from a general theorem due to Bochner that any positive semi-definite function that is continuous everywhere must have the representation of the above form.

We would like to emphasize that the use of Stieltjes integrals in the Cramèr representation and in the Wiener–Khinchin theorem is indispensable. If spectrum of the process contains a discrete component, say of frequency ω_0, then the distribution function F undergoes a discontinuous jump at frequency ω_0 and cannot be differentiable there. In the case when F is differentiable everywhere we have a purely continuous spectrum and we can introduce a function S such that

$$dF(\omega) = S(\omega) d\omega. \qquad (3.50)$$

The function S is called the *two-sided spectral density* of the stationary stochastic process. Even in this case the use of Stieltjes integrals in the

Cramér representation is unavoidable. From the Condition 2 in Theorem 1 we have $E\{|dZ(\omega)|^2\} = S(\omega)\,d\omega$, this implies that $dZ(\omega)$ is of the order $\sqrt{d\omega}$ and cannot be represented as $z(\omega)\,d\omega$ for some process $z(\omega)$; in other words the process $Z(\omega)$ is not differentiable. The proof of Theorems 1 and 2 and their extensive discussion can be found in Refs. [118, 119, 120].

Assuming that the spectral density $S(\omega)$ of the stationary stochastic process $x(t)$ exists and using the Cramèr representation and the Wiener–Khinchin theorem we have the following expression for the correlation between the Fourier components $\tilde{x}(\omega)$ of the stationary process $x(t)$,

$$E\{\tilde{x}(\omega)\,\tilde{x}(\omega')^*\} = S(\omega)\,\delta(\omega' - \omega), \tag{3.51}$$

where δ is the Dirac delta function.

Let us next consider the case of discrete-time stochastic processes, i.e. time series. Let the time series x_k, $k \in \mathbb{Z}$, be given by the formula

$$x_k := x_0\, f(k), \quad k \in \mathbb{Z}, \tag{3.52}$$

where x_0 is a complex random variable and f is some complex-valued deterministic function. Then, as in the case of continuous-time stochastic processes, the time series x_k is *stationary* if and only if it has the form

$$x_k = x_0\, e^{i\omega_0 k}, \quad k \in \mathbb{Z}. \tag{3.53}$$

The only difference is that now the index k takes only integer values and thus ω_0 is defined to within an additive constant which is an integer multiple of 2π. Consequently the Cramèr representation and the Wiener–Khinchin theorem for stationary time series take respectively the form

$$x_k = \int_{-\pi}^{\pi} e^{i\omega k}\, dZ(\omega), \quad k \in \mathbb{Z}, \tag{3.54a}$$

$$R(\ell) = \int_{-\pi}^{\pi} e^{i\omega \ell}\, dF(\omega), \quad \ell \in \mathbb{Z}, \tag{3.54b}$$

where $R(\ell)$ is the autocorrelation function of the time series,

$$R(\ell) := E\{x_k\, x_{k+\ell}\}, \quad \ell \in \mathbb{Z}. \tag{3.55}$$

If the distribution function F is differentiable and its derivative is S, we can replace Eq. (3.54b) by

$$R(\ell) = \int_{-\pi}^{\pi} e^{i\omega \ell} S(\omega)\, d\omega, \quad \ell \in \mathbb{Z}. \tag{3.56}$$

Making use of the periodic representation (11) from the Notation Section of the Dirac δ function, one can solve Eq. (3.56) with respect to the function S. The result is

$$S(\omega) = \frac{1}{2\pi} \sum_{\ell=-\infty}^{\infty} e^{-i\omega\ell} R(\ell). \tag{3.57}$$

Ergodic theorem. The practical value of stationary stochastic processes is to a considerable extent due to the fact that its mean value and its auto-correlation function can usually be calculated by using just one realization of the stochastic process. The possibility of calculating these character-istics of stationary processes from a single realization is a consequence of the fact that the *ergodic theorem* (or *law of large numbers*) is appli-cable. According to the ergodic theorem, the mathematical expectation obtained by averaging the corresponding quantities over the whole space of experimental outcomes, can be replaced by *time averages* of the same quantities.

More precisely, if x_t is a stationary time series satisfying certain very general conditions (which are almost always met in practice), then with a suitable definition of the limit of a sequence of random variables, the fol-lowing limiting relations for the mean m and the autocorrelation function R are valid:

$$m = \mathrm{E}\{x_t\} = \lim_{N\to\infty} \frac{1}{N} \sum_{t=1}^{N} x_t, \tag{3.58a}$$

$$R(\tau) = \mathrm{E}\{x_t\, x_{t+\tau}\} = \lim_{N\to\infty} \frac{1}{N} \sum_{t=1}^{N} x_t\, x_{t+\tau}. \tag{3.58b}$$

Similarly if x is a stationary continuous-time stochastic process, then

$$m = \mathrm{E}\{x(t)\} = \lim_{T\to\infty} \frac{1}{N} \int_0^T x(t)\, \mathrm{d}t, \tag{3.59a}$$

$$R(\tau) = \mathrm{E}\{x(t)\, x(t+\tau)\} = \lim_{T\to\infty} \frac{1}{N} \int_0^T x(t)\, x(t+\tau)\, \mathrm{d}t. \tag{3.59b}$$

3.3 Hypothesis testing

The basic idea behind signal detection is that the presence of the signal in the data x collected by a detector changes the statistical characteristics of the data, in particular its probability distribution. When the signal is absent the data have a probability density function (pdf) $p_0(x)$, and when the signal is present the pdf of the data is $p_1(x)$. The problem of detecting the signal in noise can be posed as a statistical *hypothesis testing* problem.

The *null hypothesis* H_0 is that the signal is absent in the data and the *alternative hypothesis* H_1 is that the signal is present. A *hypothesis test* (or a *decision rule*) δ is a partition of the observation set $\mathcal{X}$ into two subsets $\mathcal{R}$ and its complement $\mathcal{R}' := \mathcal{X} - \mathcal{R}$. If data are in $\mathcal{R}$ we accept the null hypothesis otherwise we reject it. The problem is to find a test that is in some way optimal. There are several approaches to find such a test.

The theory of hypothesis testing began with the work of Reverend Thomas Bayes [137] on "inverse" of conditional probability. Tests that minimize the chance of error were proposed by Neyman and Pearson [138]. The notion of cost and risk were introduced by Wald [139], who can be credited with the development of the theory along the lines of the theory of games.

3.3.1 Bayesian approach

In a Bayesian approach we assign *costs* to our decisions; in particular we introduce non-negative numbers C_{ij} $(i, j = 0, 1)$, where C_{ij} is the cost incurred by choosing hypothesis H_i when hypothesis H_j is true. We shall call matrix C the *cost matrix*. We define the *conditional risk* R_j $(j = 0, 1)$ of a decision rule δ for each hypothesis as

$$R_j(\delta) := C_{0j} P_j(\mathcal{R}) + C_{1j} P_j(\mathcal{R}'), \quad j = 0, 1, \tag{3.60}$$

where P_j is the probability distribution of the data when hypothesis H_j is true. Next we assign probabilities π_0 and $\pi_1 = 1 - \pi_0$ to the occurrences of hypothesis H_0 and H_1, respectively. These probabilities are called *a priori probabilities* or *priors*. We define the *Bayes risk* as the overall average cost incurred by decision rule δ:

$$r(\delta) := \pi_0 R_0(\delta) + \pi_1 R_1(\delta). \tag{3.61}$$

Finally, we define the *Bayes rule* as the rule that *minimizes*, over all decision rules δ, the Bayes risk $r(\delta)$.

Combining Eqs. (3.60) and (3.61) the Bayes risk can be written as

$$r(\delta) = \pi_0\, C_{00} + \pi_1\, C_{01}$$
$$+ \int_{\mathcal{R}'} \Big(\pi_0(C_{10} - C_{00})\, p_0(x) + \pi_1(C_{11} - C_{01})\, p_1(x) \Big)\, \mathrm{d}x. \tag{3.62}$$

Then it is not difficult to see that the Bayes rule is to accept the hypothesis H_1 if the ratio

$$\Lambda(x) := \frac{p_1(x)}{p_0(x)} \tag{3.63}$$

is greater than the *threshold* value λ given by

$$\lambda := \frac{\pi_0}{\pi_1} \frac{C_{10} - C_{00}}{C_{01} - C_{11}}. \tag{3.64}$$

The ratio Λ is called the *likelihood ratio* and as we shall see it plays a key role in signal detection theory.

Let us introduce the probabilities

$$P_{\mathrm{F}} := \int_{\mathcal{R}'} p_0(x)\,\mathrm{d}x, \tag{3.65a}$$

$$P_{\mathrm{D}} := \int_{\mathcal{R}'} p_1(x)\,\mathrm{d}x. \tag{3.65b}$$

Probabilities P_{F} and P_{D} are the probabilities that the data has the values in the region $\mathcal{R}'$ when respectively the hypothesis H_0 and H_1 is true. In terms of probabilities P_{F} and P_{D} the Bayes risk is given by

$$r(\delta) = \pi_0 C_{00} + \pi_1 C_{01} + \pi_0 (C_{10} - C_{00}) P_{\mathrm{F}} + \pi_1 (C_{11} - C_{01}) P_{\mathrm{D}}. \tag{3.66}$$

In the theory of signal detection the probabilities P_{F} and P_{D} are the probabilities of *false alarm* and *detection*, respectively.

We say that priors are equal if

$$\pi_0 = \pi_1 = \frac{1}{2}. \tag{3.67}$$

We say that costs are *uniform* if

$$C_{ij} = \begin{cases} 0 & \text{if } i = j, \\ 1 & \text{if } i \neq j. \end{cases} \tag{3.68}$$

For equal priors and uniform costs the Bayes risk reads

$$r(\delta) = \frac{1}{2}(P_{\mathrm{F}} + P_{\mathrm{FD}}), \tag{3.69}$$

where P_{FD} is the *false dismissal* probability equal to $1 - P_{\mathrm{D}}$. The threshold value λ, Eq. (3.64), of the likelihood ratio in this case is

$$\lambda = 1. \tag{3.70}$$

Example 2. *Consider the following two hypotheses concerning a real-valued observation X:*

$$H_0 : X = N, \tag{3.71a}$$

$$H_1 : X = N + \mu, \tag{3.71b}$$

where $\mu > 0$. Let us assume that N has normal distribution $N(0, \sigma^2)$, then $N + \mu$ has also normal distribution $N(\mu, \sigma^2)$.

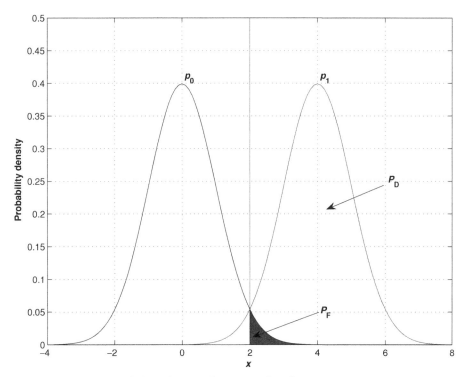

Fig. 3.1. Two probability density functions for the two alternative hypotheses considered in Example 2. The threshold value of the data x [computed from Eq. (3.73)] in the case of equal priors and uniform costs is plotted as a vertical line. The area under the curve p_0 to the right of the threshold is the false alarm probability P_{F} and the corresponding area under the curve p_1 is the detection probability P_{D}.

It is useful to introduce the quantity $d := \mu/\sigma$ that measures relative strength of the "signal" μ and noise. We shall call the quantity d the signal-to-noise ratio. *The likelihood ratio is then given by*

$$\Lambda(x) = \exp\left(\frac{x}{\sigma}d - \frac{1}{2}d^2\right). \qquad (3.72)$$

In Fig. 3.1 we have plotted the probability density functions p_0 (signal absent) and p_1 (signal present) for the two hypotheses.

Let us consider three cases for the choice of priors and costs. In all cases let us assume that $C_{00} = C_{11} = 0$.

1. Standard case. *Priors are equal and costs are uniform. Then the threshold is $\lambda = 1$.*

2. High risk case. *The prior probability π_0 is much higher than π_1. The cost C_{10} of accepting hypothesis H_1 when H_0 is true is high comparing to the cost C_{01} of accepting H_0 when H_1 is true. Let us take $\pi_0/\pi_1 = 10$ and $C_{10}/C_{01} = 10$. The threshold is $\lambda = 100$.*

3. Low risk case. *The prior π_1 is much higher than π_0 and the cost C_{01} of accepting hypothesis H_0 when H_1 is true is much higher than the cost C_{10} of asserting the opposite. We take $\pi_0/\pi_1 = 1/10$ and $C_{10}/C_{01} = 1/10$. The threshold is $\lambda = 10^{-2}$.*

An example of a high risk situation may be the case of detecting a gravitational-wave signal by a first generation detector. We expect a priori that the probability of finding such a signal is low and that the cost of erroneously announcing the detection of a gravitational wave is high. An example of a low risk situation may be the case of a radar protecting a densely populated city during the time of a high probability of a military conflict and a high cost of a successful missile strike.

As the likelihood ratio (3.72) is an increasing function of the observation x, the Bayes rule is equivalent to comparing the data x with the threshold

$$x_0 = \sigma \left(\frac{\ln \lambda}{d} + \frac{d}{2} \right). \tag{3.73}$$

When the data x exceeds the threshold value x_0 we announce the presence of the signal μ in the data. In the standard case of equal priors and uniform costs (and a normal distribution of data) the Bayes rule consists of comparing the data x with threshold $x_0 = \frac{1}{2}\mu$, i.e. with half the amplitude of the signal that we search for. In Fig. 3.2 we plot the size of the threshold x_0 as a function of the strength of the signal that we want to detect for various cases.

3.3.2 Minimax approach

Very often in practice we do not have control over or access to the mechanism generating the state of nature and we are not able to assign priors to various hypotheses. Consequently the Bayes solution cannot be found. In such a case we can consider the Bayes risk r as a function of unknown prior π_0, $r = r(\pi_0, \delta)$, where δ is the Bayes decision rule (see the preceding section). Thus for each value of the prior π_0 the decision rule δ is such that Bayes risk $r(\pi_0, \delta)$ is minimum. It can be shown that $r(\pi_0, \delta)$ is a convex function of π_0 over the interval $[0, 1]$. It then follows that r is a continuous function of π_0. One can consider the prior $\pi_0 = \pi_{\rm L}$ for which $r(\pi_0, \delta)$ attains a maximum with respect to π_0. Prior

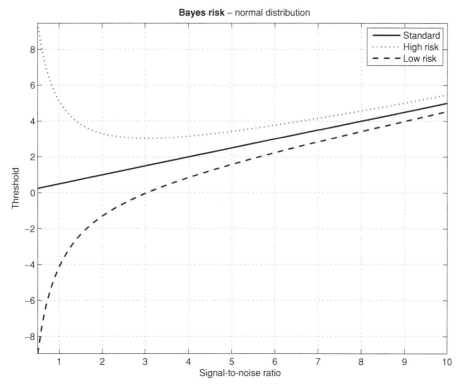

Bayes risk – normal distribution

Fig. 3.2. The threshold x_0 [computed by means of Eq. (3.73)] as a function of the signal-to-noise ratio for Bayes rules for three choices of the cost matrix considered in Example 2.

π_L is called the *least-favorable prior* because for π_L the Bayes risk is maximum. The Bayes rule corresponding to the least-favorable prior is called the *minimax rule*. Formally minimax decision rule is a decision δ_{π_L} that satisfies

$$r(\pi_L, \delta_{\pi_L}) = \min_{\delta} \max_{0 \leq \pi_0 \leq 1} r(\pi_0, \delta). \qquad (3.74)$$

We shall denote the *minimax risk* $r(\pi_L, \delta_{\pi_L})$ by r_{minmax}. In general, it may happen that the maximum of $r(\pi_0, \delta)$ is on the borders of the interval $[0, 1]$ and that the function r is not differentiable. In the latter case we need to consider *randomized decision rules*, i.e. such decisions which can be chosen randomly from a certain set. The *minimax theorem* [140, 141] says that with these extensions the minimax rule always exists.

The minimax criterion originates from game theory. John von Neumann observed that if poker players maximize their rewards, they do so by bluffing; and, more generally, that in many games it pays to be unpredictable. Formally he defined a *mixed strategy* if a player in a game chooses among two or more strategies at random according to specific probabilities. Mixed strategy in game theory corresponds to a randomized rule in theory of decisions. Von Neumann has shown that every two-person zero-sum game has a minimax solution if in the set of strategies we include both pure and mixed strategies [140, 141].

Let us consider an Observer and Nature as two opponents in a two-person, zero-sum game. Zero-sum game means that one player's gain is the other player's loss. Thus for a zero-sum game the cost matrix C of the Observer is minus the cost matrix of Nature. Nature picks one of the two hypothesis H_0 or H_1 and the Observer has to guess which was picked. If the Observer guesses H_1 when Nature has picked H_0 he or she must pay her an amount C_{10}. The minimax decision rule leads to a certain equilibrium. An Observer using the minimax rule is certain that no matter what prior probability π_0 Nature chooses his or her loss will not exceed r_{minmax}. Nature, on the other hand, is led to choose prior π_L because her gain will not be less than r_{minmax}.

The minimax rule is illustrated in Fig. 3.3, where we plot the Bayes risk as a function of the prior probability π_0. The minimax risk is given by a horizontal line tangent to r at the maximum.

3.3.3 Neyman–Pearson approach

In many problems of practical interest the imposition of a specific cost structure on the decisions made is not possible or desirable. The Neyman–Pearson approach involves a trade-off between the two types of errors that one can make in choosing a particular hypothesis. *Type I error* is choosing hypothesis H_1 when H_0 is true and *type II error* is choosing H_0 when H_1 is true. In signal detection theory probability of type I error is called *false alarm probability* whereas probability of type II error is called *false dismissal probability*. The *probability of detection* of the signal is $1-$(false dismissal probability). In hypothesis testing probability of type I error is called *significance of the test* whereas $1-$(probability of type II error) is called *power of the test*.

The Neyman–Pearson design criterion is to maximize the power of the test (probability of detection) subject to a chosen significance of the test (false alarm probability). The basic result is the *fundamental lemma of Neyman and Pearson* which we quote below as Lemma 1. In the lemma we are testing the hypothesis H_0 that the probability distribution of the data X is P_0 (no signal in the data) against the hypothesis H_1 that the data probability distribution is P_1 (signal present in the data). In the formulation of this lemma we identify the test with a function $\phi:$ $\mathcal{X} \to [0,1]$ whose values have the following interpretation: when $\phi(x)=1$ one rejects the hypothesis H_0, when $\phi(x)=0$ one accepts the hypothesis H_0, and when $\phi(x) \in (0,1)$ one uses an additional random mechanism,

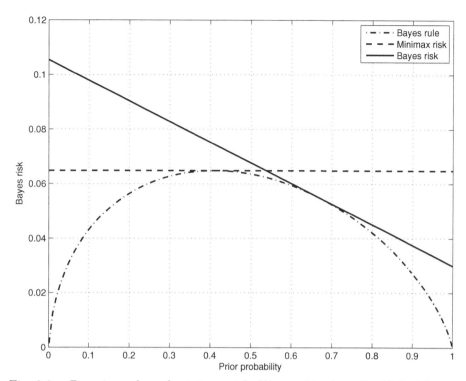

Fig. 3.3. Bayesian rule and minimax risk. We consider here the high risk case of Example 2 with a normal distribution of data and the signal-to-noise ratio $d = 4$. For each prior π_0 we apply the *Bayes rule* and obtain the corresponding Bayes risk (dashed-dotted line). The maximum of these risks is the *minimax risk*. It is shown as a horizontal line tangential to the Bayes-rule risk curve at its maximum (dashed line). The continuous line tangent to the Bayes-rule risk curve is the Bayes risk computed [by means of Eq. (3.62)] for an arbitrarily chosen prior, which, in general, does not coincide with the prior "chosen" by Nature. We see that the Bayes risk associated with a guess of the prior may be unbounded. Thus the minimax rule is a safe choice for unknown priors.

independent of the data, which rejects the hypothesis H_0 with probability $\phi(x)$. The test so constructed is called a *randomized test*.

Lemma 1. *Let P_0 and P_1 be probability distributions possessing respectively densities p_0 and p_1. Let $\alpha \in (0,1)$ be a fixed number.*

1. Existence of the test. *There exist constants λ and γ such that*

$$\phi(x) := \begin{cases} 1, & \text{when } p_1(x) > \lambda \, p_0(x), \\ \gamma, & \text{when } p_1(x) = \lambda \, p_0(x), \\ 0, & \text{when } p_1(x) < \lambda \, p_0(x) \end{cases} \qquad (3.75)$$

is a test of hypothesis H_0 against the alternative H_1 at the significance level α, i.e.

$$E_0\{\phi(X)\} = \alpha. \qquad (3.76)$$

2. Sufficient condition for the most powerful test. *If the test ϕ satisfies
   the condition (3.76) and for some constant λ it also satisfies the
   condition*

$$\phi(x) = \begin{cases} 1, & when\ p_1(x) > \lambda\, p_0(x), \\ 0, & when\ p_1(x) < \lambda\, p_0(x), \end{cases} \qquad (3.77)$$

   *then ϕ is the most powerful test for testing H_0 against H_1 at the
   significance level α.*

3. Necessary condition for the most powerful test. *If ϕ is the most
   powerful test for testing H_0 against H_1 at the significance level α,
   then for some constant λ it satisfies the condition (3.77).*

The proof of the lemma is given in Appendix B. It follows from the
Neyman–Pearson lemma that the most powerful test consists of comparing the likelihood ratio, i.e. the ratio of two pdfs p_1 and p_0 [see Eq.
(3.63)], to a threshold.

   In order to assess the performance of the likelihood ratio test based on
the Neyman–Pearson approach one usually examines power of the test
(probability of detection) as a function of significance level (false alarm
probability). For different signal-to-noise ratios d one obtains different
curves. The parametrized plot (with d being the parameter) of functions
$P_D(P_F; d)$ is called the *receiver operating characteristic* (ROC). The ROC
curves for the hypothesis problem in Example 2 (where the case of a
normal distribution of the data is considered) are shown in Fig. 3.4.

   By differentiating probabilities of detection and false alarm one immediately gets

$$\frac{dP_D}{dP_F} = \frac{p_1}{p_0} = \lambda. \qquad (3.78)$$

Thus the slope of the ROC curve for a given false alarm probability α is
equal to the threshold for Neyman–Pearson test at the level α.

   From Lemma 1 we have the following corollary.

Corollary 1. *Let β be the power of the most powerful test at the significance level α ($0 < \alpha < 1$) for a test of hypothesis H_0 against an alternative
hypothesis H_1. Then $\beta > \alpha$ unless $P_0 = P_1$.*

A consequence of the above corollary is that the ROC curves are convex.

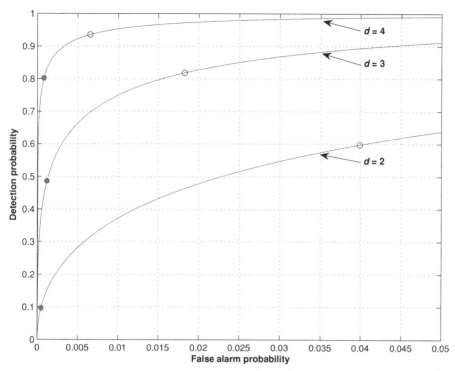

Fig. 3.4. Receiver operating characteristic (ROC). The plots are for three different signal-to-noise ratios (shown in the plot) and for the hypothesis testing problem considered in Example 2. The distribution of data is normal. The dots and circles are the Bayes and minimax solutions respectively in the high risk case of Example 2.

3.3.4 Likelihood ratio test

It is remarkable that all three very different approaches in hypothesis testing: Bayesian, minimax, and Neyman–Pearson, lead to the same test called the *likelihood ratio test* [142]. The likelihood ratio Λ is the ratio of the pdf when the signal is present to the pdf when it is absent [see Eq. (3.63)]. We accept the hypothesis H_1 if $\Lambda > \lambda$, where λ is the threshold that is calculated from the costs C_{ij}, priors π_i or the significance of the test depending on what approach is being used.

3.4 The matched filter in Gaussian noise: deterministic signal

3.4.1 Cameron–Martin formula

Let s be a *deterministic* signal we are looking for and let n be the detector's noise. For convenience we assume that the signal s is a continuous function of time t and that the noise n is a continuous stochastic process. Results for the discrete time data (that we have in practice) can then be obtained by a suitable sampling of the continuous in time expressions. Assuming that the noise n is *additive* the data x collected by the detector can be written as

$$x(t) = n(t) + s(t). \tag{3.79}$$

Suppose that the noise n is a *zero-mean* and *Gaussian* stochastic process with *autocorrelation function*

$$K(t, t') := \mathrm{E}\left\{n(t)n(t')\right\}. \tag{3.80}$$

Let the observational interval be $[0, T_o]$. Then it can be shown that the logarithm of the likelihood function is given by

$$\ln \Lambda[x] = \int_0^{T_o} q(t)x(t)\,\mathrm{d}t - \frac{1}{2}\int_0^{T_o} q(t)s(t)\,\mathrm{d}t, \tag{3.81}$$

where the function q is the solution of the integral equation

$$s(t) = \int_0^{T_o} K(t, t')q(t')\,\mathrm{d}t'. \tag{3.82}$$

The equation (3.81) for the likelihood ratio in the Gaussian noise case is called the *Cameron–Martin* formula.

From the expression (3.81) we see that the likelihood ratio test is equivalent to comparing the following quantity, which is the linear correlation of the data,

$$G := \int_0^{T_o} q(t)x(t)\,\mathrm{d}t, \tag{3.83}$$

to a threshold G_0. One easily finds that the expectation values of G when the signal s is absent or present in the data read

$$\mathrm{E}_0\left\{G\right\} = 0, \quad \mathrm{E}_1\left\{G\right\} = \int_0^{T_o} q(t)s(t)\,\mathrm{d}t, \tag{3.84}$$

and the variance of G is the same independently of whether signal is present or absent,

$$\begin{aligned}
\mathrm{Var}\{G\} &= \int_0^{T_o}\int_0^{T_o} q(t_1)q(t_2)\mathrm{E}\left\{n(t_1)n(t_2)\right\}\mathrm{d}t_1\,\mathrm{d}t_2 \\
&= \int_0^{T_o}\int_0^{T_o} q(t_1)q(t_2)K(t_1, t_2)\,\mathrm{d}t_1\,\mathrm{d}t_2 \\
&= \int_0^{T_o} q(t)s(t)\,\mathrm{d}t. \tag{3.85}
\end{aligned}$$

The quantity ρ given by

$$\rho^2 = \int_0^{T_o} q(t)s(t)\,\mathrm{d}t \qquad (3.86)$$

is called the *signal-to-noise ratio*.

Since data x are Gaussian and the statistic G (the word "statistic" means here any function applied to a set of data) is linear in x, it has a normal pdf, therefore pdfs p_0 and p_1 of the statistic G when respectively signal is absent or present in the data are given by

$$p_0(G) = \frac{1}{\sqrt{2\pi\rho^2}} \exp\left(-\frac{1}{2}\frac{G^2}{\rho^2}\right), \qquad (3.87a)$$

$$p_1(G) = \frac{1}{\sqrt{2\pi\rho^2}} \exp\left(-\frac{1}{2}\frac{(G-\rho^2)^2}{\rho^2}\right). \qquad (3.87b)$$

Probability of false alarm P_F and of detection P_D are readily expressed in terms of error functions,

$$P_F = \mathrm{erfc}\left(\frac{G_0}{\rho}\right), \qquad (3.88a)$$

$$P_D = \mathrm{erfc}\left(\frac{G_0}{\rho} - \rho\right), \qquad (3.88b)$$

where the complementary error function erfc is defined as

$$\mathrm{erfc}(x) := \frac{1}{\sqrt{2\pi}} \int_x^\infty \exp\left(-\frac{1}{2}t^2\right)\,\mathrm{d}t. \qquad (3.89)$$

We see that in the Gaussian case a single parameter, the signal-to-noise ratio ρ, determines both probabilities of false alarm and detection and consequently the receiver operating characteristic. For a given false alarm probability the greater the signal-to-noise ratio the greater the probability of detection of the signal.

3.4.2 Stationary noise

When the noise n is a *stationary* stochastic process (see Section 3.2), its autocorrelation function K depends only on the difference of times at which it is evaluated, i.e. there exists an *even* function R of one variable such that

$$\mathrm{E}\left\{n(t)\,n(t')\right\} = R(t - t'). \qquad (3.90)$$

Moreover assuming an idealized situation of observation over the infinite interval $(-\infty, \infty)$, the integral equation (3.82) takes the form

$$s(t) = \int_{-\infty}^{\infty} R(t - t')\, q(t')\, dt'. \tag{3.91}$$

This integral equation can easily be solved by taking the Fourier transform of both its sides. One then obtains that the Fourier transform $\tilde{q}$ (the tilde indicates here and below taking the Fourier transform) of the function q is given by

$$\tilde{q}(f) = \begin{cases} 2\dfrac{\tilde{s}(f)}{S_n(f)}, & \text{for } f \geq 0, \\[3mm] 2\dfrac{\tilde{s}(f)}{S_n(-f)}, & \text{for } f < 0, \end{cases} \tag{3.92}$$

where the function S_n is the *one-sided spectral density* of the detector's noise (defined only for non-negative frequencies $f \geq 0$). By Wiener–Khinchin theorem (Theorem 2) S_n is related to the Fourier transform of the noise autocorrelation function:

$$S_n(f) = 2 \int_{-\infty}^{\infty} R(\tau) \exp(-2\pi i f \tau)\, d\tau, \quad f \geq 0. \tag{3.93}$$

The one-sided spectral density S_n can also be determined through the equation for the correlation between the Fourier components of the noise,

$$\mathrm{E}\{\tilde{n}(f)\, \tilde{n}(f')^*\} = \frac{1}{2} S_n(|f|)\, \delta(f - f'), \qquad -\infty < f, f' < +\infty. \tag{3.94}$$

It is useful here to give the relation between the one-sided and *two-sided* spectral densities of the noise (or any other stationary stochastic process). Let us denote the two-sided spectral density by $\mathcal{S}_n$. It is determined for all frequencies $-\infty < f < +\infty$ and its relation with the one-sided density S_n is the following:

$$\mathcal{S}_n(f) = \begin{cases} S_n(f), & f \geq 0, \\ S_n(-f), & f < 0. \end{cases} \tag{3.95}$$

The two-sided spectral density $\mathcal{S}_n$ is just the Fourier transform of the noise autocorrelation function,

$$\mathcal{S}_n(f) = \int_{-\infty}^{\infty} R(\tau) \exp(-2\pi i f \tau)\, d\tau, \quad -\infty < f < +\infty, \tag{3.96}$$

and instead of Eq. (3.94) the following equation is fulfilled:

$$\mathrm{E}\{\tilde{n}(f)\, \tilde{n}(f')^*\} = \mathcal{S}_n(f)\, \delta(f - f'), \qquad -\infty < f, f' < +\infty. \tag{3.97}$$

Making use of the result (3.92) one can show that for the Gaussian and stationary noise the log likelihood function (3.81) can be expressed as

$$\ln \Lambda[x] = (x|s) - \frac{1}{2}(s|s),\qquad(3.98)$$

where the scalar product $(\cdot|\cdot)$ is defined by (here $\Re$ denotes the real part of a complex expression)

$$(x|y) := 4\Re \int_0^\infty \frac{\tilde{x}(f)\tilde{y}^*(f)}{S_n(f)}\,\mathrm{d}f.\qquad(3.99)$$

Using the scalar product defined above the signal-to-noise ratio ρ introduced in Eq. (3.86) can be written as

$$\rho^2 = (s|s) = 4 \int_0^\infty \frac{|\tilde{s}(f)|^2}{S_n(f)}\,\mathrm{d}f.\qquad(3.100)$$

3.4.3 Matched filtering

A *linear filter* is a map which relates its output function $O = O(t)$ to its input function $I = I(t)$ by means of equation

$$O(t) = (H * I)(t) := \int_{-\infty}^{+\infty} H(\tau)I(t-\tau)\,\mathrm{d}\tau,\qquad(3.101)$$

where $*$ denotes the *convolution*. The function H is called the *impulse response* of the filter. The Neyman–Pearson test for the zero-mean Gaussian noise consists of passing the data through a linear filter whose impulse response H is

$$H(\tau) := \begin{cases} q(T_\mathrm{o} - \tau), & 0 \le \tau \le T_\mathrm{o}, \\ 0, & \tau < 0 \ \text{ or } \ \tau > T_\mathrm{o}. \end{cases}\qquad(3.102)$$

It is easy to show that for the response function H given above it holds

$$(H * x)(T_\mathrm{o}) = G,\qquad(3.103)$$

where G is the detection statistic from Eq. (3.83). Thus the output of the filter (3.102) at the time $t = T_\mathrm{o}$ is equal to G and can be compared to a decision level G_0 to decide whether a signal is present or absent. Such a filter is called the *matched filter*.

In the case of *white noise* its autocorrelation function is proportional to the Dirac delta function,

$$R(\tau) = \frac{1}{2} S_0\,\delta(\tau),\qquad(3.104)$$

where the constant S_0 is the one-sided spectral density of noise. In this case the statistic G introduced in Eq. (3.83) is given by

$$G = \frac{2}{S_0} \int_{-\infty}^{\infty} s(t)x(t)\,dt. \tag{3.105}$$

Thus for the case of white noise the likelihood ratio test consists of correlating the data with a filter *matched* to the signal that we are searching for in the noise.

Let us consider another interpretation of the matched filter. Suppose that we pass data $x = n + s$ through linear filter F. The output signal-to-noise ratio d is defined by

$$d^2 = \frac{(s|F)^2}{\mathrm{E}\left\{(n|F)(n|F)\right\}}. \tag{3.106}$$

By Cauchy–Schwarz inequality we have $(s|F)^2 \leq (s|s)(F|F)$, where equality holds when $F(t) = A_0\,s(t)$ with A_0 being some constant. Because

$$\mathrm{E}\left\{(n|F)(n|F)\right\} = (F|F), \tag{3.107}$$

we have $d^2 \leq (s|s)$. Thus the signal-to-noise ratio d is maximum and equal to ρ [cf. Eq. (3.100)] when the filter F is proportional to the signal s. Consequently the matched filter is a filter that maximizes the signal-to-noise ratio over all linear filters [142]. This property is independent of the probability distribution of the noise in the data.

3.5 Estimation of stochastic signals

Let the data x consist of a signal s corrupted by an additive noise n,

$$x(t) = n(t) + s(t). \tag{3.108}$$

We would like to estimate the signal s given the data x. This problem is trivially solved when the signal s is deterministic and known. A non-trivial problem arises when s is a stochastic process. We consider now this case.

Let us assume that for both stochastic processes x and s the autocovariance function exists. Let us also assume for simplicity that x and s are zero-mean stochastic processes. We shall look for an estimate $\hat{s}$ of the signal s that is a linear functional of the data. Thus $\hat{s}$ has the form

$$\hat{s}(t) = \int_0^{T_o} q(t, u)x(u)\,du, \tag{3.109}$$

where, as before, $[0, T_o]$ is the observational interval. To obtain an estimate we need some optimality criterion. We shall find the best estimate in the *minimum mean squared error* sense, i.e. a $\hat{s}$ realizing the minimum

$$\min_{z \in H} \mathrm{E}\left\{ \left(z(t) - s(t) \right)^2 \right\}, \tag{3.110}$$

where H is the space of all estimators of the form (3.109). We have the following important result.

Proposition 1 (Orthogonality principle). *An estimate $\hat{s}$ of the form (3.109) realizes the minimum (3.110) if and only if*

$$\mathrm{E}\left\{ \left(\hat{s}(t) - s(t) \right) z(t') \right\} = 0 \tag{3.111}$$

for all $z \in H$.

Let us substitute data x for z in Eq. (3.111). After simple manipulations we obtain the following equation

$$C_{sx}(t, t') = \int_0^{T_o} q(t, u) C_x(u, t') \, \mathrm{d}u, \tag{3.112}$$

where C_{sx} is the cross-covariance function of the data x and the signal s and C_x is the autocovariance function of the data x. The integral equation (3.112) is called the *Wiener–Hopf equation* and its solution q is called the *Wiener–Kolmogorov filter*. Given q we have estimate $\hat{s}$ of the signal s that minimizes the mean squared error.

Let us now assume that the noise and the signal itself are both *stationary* stochastic processes and moreover that they are *uncorrelated*. Then the Wiener–Hopf equation (3.112) takes the form

$$C_s(\tau) = \int_0^{T_o} q(\tau - \tau') \, C_x(\tau') \, \mathrm{d}\tau', \tag{3.113}$$

where C_s and C_x are the autocovariances of the signal s and the data x, respectively. The Wiener–Hopf equation (3.113) can be solved by taking the Fourier transform of both its sides. One gets that the Fourier transform $\tilde{q}$ of the Wiener–Kolmogorov filter q is given by

$$\tilde{q}(f) = \frac{\mathcal{S}_s(f)}{\mathcal{S}_s(f) + \mathcal{S}_n(f)}, \tag{3.114}$$

where $\mathcal{S}_s$ and $\mathcal{S}_n$ are *two-sided* spectral densities of the signal and the noise, respectively. One can also obtain a closed-form expression for the minimum mean squared error (MMSE) [see Eq. (3.110)],

$$\mathrm{MMSE} = \int_{-\infty}^{\infty} \frac{\mathcal{S}_s(f) \, \mathcal{S}_n(f)}{\mathcal{S}_s(f) + \mathcal{S}_n(f)} \, \mathrm{d}f. \tag{3.115}$$

The matched filter for Gaussian signal in Gaussian noise. Let the data x consist of signal s superimposed on the detector's noise n. Let both s and n be zero-mean and *Gaussian* stochastic processes. We also assume that the stochastic processes s and n are *uncorrelated*. One can show that the only term in the log likelihood ratio that depends on data x has the form

$$U = \int_0^{T_o} \int_0^{T_o} x(t)\, Q(t,u)\, x(u)\, du\, dt, \qquad (3.116)$$

where $[0, T_o]$ is the observational interval and Q is given by solution of the following integral equation:

$$C_s(t,v) = \int_0^{T_o} \int_0^{T_o} C_n(t,p)\, Q(p,u)\big(C_s(u,v) + C_n(u,v)\big)\, du\, dp, \quad (3.117)$$

where C_s and C_n are the autocovariance functions of the signal and the noise, respectively.

Let us suppose that both signal and noise are *stationary* stochastic processes. Then the integral equation (3.117) takes the form

$$C_s(t-v) = \int_0^{T_o} \int_0^{T_o} C_n(t-p)\, Q(p-u)\big(C_s(u-v) + C_n(u-v)\big)\, du\, dp.$$

$$(3.118)$$

When the observation time T_o is much larger than the correlation time for both the signal s and the noise n we can approximately solve the above equation by means of the Fourier transform technique. We find that the Fourier transform $\tilde{Q}$ of the filter Q is given by

$$\tilde{Q}(f) = \frac{S_s(f)}{S_n(f)\,(S_s(f) + S_n(f))}. \qquad (3.119)$$

We see that the optimal filter given by Eq. (3.116) is quadratic in the data. The optimal filtering to detect the stochastic signal s also has the following interesting interpretation. We can consider it as a two-stage procedure. The first step is an estimation of the signal s. To obtain the estimate $\hat{s}$ of the signal we use the optimal (in the mean squared error sense) filtering procedure given by Eqs. (3.109) and (3.112). We then apply matched filter defined by Eqs. (3.81) and (3.82) to detect the signal $\hat{s}$ in data x. Thus we can interpret the optimal statistic U as an *estimator–correlator*.

Suppose that we have two detectors measuring the same stochastic signal s, i.e. we have two data streams x_1 and x_2:

$$x_1(t) = n_1(t) + r_1(t), \qquad (3.120a)$$
$$x_2(t) = n_2(t) + r_2(t), \qquad (3.120b)$$

where r_1 and r_2 are the responses of respectively the first and the second detector to the same stochastic signal s. The two responses are in general different for detectors located in different places on Earth. This leads to the following optimal statistic

$$U_{12} = \int_0^{T_o} \int_0^{T_o} x_1(t) \, Q_{12}(t, u) \, x_2(u) \, du \, dt. \qquad (3.121)$$

Using the estimator–correlator concept introduced above we can interpret the statistic U_{12} as an estimate of the signal s from the data of the first detector followed by a matched filter to detect the signal in the data of the second detector. In Section 6.4 we shall derive the form of the filter Q_{12} that maximizes the signal-to-noise ratio.

3.6 Estimation of parameters

Very often we know the form of the signal that we are searching for in the data in terms of a finite number of unknown parameters. We would like to find optimal procedures of estimating these parameters. An estimator of a parameter θ is a function $\hat{\theta}(x)$ that assigns to each data x the "best" guess of the true value of θ. Note that as $\hat{\theta}(x)$ depends on the random data an estimator is always a random variable.

In this section we consider a *statistical model* $(\mathcal{X}, \mathcal{O}, \{P_\theta, \boldsymbol{\theta} \in \Theta\})$, where $\{P_\theta, \boldsymbol{\theta} \in \Theta\}$ is a family of probability distributions (defined on some σ-algebra $\mathcal{O}$ of events) parametrized by parameters $\boldsymbol{\theta} = (\theta_1, \dots, \theta_K)$ which take their values from the set $\Theta \subset \mathbb{R}^K$. Functions $T : \mathcal{X} \to \mathbb{R}^k$ are called (k-dimensional) *statistics*. We also introduce n-dimensional random variable $\mathbf{X} = (X_1, \dots, X_n)$ called a *sample* drawn from a probability distribution P_θ (for some $\boldsymbol{\theta} \in \Theta$), where X_i ($i = 1, \dots, n$) are independent and identically distributed random variables with probability distribution P_θ.

3.6.1 Uniformly minimum variance unbiased estimators

Ideally we would like our estimator to be (i) *unbiased*, i.e. its expectation value to be equal to the true value of the parameter, and (ii) *of minimum variance*. Also we would like that our estimator is of minimum variance and unbiased independently of the value of the parameter. Such estimators are termed *uniformly minimum variance unbiased* estimators. In this section we consider a general structure of such estimators. We begin by introducing several basic concepts of estimation theory.

Definition 1. *A statistic T is said to be a* sufficient *statistic for the parameter θ if the conditional distribution $P_\theta(\,\cdot\,|T=t)$ is independent of θ.*

For a sufficient statistic T, knowledge of $T(\mathbf{X})$ removes any further dependence of the distribution of $\mathbf{X}$ on θ. Thus a sufficient statistic $T(\mathbf{X})$ contains all the information in $\mathbf{X}$ about the parameter θ.

Definition 2. *A statistic T sufficient for the parameter θ is said to be a* minimal sufficient *statistic for θ if for any statistic S which is sufficient for θ there exists a function f such that $T=f(S)$.*

The following theorem characterizes a sufficient statistic uniquely.

Theorem 3 (The factorization criterion). *A statistic T is sufficient if and only if the pdf p_θ of the sample $\mathbf{X}$ has the form*

$$p_\theta(\mathbf{x}) = g_\theta\big(T(\mathbf{x})\big)\,h(\mathbf{x}), \tag{3.122}$$

where the function h is independent of θ and the function g_θ depends on $\mathbf{x}$ only through the value of the statistic T.

The above theorem is useful in extracting sufficient statistics from probability distributions.

Definition 3 (Completeness). *A family of distributions P_θ is said to be* complete *if condition $\mathrm{E}_\theta\{f(\mathbf{X})\}=0$ for all θ implies $f=0$. A statistic T is complete if its family of distributions is complete.*

For a complete statistic there is no (not being a constant) function f of the statistic that has the expectation value independent of θ.

Example 3. *Let $\mathbf{X}=(X_1,\ldots,X_n)$ be a sample drawn from a Poisson distribution $P(\tau)$. Then the joint pdf of $\mathbf{X}$ is given by*

$$p(\mathbf{x};\tau) = \frac{\tau^{x_1+\ldots+x_n}}{x_1!\cdots x_n!}\mathrm{e}^{-\tau n} = g_\tau(T(x))\,h(x), \tag{3.123}$$

where

$$T(x) := \sum_{i=1}^{n} x_i, \quad g_\tau(T) := \mathrm{e}^{-\tau n}\tau^T, \quad h(x) := \prod_{i=1}^{n}\frac{1}{x_i!}. \tag{3.124}$$

Thus T is a sufficient statistic. It is also a complete sufficient statistic.

A general relation between completeness and sufficiency is given by the following theorem.

Theorem 4. *If T is a complete sufficient statistic then it is a minimal sufficient statistic.*

An effective construction of minimum variance unbiased estimators is based on the following two theorems.

Theorem 5 (Rao–Blackwell theorem). *Let P_θ be a family of distributions and let $g(\theta)$ be a given function. If $\hat{g}$ is an unbiased estimator of function g and if T is a sufficient statistic for θ then $E_\theta\{(\hat{g}|T)\}$ is also an unbiased estimator and its variance is uniformly not greater than variance of $\hat{g}$.*

Theorem 6. *If T is a complete sufficient statistic and if for a given function g there exists another function $\hat{g}$ such that*

$$E_\theta\{\hat{g}(T)\} = g(\theta), \qquad (3.125)$$

then $\hat{g}(T)$ is the unbiased minimum variance estimator of $g(\theta)$.

From the two theorems above it follows that if $\tilde{g}(\theta)$ is an arbitrary unbiased estimator of function $g(\theta)$ and if T is a complete sufficient statistic then $\hat{g}(T) = E_\theta\{(\tilde{g}|T)\}$ is the minimum variance unbiased estimator of $g(\theta)$.

Example 4. *We assume that distribution of the number of certain events follows the Poisson distribution $P(\tau)$. We want to estimate from a sample $\mathbf{X} = (X_1, X_2, \ldots, X_n)$ of the distribution the probability $\lambda := \exp(-\tau)$ that in a certain interval no event will occur. A simple estimator $\tilde{\lambda}$ is given by*

$$\tilde{\lambda} := \begin{cases} 1, & when\ x_1 = 0, \\ 0, & otherwise. \end{cases} \qquad (3.126)$$

As $T(\mathbf{X}) := \sum_{i=1}^{n} X_i$ is a complete sufficient statistic, the estimator $\hat{\lambda} = E\{(\tilde{\lambda}|T)\}$ is the desired minimum variance unbiased estimator. One can show that

$$\hat{\lambda} = \left(1 - \frac{1}{n}\right)^T. \qquad (3.127)$$

Thus from a crude unbiased estimator by Rao–Blackwell theorem we can obtain the minimum variance unbiased estimator.

3.6.2 Exponential families of probability distributions

It is useful to introduce a general family of *exponential* probability distributions. Let $\boldsymbol{\theta}$ be a vector consisting of K parameters $(\theta_1, \theta_2, \ldots, \theta_K) \in \Theta \subset \mathbb{R}^K$. Thus the parameters $\boldsymbol{\theta}$ span a subset Θ of a K-dimensional Euclidean space $\mathbb{R}^K$.

Definition 4. *A family of probability distributions is* exponential *if their pdfs have the form*

$$p(x; \boldsymbol{\theta}) = C(\boldsymbol{\theta}) \exp \left(\sum_{i=1}^{K} Q_i(\boldsymbol{\theta}) \, T_i(x) \right) h(x). \qquad (3.128)$$

Often it is enough to consider an exponential family of pdfs of the following simpler form:

$$p(x; \boldsymbol{\theta}) = C(\boldsymbol{\theta}) \exp \left(\sum_{i=1}^{K} \theta_i \, T_i(x) \right) h(x). \qquad (3.129)$$

Theorem 7. *If $p(\cdot\,; \boldsymbol{\theta})$ is an exponential family of pdfs given by equation (3.129) then $(T_1(X), T_2(X), \ldots, T_K(X))$ is a K-dimensional minimal and complete sufficient statistic.*

Under certain regularity conditions $(T_1(X), T_2(X), \ldots, T_K(X))$ is a K-dimensional minimal and complete sufficient statistic also for an exponential family given by Eq. (3.128).

One can prove that for the family of pdfs given in Eq. (3.129) the following formula holds:

$$\mathrm{E}_{\boldsymbol{\theta}} \left\{ \prod_{i=1}^{K} T_i^{l_i}(X) \right\} = C(\boldsymbol{\theta}) \, \frac{\partial^l}{\partial \theta_1^{l_1} \cdots \partial \theta_K^{l_K}} \frac{1}{C(\boldsymbol{\theta})}, \qquad (3.130)$$

where $l = l_1 + l_2 + \cdots + l_K$. In particular,

$$\mathrm{E}_{\boldsymbol{\theta}}\{T_i(X)\} = \frac{\partial \ln C(\boldsymbol{\theta})}{\partial \theta_i}, \qquad i = 1, \ldots, K. \qquad (3.131)$$

Example 5. *Let a sample $\mathbf{X} = (X_1, \ldots, X_n)$ be drawn from a normal distribution $N(\mu, \sigma^2)$, whose pdf can be written as [here $\boldsymbol{\theta} = (\mu, \sigma)$]*

$$p(x; \mu, \sigma) = \exp \left(-\frac{1}{2\sigma^2} x^2 + \frac{\mu}{\sigma^2} x - \left(\frac{\mu^2}{2\sigma^2} + \ln \left(\sigma \sqrt{2\pi} \right) \right) \right). \qquad (3.132)$$

Normal distribution is thus an exponential distribution and $T_1(\mathbf{X}) := \sum_{i=1}^{n} X_i$, $T_2(\mathbf{X}) := \sum_{i=1}^{n} X_i^2$ *are complete sufficient statistics. Let us consider the following two functions of the statistics* T_1 *and* T_2:

$$\hat{g}_1(T_1) := \frac{T_1}{n}, \qquad \hat{g}_2(T_1, T_2) := \frac{1}{n-1}\left(T_2 - \frac{1}{n}T_1^2\right). \qquad (3.133)$$

It is easy to check that $\mathrm{E}_{\boldsymbol{\theta}}\{\hat{g}_1\} = \mu$ *and* $\mathrm{E}_{\boldsymbol{\theta}}\{\hat{g}_2\} = \sigma^2$. *It then follows from Theorem 6 that* $\hat{g}_1$ *and* $\hat{g}_2$ *are minimum variance unbiased estimators of the mean and variance of the normal distribution. The estimator* $\hat{g}_1$ *has the normal distribution* $N(\mu, \sigma^2/n)$, *whereas* $(n-1)\hat{g}_2/\sigma^2$ *has the* χ^2 *distribution with* $n-1$ *degrees of freedom. The variances of the estimators* $\hat{g}_1$ *and* $\hat{g}_2$ *are* σ^2/n *and* $\sigma^4/(n-1)$ *respectively. Thus as sample size* n *increases these variances tend to 0. Estimators with this property are called consistent.*

3.6.3 Fisher information

Let $p(x; \boldsymbol{\theta})$ be a pdf of a random variable X that depends on the parameters $\boldsymbol{\theta} = (\theta_1, \theta_2, \dots, \theta_K)$. The square $K \times K$ matrix $\Gamma(\boldsymbol{\theta})$ with components

$$\Gamma_{ij}(\boldsymbol{\theta}) := \mathrm{E}_{\boldsymbol{\theta}}\left\{ \frac{\partial \ln p(x; \boldsymbol{\theta})}{\partial \theta_i} \frac{\partial \ln p(x; \boldsymbol{\theta})}{\partial \theta_j} \right\}, \quad i, j = 1, \dots, K, \quad (3.134)$$

is called the *Fisher information matrix* about parameters $\boldsymbol{\theta}$ based on the random variable X. The random vector with coordinates $\partial \ln p(x; \boldsymbol{\theta})/\partial \theta_i$ $(i = 1, \dots, K)$ is called the *score function*. We have the following results

$$\mathrm{E}_{\boldsymbol{\theta}}\left\{ \frac{\partial \ln p(x; \boldsymbol{\theta})}{\partial \theta_i} \right\} = 0, \quad i = 1, \dots, K, \qquad (3.135\text{a})$$

$$\mathrm{E}_{\boldsymbol{\theta}}\left\{ \frac{\partial^2 \ln p(x; \boldsymbol{\theta})}{\partial \theta_i \partial \theta_j} \right\} = -\Gamma_{ij}(\boldsymbol{\theta}), \quad i, j = 1, \dots, K. \qquad (3.135\text{b})$$

In the case of an exponential family of pdfs given in Eq. (3.129), the Fisher matrix $\Gamma(\boldsymbol{\theta})$ can easily be calculated using Eq. (3.135b). The result is

$$\Gamma_{ij}(\boldsymbol{\theta}) = -\frac{\partial^2}{\partial \theta_i \partial \theta_j} \ln C(\boldsymbol{\theta}), \quad i, j = 1, \dots, K. \qquad (3.136)$$

Example 6. *1. For the binomial distribution* $b(p, n)$ *the Fisher information* $\Gamma(p)$ *is given by*

$$\Gamma(p) = \frac{n}{p(1-p)}. \qquad (3.137)$$

2. *For the Poisson distribution* $P(\tau)$ *the Fisher information* $\Gamma(\tau)$ *equals*

$$\Gamma(\tau) = \frac{1}{\tau}. \tag{3.138}$$

3. *For the normal distribution* $N(\mu, \sigma^2)$ *the two-dimensional Fisher matrix* $\Gamma(\boldsymbol{\theta})$, *where* $\boldsymbol{\theta} = (\mu, \sigma)$, *is given by*

$$\Gamma(\boldsymbol{\theta}) = \begin{pmatrix} \dfrac{1}{\sigma^2} & 0 \\ 0 & \dfrac{2}{\sigma^2} \end{pmatrix}. \tag{3.139}$$

3.6.4 Kullback–Leibler divergence

Kullback and Leibler [143] define the *divergence* $d_{\mathrm{KL}}(p, q)$ between two pdfs p and q as

$$d_{\mathrm{KL}}(p, q) := E_p\left\{\ln \frac{p}{q}\right\} + E_q\left\{\ln \frac{q}{p}\right\}. \tag{3.140}$$

The Kullback–Leibler divergence measures how far apart two distributions are in the sense of likelihood. That is, if an observation were to come from one of the distributions, how likely is it that the observation did not come from the other distribution?

Let us consider the case of detection of a signal in noise. Then we have two pdfs: p_0 when the signal is absent and p_1 when the signal is present. Let $L(x) := \ln\big(p_1(x)/p_0(x)\big)$ be the log likelihood ratio. Then the Kullback–Leibler divergence can be expressed as

$$d_{\mathrm{KL}}(p_0, p_1) = E_1\{L\} - E_0\{L\}, \tag{3.141}$$

where E_1, E_0 are expectation values when respectively signal is present and absent. Let us assume that the noise in the detector is zero-mean, Gaussian, and stationary. Then the log likelihood ratio is given by

$$L = (x|s) - \frac{1}{2}(s|s), \tag{3.142}$$

where x are the data, s is the signal we are looking for, and the scalar product $(\cdot|\cdot)$ is defined in Eq. (3.99). Moreover we assume that the noise n in the data is additive. Then we immediately have

$$d_{\mathrm{KL}}(p_0, p_1) = (s|s). \tag{3.143}$$

Thus in this case the Kullback–Leibler divergence is precisely equal to the signal-to-noise ratio squared ρ^2 [see Eq. (3.100)].

Suppose now that we consider two pdfs that belong to the same K-dimensional family $p(x; \boldsymbol{\theta})$ of pdfs parametrized by the parameters

$\boldsymbol{\theta} = (\theta_1, \ldots, \theta_K)$. We shall denote the KL-divergence $d_{\mathrm{KL}}\big(p(x;\boldsymbol{\theta}), p(x;\boldsymbol{\theta}')\big)$ by $d_{\mathrm{KL}}(\boldsymbol{\theta}, \boldsymbol{\theta}')$. Then it is not difficult to find the following relation between the KL-divergence and the Fisher matrix Γ:

$$\frac{\partial^2 d_{\mathrm{KL}}(\boldsymbol{\theta}, \boldsymbol{\theta}')}{\partial \theta'_i \partial \theta'_j}\bigg|_{\boldsymbol{\theta}'=\boldsymbol{\theta}} = 2\,\Gamma_{ij}(\boldsymbol{\theta}), \quad i, j = 1, \ldots, K. \tag{3.144}$$

One also easily finds that

$$d_{\mathrm{KL}}(\boldsymbol{\theta}, \boldsymbol{\theta}) = 0, \qquad \frac{\partial d_{\mathrm{KL}}(\boldsymbol{\theta}, \boldsymbol{\theta}')}{\partial \theta'_i}\bigg|_{\boldsymbol{\theta}'=\boldsymbol{\theta}} = 0, \quad i = 1, \ldots, K. \tag{3.145}$$

When $\boldsymbol{\theta}' = \boldsymbol{\theta} + \delta\boldsymbol{\theta}$ where $\delta\boldsymbol{\theta}$ is small we can perform the Taylor expansion of d_{KL}. Keeping only the terms up to the second order and using Eqs. (3.144) and (3.145) above we find that

$$d_{\mathrm{KL}}(\boldsymbol{\theta}, \boldsymbol{\theta} + \delta\boldsymbol{\theta}) \cong \sum_{i,j=1}^{K} \Gamma_{ij}(\boldsymbol{\theta})\, \delta\theta_i\, \delta\theta_j. \tag{3.146}$$

The above equation gives the interpretation of the Fisher matrix as a measure of distance between the pdfs of a parametric family. The advantage of the KL-divergence over the Fisher matrix as a measure of distance between the pdfs is that it can be used even if the pdfs under consideration are not all members of the same parametric family. The disadvantage of the KL-divergence is that it does not fulfill the triangle inequality and therefore it is not a true distance.

3.6.5 L_1-norm distance

We can define the following L_1-norm distance d_L between the pdfs p and q defined on the same observation space X [144]

$$d_L(p, q) := \frac{1}{2} \int_X |p(x) - q(x)|\, \mathrm{d}x. \tag{3.147}$$

The d_L distance is a true metric, i.e. $d_L(p, q) = 0$ if and only if $p = q$, $d_L(p, q) = d_L(q, p)$, and the triangle inequality holds: $d_L(p, q) + d_L(q, r) \geq d_L(p, r)$. Moreover the distance d_L has a probabilistic interpretation. It is related to Bayes risk r for equal priors and uniform costs [see Eq. (3.69)] by the following formula (see Section 3.1 of [144]):

$$r = \frac{1}{2}(1 - d_L). \tag{3.148}$$

Consequently for the case of detection of a known signal in noise we have

$$d_L = P_{\mathrm{D}} - P_{\mathrm{F}}, \tag{3.149}$$

where P_D and P_F are probability of detection and false alarm, respectively. Thus the two approaches: the Bayesian one that seeks to minimize Bayes risk and the Neyman–Pearson that seeks to maximize probability of detection subject to a given false alarm probability, lead to maximization of the L_1-norm distance.

In the Gaussian case the distance d_L between the pdf when signal s is present and the pdf when the signal is absent (see [144]) is given by

$$d_L = 1 - 2\,\mathrm{erfc}\left(\frac{1}{2\sqrt{2}}\sqrt{(s|s)}\right), \qquad (3.150)$$

where erfc is the complementary error function defined in Eq. (3.89).

3.6.6 Entropy

We define the *entropy* H of a pdf p as

$$H := -\mathrm{E}\left\{\ln(p(x))\right\}. \qquad (3.151)$$

The quantity $I := -H$ is called *Shannon information* [145] associated with the pdf p.

Maximum entropy principle. Suppose we have certain data x and we would like to determine what is the probability distribution of x. Suppose we have certain additional information about data that can be expressed in terms of K constraints:

$$\mathrm{E}\left\{T_i(x)\right\} = \lambda_i \quad \text{for } i = 1, \ldots, K. \qquad (3.152)$$

The *maximum entropy principle* of Jaynes [146] states that we should choose p so that its entropy is maximum. By the Boltzmann theorem the distribution that satisfies the maximum entropy principle subject to the constraints (3.152) is the exponential distribution given by Eq. (3.129).

3.6.7 Bayesian estimation

We assign a cost function $C(\theta', \theta)$ of estimating the true value of θ as θ'. We then associate with an estimator $\hat{\theta}$ a conditional risk or cost averaged over all realizations of data x for each value of the parameter θ:

$$R_\theta(\hat{\theta}) = \mathrm{E}_\theta\{C(\hat{\theta}, \theta)\} = \int_X C(\hat{\theta}(x), \theta)\, p_\theta(x)\, \mathrm{d}x, \qquad (3.153)$$

where X is the set of observations and $p_\theta(x)$ is the joint probability distribution of data x and parameter θ. We further assume that there is a certain a priori probability distribution $\pi(\theta)$ of the parameter θ. We then define the *Bayes estimator* as the estimator that minimizes the average risk defined as

$$r(\hat{\theta}) = \mathrm{E}\{R_\theta(\hat{\theta})\} = \int_X \int_\Theta C(\hat{\theta}(x), \theta)\, p_\theta(x)\, \pi(\theta)\, \mathrm{d}\theta\, \mathrm{d}x, \qquad (3.154)$$

where E is the expectation value with respect to a priori distribution π and Θ is the set of observations of the parameter θ.

It is not difficult to show that for a commonly used cost function

$$C(\theta', \theta) := (\theta' - \theta)^2, \qquad (3.155)$$

the Bayesian estimator is the conditional mean of the parameter θ given data x, i.e.

$$\hat{\theta}(x) = \mathrm{E}\{\theta | x\} = \int_\Theta \theta\, p(\theta | x)\, \mathrm{d}\theta, \qquad (3.156)$$

where $p(\theta | x)$ is the conditional probability density of parameter θ given the data x.

3.6.8 Maximum a posteriori probability estimation

Suppose that in a given estimation problem we are not able to assign a particular cost function $C(\theta', \theta)$. Then a natural choice is a uniform cost function equal to 0 over a certain interval I_θ of the parameter θ [131]. From Bayes theorem [137] we have

$$p(\theta | x) = \frac{p_\theta(x) \pi(\theta)}{p(x)}, \qquad (3.157)$$

where $p(x)$ is the pdf of data x. Then from Eq. (3.154) one can deduce that for each data x the Bayes estimate is any value of θ that maximizes conditional probability $p(\theta | x)$. The pdf $p(\theta | x)$ is also called *a posteriori* pdf of the parameter θ and the estimator that maximizes $p(\theta | x)$ is called the *maximum a posteriori* (MAP) estimator. It is denoted by $\hat{\theta}_{\mathrm{MAP}}$. The MAP estimators are thus solutions of the following equation

$$\frac{\partial \ln p(x, \theta)}{\partial \theta} = -\frac{\partial \ln \pi(\theta)}{\partial \theta}, \qquad (3.158)$$

called the *MAP equation*.

3.6.9 Maximum-likelihood estimation

Often we do not know a priori the probability density of a given parameter and we simply assign to it a uniform probability. In such a case the maximization of a posteriori probability is equivalent to the maximization of the probability density $p(x, \theta)$ treated as a function of θ. We call the function $l(\theta, x) := p(x, \theta)$ the *likelihood function* and the value of the parameter θ that maximizes $l(\theta, x)$ the *maximum likelihood* (ML) estimator. Instead of function l we can use the function $\Lambda(\theta, x) := l(\theta, x)/p(x)$ (assuming that $p(x) > 0$). Λ is then equivalent to the likelihood ratio [see Eq. (3.63)] when the parameters of the signal are known. Then ML estimators are obtained by solving the equation

$$\frac{\partial \ln \Lambda(\theta, x)}{\partial \theta} = 0, \qquad (3.159)$$

called the *ML equation*.

A very important property of the ML estimators is that asymptotically (i.e. for signal-to-noise ratio tending to infinity) they are (i) unbiased, and (ii) they have a Gaussian distribution with the covariance matrix equal to the inverse of the Fisher information matrix.

3.6.10 Lower bounds on the variance of estimators

Let first consider the case of one parameter θ. Assuming a number of mild conditions on the pdf $p(x; \theta)$ we have the following theorem [147, 148].

Theorem 8 (Cramèr–Rao lower bound). *Let $\Gamma(\theta)$ be the Fisher information. Suppose that $\Gamma(\theta) > 0$, for all θ. Let $\phi(X)$ be a one-dimensional statistic with $\mathrm{E}\{\phi(X)\} < \infty$ for all θ. Then*

$$\mathrm{Var}\{\phi(X)\} \geq \frac{\left(\dfrac{\mathrm{d}}{\mathrm{d}\theta} \mathrm{E}\{\phi(X)\}\right)^2}{\Gamma(\theta)}. \qquad (3.160)$$

Proof. We differentiate the normalization condition for the pdf p,

$$\int p(x; \theta) \, \mathrm{d}x = 1, \qquad (3.161)$$

with respect to the parameter θ. Assuming that we can differentiate under the integral sign, we get

$$0 = \int \frac{\partial p(x; \theta)}{\partial \theta} \, \mathrm{d}x = \int \frac{\partial \ln p(x; \theta)}{\partial \theta} p(x; \theta) \, \mathrm{d}x = \mathrm{E}\left\{\frac{\partial \ln p(x; \theta)}{\partial \theta}\right\}. \qquad (3.162)$$

Let us also differentiate with respect to θ the mean value of the statistic ϕ,

$$\frac{\mathrm{d}}{\mathrm{d}\theta} \mathrm{E}\{\phi(X)\} = \int \phi(x) \frac{\partial p(x; \theta)}{\partial \theta} \, \mathrm{d}x = \mathrm{E}\left\{\phi(X) \frac{\partial \ln p(x; \theta)}{\partial \theta}\right\}$$

$$= \mathrm{E}\left\{\left(\phi(X) - \mathrm{E}\{\phi(X)\}\right) \frac{\partial \ln p(x; \theta)}{\partial \theta}\right\}. \qquad (3.163)$$

Now taking the absolute value and using the Cauchy–Schwarz inequality (which states that $|\mathrm{E}\{\phi\psi\}| \leq \sqrt{\mathrm{E}\{\phi^2\}}\sqrt{\mathrm{E}\{\psi^2\}}$ for any two random variables ϕ and ψ) we obtain

$$\left|\frac{\mathrm{d}}{\mathrm{d}\theta}\mathrm{E}\left\{\phi(X)\right\}\right| \leq \sqrt{\mathrm{E}\left\{\left(\phi(X) - \mathrm{E}\left\{\phi(X)\right\}\right)^2\right\}}\sqrt{\mathrm{E}\left\{\left(\frac{\partial \ln p(x;\theta)}{\partial \theta}\right)^2\right\}}.$$

$$(3.164)$$

By squaring this and using Eq. (3.134) for Fisher information (for one-dimensional case) we obtain the inequality (3.160). $\qquad\square$

If the variance of an unbiased estimator of the parameter θ attains the lower Cramèr–Rao bound the estimator is called *efficient*.

In the multiparameter case where the vector $\boldsymbol{\theta} = (\theta_1,\ldots,\theta_K)$ collects all the K parameters, we have the following lower bound.

Theorem 9 (Multiparameter Cramèr–Rao lower bound). *Let* $\Gamma(\boldsymbol{\theta})$ *be the Fisher information matrix. Suppose that* $\Gamma(\boldsymbol{\theta})$ *is positive definite for all* $\boldsymbol{\theta}$. *Let* $\phi(X)$ *be a statistic with* $\mathrm{E}\left\{\phi(X)\right\} < \infty$ *for all* $\boldsymbol{\theta}$ *and let*

$$\gamma(\boldsymbol{\theta}) := \left(\frac{\partial \mathrm{E}\left\{\phi(X)\right\}}{\partial \theta_1},\ldots,\frac{\partial \mathrm{E}\left\{\phi(X)\right\}}{\partial \theta_K}\right)^{\mathsf{T}},$$

$$(3.165)$$

where T *denotes the matrix transposition. Then*

$$\mathrm{Var}\{\phi(X)\} \geq \gamma(\boldsymbol{\theta})^{\mathsf{T}} \cdot \Gamma(\boldsymbol{\theta})^{-1} \cdot \gamma(\boldsymbol{\theta}).$$

$$(3.166)$$

For an unbiased estimator $\phi(X)$ of the parameter θ_i the diagonal (i,i) component of the inverse of the Fisher matrix is the smallest possible variance.

One can improve the Cramèr–Rao bound considering higher (than the first one) derivatives of the pdf. This leads to Bhattacharyya lower bounds [149]. In the following theorem we shall present the Bhattacharyya lower bound for the case of one parameter. Let us introduce the following quantities:

$$\gamma_a(\theta) := \frac{\mathrm{d}^a \mathrm{E}\left\{\phi(X)\right\}}{\mathrm{d}\theta^a}, \quad a = 1,\ldots,\mathcal{N}, \tag{3.167a}$$

$$\gamma(\theta) := \left(\gamma_1(\theta),\ldots,\gamma_\mathcal{N}(\theta)\right)^{\mathsf{T}}, \tag{3.167b}$$

$$\psi_a(x;\theta) := \frac{1}{p(x;\theta)}\frac{\mathrm{d}^a p(x;\theta)}{\mathrm{d}\theta^a}, \quad a = 1,\ldots,\mathcal{N}, \tag{3.167c}$$

$$J_{ab}(\theta) := \mathrm{E}\left\{\psi_a(x;\theta)\,\psi_b(x;\theta)\right\}, \quad a,b = 1,\ldots,\mathcal{N}. \tag{3.167d}$$

Theorem 10 (Bhattacharyya system of bounds). *Let $\phi(X)$ be a one-dimensional statistic with $\mathrm{E}\{\phi(X)\} < \infty$ for all θ and let the square $N \times N$ matrix J (with the components J_{ab} defined above) be non-singular. Then*

$$\mathrm{Var}\{\phi(X)\} \geq \gamma(\theta)^{\mathsf{T}} \cdot J(\theta)^{-1} \cdot \gamma(\theta). \qquad (3.168)$$

There exist more sophisticated bounds called Weinstein and Ziv–Zakai bounds that capture the global structure of the likelihood function $p(\theta, x)$. These bounds are discussed in the context of gravitational-wave data analysis by Nicholson and Vecchio [150].

3.6.11 Jeffreys' prior

Fisher information pays a role in a method for choosing a prior density π with the property that when the parameter θ is transformed by a function g (which is one-to-one and differentiable) then also prior density is transformed in the same way. In turns out that a prior probability density $\pi(\theta)$ with this property is given by

$$\pi(\theta) = \pi_0 \sqrt{\Gamma_X(\theta)}, \qquad (3.169)$$

where $\Gamma_X(\theta)$ is the Fisher information about θ based on random variable (i.e. data) X and π_0 is a constant so that the integral of $\pi(\theta)$ is 1.

3.7 Non-stationary stochastic processes

3.7.1 Karhunen–Loéve expansion

Suppose that $x(t)$, $0 \leq t \leq T_o$ is a zero-mean stochastic process defined on a finite interval $[0, T_o]$ with autocovariance function $C(t, u)$ continuous on the square $[0, T_o] \times [0, T_o]$. Then C can be expanded in the uniformly and absolutely convergent series

$$C(t, u) = \sum_{k=1}^{\infty} \lambda_k \, \psi_k(t) \, \psi_k(u), \qquad (3.170)$$

where λ_k and ψ_k are the *eigenvalues* and corresponding *orthonormal eigenfunctions* of C, i.e. λ_k and ψ_k are solutions of the integral equation

$$\lambda_k \, \psi_k(t) = \int_0^{T_o} C(t, u) \, \psi_k(u) \, \mathrm{d}u, \qquad (3.171)$$

and the following orthonormality condition is fulfilled:

$$\int_0^{T_o} \psi_k(t)\,\psi_l(t)\,\mathrm{d}t = \delta_{kl}.$$

Furthermore, the stochastic process x can be represented by the following mean-squared convergent series:

$$x(t) = \sum_{k=1}^{\infty} z_k\,\psi_k(t), \quad 0 \le t \le T_o, \tag{3.172}$$

where the coefficients z_k are random variables defined as

$$z_k := \int_0^{T_o} x(t)\,\psi_k(t)\,\mathrm{d}t, \quad k = 1, 2, \dots . \tag{3.173}$$

The representation (3.170) is known as *Mercer's theorem* and the expansion (3.172) is known as the *Karhunen–Loéve expansion*. If the stochastic process x is Gaussian then the coefficients z_k are independent Gaussian random variables.

3.7.2 Wiener processes

Definition 5 (Wiener process). *A standard Wiener process (also called a standard Brown process) is a stochastic process $w(t)$, $t \ge 0$, with the following properties:*

1. *$w(0) = 0$;*

2. *$w(t)$ has continuous sample functions;*

3. *$w(t)$ has independent increments [i.e. for any t_i, $i = 1, \dots, 4$, such that $0 \le t_1 \le t_2 \le t_3 \le t_4$, the differences $w(t_2) - w(t_1)$ and $w(t_4) - w(t_3)$ of random variables are independent];*

4. *for any times t_1, t_2 the increment $w(t_2) - w(t_1)$ has Gaussian distribution $N(0, |t_2 - t_1|)$.*

Wiener process is a mathematical model of random motions of pollen particles first observed by botanist Robert Brown. Norbert Wiener gave first rigorous mathematical proof of the existence of the stochastic process defined above. The Wiener process is important for several reasons. A large class of processes can be transformed into Wiener process and a large class of processes can be generated by passing the Wiener process through linear or non-linear systems.

The autocorrelation function K of the Wiener process is given by

$$K(t_1, t_2) = \mathrm{E}\left\{w(t_1)w(t_2)\right\}$$

$$= \mathrm{E}\left\{w(t_1)\big((w(t_2) - w(t_1)) + w(t_1)\big)\right\}$$

$$= \mathrm{E}\left\{w(t_1)^2\right\} + \mathrm{E}\left\{(w(t_1) - w(0))(w(t_2) - w(t_1))\right\}. \quad (3.174)$$

If $t_2 > t_1$ then from the property (4) of Definition (5) the second term of the expression above is zero and $K(t_1, t_2) = t_1$. In general

$$K(t_1, t_2) = \min(t_1, t_2). \qquad (3.175)$$

The autocorrelation function is not of the form (3.37) and thus Wiener process is not stationary even in the wide sense.

Another non-stationary process related to Wiener process is the *Brownian bridge* $B(t)$. It has the property that both $B(0) = 0$ and $B(1) = 0$ (for Wiener process only $B(0) = 0$). Brownian bridge is obtained from standard Wiener process by $B(t) = W(t) - tW(1)$. Unlike Wiener process the increments of the Brownian bridge are not independent. The mean of $B(t)$ is 0 and its auotocorrelation function $K(t_1, t_2)$ is given by

$$K(t_1, t_2) = \min(t_1, t_2) - t_1 t_2. \qquad (3.176)$$

It is not difficult to obtain the Karhunen–Loéve expansion [introduced in Eq. (3.172)] of the Wiener process. We have (see [128])

$$w(t) = \sqrt{2} \sum_{n=0}^{\infty} w_n \frac{\sin\left((n + \tfrac{1}{2})\pi t\right)}{(n + \tfrac{1}{2})\pi}, \qquad (3.177)$$

where w_n are independent random variables with Gaussian distribution $N(0, 1)$. Let us consider the derivative $\dot{w}(t)$ of the Wiener process and let us denote it by $n(t)$. We obtain the process $n(t)$ by formally differentiating its Karhunen–Loéve expansion (3.177) term by term. We have

$$n(t) = \sqrt{2} \sum_{n=0}^{\infty} w_n \cos\left((n + \frac{1}{2})\pi t\right). \qquad (3.178)$$

The expansion above formally is a Karhunen–Loéve expansion of a process with all eigenvalues equal to one. Because in such a case the series (3.178) does not converge the derivative of the Wiener process does not exist in a strict mathematical sense. However we can still obtain formally the autocovariance function of the process $n(t)$

$$K(t, u) = \frac{\partial^2}{\partial t\, \partial u}\big(\min(t, u)\big) = \delta(t - u), \qquad (3.179)$$

where δ is the Dirac delta function. Thus the variance of the process $n(t)$ is infinite. The stochastic process $n(t)$ is called Gaussian *white noise*. If we take (formally) the Fourier transform of the autocorrelation function of $n(t)$ we obtain that the spectrum of the white noise process is equal to 1 for all frequencies. White noise process is an artifice useful in many models. In rigorous mathematical treatment we consider the integral of white noise process

$$w(t) = \int_0^t n(u)\,du, \tag{3.180}$$

which is a well-defined Wiener process. We then state the basic signal detection problem as a hypothesis testing problem,

$$\text{signal absent:} \quad y(t) = w(t), \tag{3.181a}$$

$$\text{signal present:} \quad y(t) = \int_0^t s(u)\,du + w(t). \tag{3.181b}$$

This allows a rigorous mathematical proof of formulae such as the Cameron–Martin formula (3.81).

3.7.3 Resampled stochastic processes

Let $x(t)$ be a stochastic process and let $t_r = t + k(t; \theta)$ be a smooth one-to-one function both of the index t and the parameter set θ. A *resampled stochastic process* $y(t_r)$ is defined as the stochastic process $x(t)$ taken at time t_r, i.e. $y(t_r) = x(t)$. In the following for simplicity we assume that the process $x(t)$ has zero mean. It immediately follows that the resampled process $y(t_r)$ also has zero mean. Suppose that the original process is stationary. Let us first convince ourselves that the resampled process is, in general, *non-stationary*. Let $C(t_r', t_r) := \mathrm{E}\{y(t_r')y(t_r)\}$ be the auto-covariance function of the resampled process. By definition of the resampled process we have that $C(t_r', t_r) = \mathrm{E}\{x(t')x(t)\}$ and by stationarity of $x(t)$ we have $C(t_r', t_r) = R(t' - t)$. By implicit function theorem we have that there exists a smooth function $t = t_r + g(t_r; \theta)$. In order that the resampled process be stationary the function R must depend only on the difference $t_r' - t_r$. This is the case if and only if t is a linear function of t_r, i.e. $t = t_r + a(\theta)t_r + b(\theta)$. Thus when the resampling transformation is non-linear the resulting resampled process is non-stationary.

In the linear resampling case the Fourier transform $\tilde{y}(\omega_r)$ of the resampled process at frequency ω_r is related to the Fourier transform $\tilde{x}(\omega)$ of the original process at frequency ω by the following formula:

$$\tilde{y}(\omega_r) = \frac{\exp(i\omega b)}{1+a}\tilde{x}(\omega), \tag{3.182}$$

where $\omega = \omega_r/(1+a)$.

Let us consider the autocovariance function $C(t'_r, t_r)$ of the resampled process $y(t_r)$. It can be written as

$$C(t'_r, t_r) = \int_{-\infty}^{\infty} \phi^*_{t'_r}(\omega)\,\phi_{t_r}(\omega)\,\mathrm{d}F(\omega), \qquad (3.183)$$

where F is the integrated spectrum of the stationary process $x(t)$ and we have introduced a set of functions

$$\phi_{t_r}(\omega) = \exp\left(\mathrm{i}\omega\big(t_r + g(t_r)\big)\right). \qquad (3.184)$$

We have the following important result [120, 151].

Theorem 11 (General orthogonal expansions). *Let $y(t_r)$ be a continuous parameter zero-mean stochastic process (not necessarily stationary) with autocovariance function $C(t_r, s_r) = \mathrm{E}\{y(t_r)y(s_r)\}$. If there exists a family of functions, $\{\phi_{t_r}(\omega)\}$, defined on the real line and indexed by suffix t_r, and a measure, $\mu(\omega)$, on the real line such that for each t_r, $\phi_{t_r}(\omega)$ is quadratically integrable with respect to the measure μ, i.e.*

$$\int_{-\infty}^{\infty} |\phi_{t_r}(\omega)|^2\,\mathrm{d}\mu(\omega) < \infty, \qquad (3.185)$$

and $C(t_r, s_r)$ admits for all t_r, s_r a representation of the form

$$C(t'_r, t_r) = \int_{-\infty}^{\infty} \phi^*_{t'_r}(\omega)\,\phi_{t_r}(\omega)\,\mathrm{d}\mu(\omega), \qquad (3.186)$$

then the process $y(t_r)$ admits a representation of the form

$$y(t_r) = \int_{-\infty}^{\infty} \phi_{t_r}(\omega)\,\mathrm{d}Z(\omega), \qquad (3.187)$$

where $Z(\omega)$ is an orthogonal process with

$$\mathrm{E}\left\{|\mathrm{d}Z(\omega)|^2\right\} = \mathrm{d}\mu(\omega). \qquad (3.188)$$

Conversely if $y(t_r)$ admits a representation of the form (3.187) with an orthogonal process satisfying (3.188), then $C(t'_r, t_r)$ admits a representation of the form (3.186).

The formula (3.186) is called *generalized Wiener–Khinchin* relation and formula (3.187) is called *generalized Cramèr representation* of the stochastic process. In our case the generalized Cramèr representation reads

$$y(t_r) = \int_{-\infty}^{\infty} \exp\left(\mathrm{i}\omega\big(t_r + g(t_r)\big)\right)\,\mathrm{d}Z(\omega). \qquad (3.189)$$

This representation clearly shows that the resampled process is in general non-stationary because the choice of basis function

$$\phi_{t_r}(\omega) = \exp(i\omega t_r) \tag{3.190}$$

is not in general possible. The generalized Cramèr representation is also immediate from the Cramèr representation for the original stationary process and its transformation to resampled time. It also follows that the measure μ coincides with the integrated spectrum $H(\omega)$ of the original stationary process. However $H(\omega)$ cannot be interpreted as the spectrum of the resampled process. Indeed for the resampled process which is non-stationary the concept of spectrum is not mathematically well defined.

The general orthogonal expansion theorem has already been used by M. B. Priestley [120] to develop the theory of so called *evolutionary spectra*. This theory describes a very important class of non-stationary processes often occurring in practice for which the amplitude of the Fourier transform slowly changes with time.

In the continuous case it is instructive to calculate the correlation function for the Fourier frequency components of the resampled (non-stationary) process. Using Eq. (3.187) the Fourier transform $\tilde{y}(\omega)$ of the resampled process $y(t_r)$ can be written as

$$\tilde{y}(\omega) = \int_{-\infty}^{\infty} Q(\omega_1, \omega) \, dZ(\omega_1), \tag{3.191}$$

where the kernel Q is given by

$$Q(\omega_1, \omega) = \int_{-\infty}^{\infty} \phi_{t_r}(\omega_1) \exp(-i\omega t_r) \, dt_r. \tag{3.192}$$

The correlation between two Fourier components of the resampled process takes the form

$$E\{\tilde{y}(\omega)\,\tilde{y}(\omega')^*\} = \int_{-\infty}^{\infty} \int_{-\infty}^{\infty} Q(\omega_1, \omega')^* \, Q(\omega_2, \omega) \, E\{dZ(\omega_1)^* dZ(\omega_2)\}$$

$$= \int_{-\infty}^{\infty} Q(\omega_1, \omega')^* \, Q(\omega_1, \omega) \, S(\omega_1) \, d\omega_1. \tag{3.193}$$

This formula should be compared with Eq. (3.51) for stationary processes. We see that for a resampled stochastic process the Fourier components at different frequencies are correlated. This is another manifestation of the non-stationarity of the process.

Let us now consider the example of white noise for which the spectral density $S(\omega)$ is independent of the frequency ω. It is straightforward to show that

$$\int_{-\infty}^{\infty} Q(\omega_1, \omega')^* \, Q(\omega_1, \omega) \, d\omega_1 = \delta(\omega' - \omega). \qquad (3.194)$$

Thus for the case of white noise we have

$$E\{\tilde{y}(\omega)\,\tilde{y}(\omega')^*\} = S(\omega)\,\delta(\omega' - \omega), \qquad (3.195)$$

and consequently in this case the noise remains stationary (and white) after resampling. It is also easy to see that the Fourier components at different frequencies will be uncorrelated if the spectral density is constant over the bandwidth of the kernel Q.

3.7.4 Cyclostationary processes

Let $x(t)$ $(t \in \mathbb{R})$ be a continuous-time stochastic process having the finite first- and second-order moments

$$m(t) := E\{x(t)\}, \quad K(t, t') := E\{x(t)x(t')\}. \qquad (3.196)$$

Process x is said to be *cyclostationary* with period T if the moments m and K are periodic functions of t and t' with period T, i.e. for every $(t, t') \in \mathbb{R} \times \mathbb{R}$ we have

$$m(t + T) = m(t), \quad K(t + T, t' + T) = K(t, t'). \qquad (3.197)$$

For simplicity from now on we shall assume that $m(t) = 0$.

If x is cyclostationary, then the function $B(t, \tau) := K(t + \tau, t)$ for a given $\tau \in \mathbb{R}$ is a periodic function of t with period T. It can thus be represented by the Fourier series,

$$B(t, \tau) = \sum_{r=-\infty}^{\infty} B_r(\tau) \exp\left(2\pi i \frac{rt}{T}\right), \qquad (3.198)$$

where the functions B_r are given by

$$B_r(\tau) = \frac{1}{T} \int_0^T B(t, \tau) \exp\left(-2\pi i \frac{rt}{T}\right) dt. \qquad (3.199)$$

The Fourier transforms g_r of B_r are the so called *cyclic spectra* of the cyclostationary process x [152],

$$g_r(f) := \int_{-\infty}^{\infty} B_r(\tau) \exp\left(-2\pi i f \tau\right) d\tau. \qquad (3.200)$$

If a cyclostationary process is real, the following relationships between the cyclic spectra hold

$$B_{-r}(\tau) = B_r^*(\tau), \qquad (3.201a)$$

$$g_{-r}(-f) = g_r^*(f). \qquad (3.201b)$$

This implies that for a real cyclostationary process, the cyclic spectra with $r \geq 0$ contain all the information needed to characterize the process itself.

The function $\sigma^2(t) := B(t, 0)$ is the variance of the cyclostationary process x, and it can be written as a Fourier decomposition as a consequence of Eq. (3.199)

$$\sigma^2(t) = \sum_{r=-\infty}^{\infty} H_r \exp\left(2\pi i \frac{rt}{T}\right), \qquad (3.202)$$

where $H_r := B_r(0)$ are harmonics of the variance σ^2. From Eq. (3.201a) it follows that $H_{-r} = H_r^*$.

For a discrete, finite, and real time series x_t, $t = 1, \ldots, N$, we can estimate the cyclic spectra by generalizing standard methods of spectrum estimation used with stationary processes. Assuming again the mean value of the time series x_t to be zero, the cyclic autocorrelation sequences are defined as

$$s_l^r := \frac{1}{N} \sum_{t=1}^{N-|l|} x_t\, x_{t+|l|} \exp\left(-2\pi i \frac{r(t-1)}{T}\right). \qquad (3.203)$$

It has been shown [152] that the cyclic autocorrelations are asymptotically (i.e. for $N \to \infty$) unbiased estimators of the functions $B_r(\tau)$. The Fourier transforms of the cyclic autocorrelation sequences s_l^r are estimators of the cyclic spectra $g_r(f)$. These estimators are asymptotically unbiased and inconsistent, i.e. their variances do not asymptotically tend to zero. In the case of Gaussian processes [152] consistent estimators can be obtained by first applying a lag-window to the cyclic autocorrelation and then performing a Fourier transform. This procedure represents a generalization of the well-known technique for estimating the spectra of stationary stochastic processes (see Section 4.2.1).

An alternative procedure for identifying consistent estimators of the cyclic spectra is first to take the Fourier transform $\tilde{x}(f)$ of the time series x_t,

$$\tilde{x}(f) = \sum_{t=1}^{N} x_t \exp\left(-2\pi i f(t-1)\right), \qquad (3.204)$$

and then estimate the cyclic periodograms $g_r(f)$,

$$g_r(f) = \frac{1}{N}\, \tilde{x}(f)\, \tilde{x}\left(f - \frac{r}{T}\right)^*. \qquad (3.205)$$

By finally smoothing the cyclic periodograms, consistent estimators of the spectra $g_r(f)$ are then obtained.

The estimators of the harmonics H_r of the variance σ^2 of a cyclostation-ary stochastic process can be obtained by first forming a sample variance of the time series x_t. The sample variance is obtained by dividing the time series x_t into contiguous segments of length τ_0 such that τ_0 is much smaller than the period T of the cyclostationary process, and by calculating the variance σ_I^2 over each segment. Estimators of the harmonics are obtained either by Fourier analyzing the series σ_I^2 or by making a least-square fit to σ_I^2 with the appropriate number of harmonics. Note that the definitions of (i) zero order ($r = 0$) cyclic autocorrelation, (ii) periodogram, and (iii) zero order harmonic of the variance, coincide with those usually adopted for stationary stochastic processes. Thus, even though a cyclostationary time series is not stationary, the ordinary spectral analysis can be used for obtaining the zero order spectra. Note, however, that cyclostationary stochastic processes provide more spectral information about the time series they are associated with due to the existence of cyclic spectra with $r > 0$.

As an important and practical application, let us consider a time series y_t consisting of the sum of a stationary stochastic process n_t and a cyclo-stationary one x_t (i.e. $y_t = n_t + x_t$). Let the variance of the stationary time series n_t be ν^2 and its spectral density be $\mathcal{E}(f)$. It is easy to see that the resulting process is also cyclostationary. If the two processes are uncorrelated, then the zero order harmonic Σ_0^2 of the variance of the combined processes is equal to

$$\Sigma_0^2 = \nu^2 + \sigma_0^2, \tag{3.206}$$

and the zero order spectrum $G_0(f)$ of y_t is

$$G_0(f) = \mathcal{E}(f) + g_0(f). \tag{3.207}$$

The harmonics of the variance as well as the cyclic spectra of y_t with $r > 0$ coincide instead with those of x_t. In other words, the harmonics of the variance and the cyclic spectra of the process y_t with $r > 0$ contain information only about the cyclostationary process x_t, and are not "contaminated" by the stationary process n_t.

4

Time series analysis

This chapter is an introduction to the theory of time series analysis. In Section 4.1 we discuss the estimators of the sample mean and the correlation function of a time series. In Section 4.2 we introduce non-parametric methods of the spectral analysis of time series, including the multitapering method. A detailed discussion of the time series spectral analysis can be found in Refs. [153, 154, 155, 156].

In Sections 4.3–4.5 we discuss useful tests of the time series. One type of test is for the presence of periodicities in the data, which we discuss in Section 4.3. In Section 4.4 we introduce two goodness-of-fit tests describing whether the data come from a given probability distribution: Pearson's χ^2 test and Kolmogorov–Smirnov test. Other types of tests are tests for Gaussianity and linearity of the data, which are discussed in Section 4.5. Both tests use higher-order spectra of time series, which are also introduced in Section 4.5.

4.1 Sample mean and correlation function

We assume that we have N contiguous data samples x_k $(k = 1, \ldots, N)$ of the stochastic process. We also assume that the underlying process is stationary and ergodic (i.e. satisfying the ergodic theorem, see Section 3.2). We immediately see that the N samples of the stochastic process that constitute our observation cannot be considered as a stationary process. They would be a stationary sequence only asymptotically as we extend the number of samples N to infinity. As we shall see this has profound consequences on the statistical properties of the estimators of the spectrum.

Let us first consider estimators of the mean and the correlation function of the random sequence. The mean value m and the correlation function $R(l)$ of the random sequence x_k are defined as

$$m := \mathrm{E}\{x_k\}, \tag{4.1a}$$

$$R(l) := \mathrm{E}\{x_k x_{k-l}\}, \quad l = -N+1, \ldots, N-1. \tag{4.1b}$$

It is also useful to define the *covariance sequence* $C(l)$ of the random sequence,

$$C(l) := \mathrm{E}\{(x_k - m)(x_{k-l} - m)\}, \quad l = -N+1, \ldots, N-1. \tag{4.2}$$

Note that because of the stationarity of the stochastic process the mean, the correlation, and the covariance functions are independent of the sample number k. The first component $C(0)$ of the covariance sequence is just the variance σ^2 of the stochastic process.

The $N \times N$ *correlation matrix* R has components defined as:

$$R_{kl} := \mathrm{E}\{x_k x_l\}, \quad k, l = 1, \ldots, N. \tag{4.3}$$

For the stationary time series x_k the correlation matrix R has the following structure

$$R = \begin{pmatrix} R(0) & R(-1) & R(-2) & \cdots & R(-N+1) \\ R(1) & R(0) & R(-1) & \cdots & R(-N+2) \\ R(2) & R(1) & R(0) & \cdots & R(-N+3) \\ \cdots & \cdots & \cdots & \cdots & \cdots \\ R(N-1) & R(N-2) & R(N-3) & \cdots & R(0) \end{pmatrix}. \tag{4.4}$$

The correlation matrix has the property that all elements on the diagonals are equal. A matrix with this property is called *Toeplitz matrix*. We can similarly define the $N \times N$ *covariance matrix* C with components

$$C_{kl} := \mathrm{E}\{(x_k - m)(x_l - m)\}, \quad k, l = 1, \ldots, N. \tag{4.5}$$

The covariance matrix of the stationary process is also a Toeplitz matrix.

An estimator of the mean m of the process is the *sample mean* $\hat{m}$ defined by

$$\hat{m} := \frac{1}{N} \sum_{k=1}^{N} x_k. \tag{4.6}$$

The expectation value of the sample mean is easily calculated:

$$\mathrm{E}\{\hat{m}\} = \frac{1}{N} \sum_{k=1}^{N} \mathrm{E}\{x_k\} = \frac{1}{N} \sum_{k=1}^{N} m = m. \tag{4.7}$$

Thus we see that the sample mean is an unbiased estimator of the mean of the process. The variance of the sample mean is given by

$$\text{Var}\{\hat{m}\} = \frac{1}{N^2} \sum_{k=1}^{N} \sum_{l=1}^{N} \text{E}\{x_k x_l\} - m^2 = \frac{1}{N^2} \sum_{k=1}^{N} \sum_{l=1}^{N} C_{kl}$$

$$= \frac{1}{N^2} \sum_{l=-N}^{N} (N - |l|) C(l) = \frac{1}{N} \sum_{l=-N}^{N} \left(1 - \frac{|l|}{N}\right) C(l). \quad (4.8)$$

The second line of the above derivation follows from the fact that C is a Toeplitz matrix and thus in the double sum we have N terms of value $C(0)$, $N - 1$ terms of value $C(1)$ and $C(-1)$, and so on. Since variance of the sample mean approaches zero as N gets large, the sample mean is a consistent estimator.

The estimator of the correlation function known as the *sample correlation function* $\hat{R}'(l)$ is defined by

$$\hat{R}'(l) := \frac{1}{N - |l|} \sum_{n=1}^{N-|l|} x_{n+|l|} x_n, \quad |l| < N. \quad (4.9)$$

It is easy to show that

$$\text{E}\{\hat{R}'(l)\} = \frac{1}{N - |l|} \sum_{n=1}^{N-|l|} \text{E}\{x_{n+|l|} x_n\} = \frac{1}{N - |l|} \sum_{n=1}^{N-|l|} R(l) = R(l),$$

$$(4.10)$$

and thus estimator $\hat{R}'(l)$ is an unbiased estimator. One can also show that the sample correlation function is a consistent estimator. Although the sample correlation function (4.9) is unbiased and consistent, it has one major problem. This estimator is not always positive semi-definite. It turns out that the following biased estimator of the correlation function is positive semi-definite,

$$\hat{R}(l) := \frac{1}{N} \sum_{n=1}^{N-|l|} x_{n+|l|} x_n, \quad |l| < N. \quad (4.11)$$

The estimator $\hat{R}(l)$ is also a consistent estimator of the correlation function.

4.2 Power spectrum estimation

In this section we shall review methods to estimate the power spectrum density of a time series.

4.2.1 Periodogram

By Wiener–Khinchin theorem the power spectrum of a stationary stochastic process is the Fourier transform of the autocorrelation function. Therefore it seems natural to take as the estimator $\hat{S}(\omega)$ of the power spectrum $S(\omega)$ the Fourier transform of the sample correlation function $\hat{R}(l)$. Thus we have

$$\hat{S}(\omega) := \sum_{l=-N}^{N} \hat{R}(l)\,e^{-i\omega l} = \frac{1}{N}\sum_{l=-N}^{N}\sum_{n=1}^{N-|l|} x_{n+|l|}\,x_n\,e^{-i\omega l}$$

$$= \frac{1}{N}\sum_{j=1}^{N}\sum_{k=1}^{N} x_j\,x_k\,e^{-i\omega(k-j)} = \frac{1}{N}\left|\sum_{k=1}^{N} x_k\,e^{-i\omega k}\right|^2, \qquad (4.12)$$

after a change of variables in the double summation. The estimator $\hat{S}$ is called the *periodogram*. The periodogram is thus the absolute value squared of the discrete Fourier transform of the data divided by the number of data points.

Let us consider some basic statistical properties of the periodogram. Let us first consider its expectation value,

$$\mathrm{E}\{\hat{S}(\omega)\} = \sum_{l=-N}^{N} \mathrm{E}\{\hat{R}(l)\}\,e^{-i\omega l} = \sum_{l=-N}^{N} \frac{N-|l|}{N} R(l)\,e^{-i\omega l}. \qquad (4.13)$$

It is convenient to write the above expectation value in the following form

$$\mathrm{E}\{\hat{S}(\omega)\} = \sum_{l=-\infty}^{\infty} w_l^B R(l)\,e^{-i\omega l}, \qquad (4.14)$$

where

$$w_l^B := \begin{cases} \dfrac{N-|l|}{N}, & |l| < N, \\ 0, & \text{otherwise.} \end{cases} \qquad (4.15)$$

Making use of the formula (3.56) (which follows from the Wiener–Khinchin theorem) we can rewrite the expectation value (4.14) in the form

$$\mathrm{E}\{\hat{S}(\omega)\} = \int_{-\pi}^{\pi} S(\omega')\,F_N(\omega-\omega')\,d\omega', \qquad (4.16)$$

where F_N is the *Fejér kernel* defined by

$$F_N(u) := \frac{1}{N}\frac{\sin^2(Nu/2)}{\sin^2(u/2)}. \qquad (4.17)$$

Thus the expectation value of the periodogram is the convolution of the true spectral density of the time series with the Fejér kernel.

Let us also obtain the expectation value of the Fourier transform of the *unbiased* estimator $\hat{R}'(l)$ of the correlation function [we shall denote this Fourier transform by $\hat{S}'(\omega)$],

$$\mathrm{E}\{\hat{S}'(\omega)\} = \sum_{l=-N}^{N} \mathrm{E}\{\hat{R}'(l)\}\,\mathrm{e}^{-i\omega l} = \sum_{l=-N}^{N} R(l)\,\mathrm{e}^{-i\omega l}. \tag{4.18}$$

It is convenient to write the above expectation value in the following form

$$\mathrm{E}\{\hat{S}'(\omega)\} = \sum_{l=-\infty}^{\infty} w_l\, R(l)\,\mathrm{e}^{-i\omega l}, \tag{4.19}$$

where

$$w_l := \begin{cases} 1, & |l| < N, \\ 0, & \text{otherwise.} \end{cases} \tag{4.20}$$

Using the Wiener–Khinchin theorem again we have

$$\mathrm{E}\{\hat{S}'(\omega)\} = \int_{-\pi}^{\pi} S(\omega')\,\delta_N(\omega - \omega')\,\mathrm{d}\omega', \tag{4.21}$$

where δ_N is the *Dirichlet kernel* defined by

$$\delta_N(u) := \frac{\sin\left((N+1/2)u\right)}{\sin(u/2)}. \tag{4.22}$$

The essential difference between (4.16) and (4.21) is that while the Fejér kernel is non-negative for every value of ω, the transform of the rectangular window is *negative* in certain regions. As a result, the expectation value of the spectral estimator using the unbiased sample correlation function $\hat{R}'(l)$ can also be negative. The unbiased correlation function is therefore rather undesirable for spectral estimation.

From Eq. (4.16) we see that unless the spectrum of the stationary process is white the periodogram is a biased estimator of the spectrum. The bias is the greater the greater the dynamic range of the spectrum of the process. This bias is the consequence of the finite size N of the observed samples of the process. The bias of the periodogram is sometimes called *spectral leakage*. This is because as a result of convolution with a window function some power in a given frequency of the spectrum is transferred to all other frequencies of the estimator.

The covariance of the periodogram estimator is given by

$$\mathrm{E}\{\hat{S}(\omega_1)\hat{S}(\omega_2)\} = S(\omega_1)S(\omega_2)\frac{F_N(\omega_1 + \omega_2) + F_N(\omega_1 - \omega_2)}{N}. \tag{4.23}$$

Asymptotically, as $N \to \infty$, and subject to finiteness of certain higher-order moments of the time series, the periodogram has the following statistical properties (see Ref. [157], Section 5.2):

$$\hat{S}(\omega) = \begin{cases} S(\omega)\,\chi_2^2/2, & \text{for } \omega \in (0, \pi), \\ S(\omega)\,\chi_1^2, & \text{for } \omega \in \{0, \pi\}, \end{cases} \qquad (4.24)$$

where χ_n^2 denotes a random variable having a χ^2 distribution with n degrees of freedom. From the properties of the χ^2 distribution it follows that asymptotically the mean and the variance of the periodogram are given by

$$\mathrm{E}\{\hat{S}(\omega)\} = S(\omega), \qquad \text{for } \omega \in [0, \pi]; \qquad (4.25\mathrm{a})$$

$$\mathrm{Var}\{\hat{S}(\omega)\} = S(\omega)^2, \qquad \text{for } \omega \in (0, \pi),$$
$$\mathrm{Var}\{\hat{S}(\omega)\} = 2S(\omega)^2, \quad \text{for } \omega \in \{0, \pi\}. \qquad (4.25\mathrm{b})$$

Thus the periodogram has the advantage that asymptotically it is an unbiased estimator of the spectrum. However, it has the disadvantage that it is an inconsistent estimator as its variance does not decrease to zero as $N \to \infty$. It is not clear what size N of the sample of stochastic process is enough for the bias of the periodogram to be sufficiently small and no matter what is the sample size the variance of the periodogram does not decrease. Thus the periodogram is not a satisfactory estimator of the spectral density. There are several ways to improve the periodogram to reduce its bias and obtain a consistent estimator of the spectrum.

4.2.2 Averaging

A simple way to reduce the variance of the periodogram is to average a number of different periodograms. We take a long data segment, break it into smaller data segments of a fixed length and average periodograms of the smaller data segments (this is the Bartlett procedure [158]). Let $\hat{S}^{(k)}(\omega)$ be the periodogram of the kth data segment. Then Bartlett's estimator is defined by

$$\hat{S}_{\mathrm{B}}(\omega) := \frac{1}{K} \sum_{k=1}^{K} S^{(k)}(\omega). \qquad (4.26)$$

In this way the variance of the resulting estimator decreases by a factor of $1/K$ where K is the number of data segments (assuming that the data segments are uncorrelated). Clearly we obtain a consistent estimator since as number of data N increases so does the number of data segments.

4.2.3 Windowing and smoothing

Another way to improve the spectral estimator is to apply a *window* to the estimated sample correlation function before doing the Fourier transform. This method was promoted by Blackman and Tuckey [159]. The spectral estimator takes then the form

$$\hat{S}_{\mathrm{BT}}(\omega) := \sum_{l=-L}^{L} w_l \, \hat{R}(l) \, \mathrm{e}^{-\mathrm{i}\omega l} \quad \text{for} \quad |l| < N. \tag{4.27}$$

The window w_l is sometimes called the *lag-window*. The resulting estimator is equivalent to smoothing the periodogram across a certain frequency interval. When computed in this way it is called the *Daniell periodogram* [160]. The variance of the estimator is approximately reduced by the factor

$$\frac{1}{N} \sum_{l=-L}^{L} w_l^2 = \frac{1}{N} \int_{-\pi}^{\pi} |\tilde{w}(\omega)|^2 \, \mathrm{d}\omega. \tag{4.28}$$

4.2.4 Combined windowing and averaging

A final strategy is to use a combination of windowing and averaging. This procedure was studied by Welch [161]. In Welch's procedure the original data sequence is divided into a number K of possibly overlapping segments. A window (taper) is applied to these segments and the resulting modified periodograms are averaged. Let $x_n^{(k)}$ represent the kth data segment (of length N) and let w_n be normalized window, then the modified periodograms are defined as

$$\tilde{S}^{(k)}(\omega) := \frac{1}{N} \left| \sum_{n=1}^{N} w_n x_n^{(k)} \mathrm{e}^{-\mathrm{i}\omega n} \right|^2, \tag{4.29}$$

where w_n has the property

$$\frac{1}{N} \sum_{n=1}^{N} w_n^2 = 1. \tag{4.30}$$

The spectral estimator is then taken as

$$\hat{S}_{\mathrm{W}}(\omega) := \frac{1}{K} \sum_{k=1}^{K} \tilde{S}^{(k)}(\omega). \tag{4.31}$$

When the segments are non-overlapping the variance of the above estimator is approximately the same as that of Bartlett. In Welch's procedure the use of appropriate window reduces the bias of the estimator whereas averaging of periodograms of the segments reduces its variance.

4.2.5 Multitapering

Multitaper method (MTM) was introduced by Thomson ([153], see also [156] for a detailed exposition of the method and [162] for application to gravitational-wave data analysis). MTM attempts to reduce the variance of spectral estimators by using a small set of *tapers* rather than the unique data taper or spectral window used by Blackman–Tukey methods. A set of independent estimators of the power spectrum is computed, by pre-multiplying the data by orthogonal tapers which are constructed to minimize the spectral leakage due to the finite length of the data set. The optimal tapers or "eigentapers" belong to a family of functions known as *discrete prolate spheroidal sequences* (DPSS) [163]. Averaging over this (small) ensemble of spectra yields a better and more stable estimator (i.e. one with lower variance) than do single-taper methods. Note that the case $n = 1$ reduces to the Blackman–Tukey case. Because the windowing functions or eigentapers are the specific solution to an appropriate variational problem, this method is less heuristic than traditional non-parametric techniques presented above. Detailed algorithms for the calculation of the eigentapers are readily available [156].

The tapers are the discrete set of n eigenfunctions which solve the variational problem of minimizing leakage outside of a frequency band of half-bandwidth $W := nw$, where $w := 1/N$. Consider a sequence w_t of length N and let $U(\omega)$ be its Fourier transform. Then we pose the problem of finding sequences w_t so that their power spectra are maximally concentrated in the interval $[-o, o]$, i.e.

$$\int_{-o}^{o} |U(\omega)|^2 \, d\omega \tag{4.32}$$

subject to normalization

$$\sum_{t=1}^{N} w_t^2 = 1. \tag{4.33}$$

It can be shown that the optimality condition leads to a matrix eigenvalue equation for $w_t(o, N)$

$$\sum_{t'=1}^{N} \frac{\sin\left(o(t - t')\right)}{\pi(t - t')} w_{t'}^{(k)} = \lambda^{(k)} w_t^{(k)}, \tag{4.34}$$

where index (k) numbers the eigenfunctions $w_t^{(k)}(o, N)$ and eigenvalues $\lambda^{(k)}$. The eigenvectors of this equations are DPSS. The remarkable fact is that the first $2NW$ eigenvalues $\lambda^{(k)}$ (sorted in descending order) are each approximately equal to one, while the remainder are approximately zero. Thus, the number K of tapers used should be less than $2NW$ in any application of MTM. The choice of the bandwidth and the number of tapers K thus represents the classical tradeoff between spectral resolution and the stability or "variance" properties of the spectral estimator [153]. Longer data sets can admit the use of a greater number K of tapers while maintaining a desired frequency resolution, and the optimal choice of K and NW is, in general, most decidedly application specific.

Once the tapers are computed for a chosen frequency bandwidth, the total power spectrum can be estimated by averaging the individual spectra given by each tapered version of the data set. We call $|\tilde{x}^{(k)}(\omega)|^2$ the kth eigenspectrum, where $\tilde{x}^{(k)}(\omega)$ is the discrete Fourier transform of $w_t^{(k)} x_t$. The high-resolution multitaper spectrum is a simple weighted sum of the K eigenspectra,

$$S_{\mathrm{MT}}(\omega) := \frac{1}{K} \sum_{k=1}^{K} |\tilde{x}^{(k)}(\omega)|^2, \tag{4.35}$$

where

$$\tilde{x}^{(k)}(\omega) = \sum_{l=1}^{N} w_t^{(k)} x_t \, e^{-i\omega t}. \tag{4.36}$$

Its resolution in frequency is K/N which means that "line" components will be detected as peaks or bumps of width K/N. For a white noise the MTM spectrum estimator is χ^2 distributed with $2K$ degrees of freedom.

4.3 Tests for periodicity

In this section we shall consider tests for the presence of periodic signals in the noise. Let us consider the following data model

$$x_t = \mu + \sum_{l=1}^{L} A_l \cos(\omega_l t + \phi_l) + n_t, \tag{4.37}$$

where μ is a constant and n_t is a Gaussian stochastic process with zero mean and variance σ^2. Thus data x_t is a superposition of L monochromatic signals added to white Gaussian noise. Each monochromatic signal is characterized by three parameters: amplitude A_l, phase ϕ_l and angular frequency ω_l. For mathematical convenience let us assume that the number N of data points is odd. To test for periodicity in time series x_t we calculate the periodogram $\hat{S}(\omega_k)$ at Fourier frequencies $\omega_k := 2\pi k/N$. Under the null hypothesis

$$A_1 = \ldots = A_L = 0 \qquad (4.38)$$

the quantities

$$\frac{2\hat{S}(\omega_k)}{\sigma^2}, \quad k = 1, \ldots, m, \qquad (4.39)$$

where $m := (N-1)/2$, have a χ^2 distribution with two degrees of freedom. The test introduced by Schuster [164] is to compare the maximum γ of the quantities (4.39),

$$\gamma := \max_{1 \leq k \leq m} \frac{2\hat{S}(\omega_k)}{\sigma^2}, \qquad (4.40)$$

against a threshold u_0. The false alarm probability α in this case, i.e. the probability that the maximum of m independently distributed χ^2 quantities with two degrees of freedom exceeds u_0, is given by

$$\alpha_{\text{Sch}} = 1 - \left(1 - e^{-u_0/2}\right)^m. \qquad (4.41)$$

The disadvantage of the above test is that we must know the variance σ^2 of the noise exactly whereas in practice we can only estimate it. Consequently Fisher [165] proposed another test based upon the statistic

$$g := \frac{\displaystyle\max_{1 \leq k \leq m} \hat{S}(\omega_k)}{\displaystyle\sum_{j=1}^{m} \hat{S}(\omega_j)}. \qquad (4.42)$$

The false alarm probability α_F for the statistic g introduced above is given by

$$\alpha_F = \sum_{j=1}^{M} (-1)^{j-1} \binom{m}{j} (1 - j\, g_0)^{m-1}, \qquad (4.43)$$

where M is the largest integer satisfying both $M < 1/g_0$ and $M \leq m$. Anderson [166] has shown that the Fisher's test is the most powerful test for the case of a single monochromatic signal present in the white noise.

Siegel [167] has studied the problem when there are several periodicities present in the data and he has found a test based on large values of the rescaled periodogram

$$\tilde{S}(\omega_k) := \frac{\hat{S}(\omega_k)}{\sum\limits_{j=1}^{m} \hat{S}(\omega_j)}, \tag{4.44}$$

instead of only their maximum. In addition to threshold g_0 Siegel introduces a second parameter λ $(0 < \lambda \leq 1)$ and considers the following test statistic T_λ:

$$T_\lambda := \sum_{k=1}^{m} \left(\tilde{S}(\omega_k) - \lambda g_0 \right)_+, \tag{4.45}$$

where $(a)_+ := \max(a, 0)$ and g_0 is the threshold for the Fisher's test. Thus the test consists of *summing* the differences of the rescaled periodogram and a predetermined threshold. Siegel has shown that the false alarm probability for his test, i.e. the probability that T_λ crosses a certain threshold t_0, is given by

$$\alpha_S = \sum_{j=1}^{m} \sum_{k=0}^{j-1} (-1)^{j+k+1} \binom{m}{j} \binom{j-1}{k} \binom{m-1}{k} t_0^k \left(1 - j\lambda g_0 - t_0\right)_+^{m-k-1}. \tag{4.46}$$

Siegel has found that for single periodicities his test is only slightly less powerful than the Fisher's test whereas for compound periodicities it outperforms the Fisher's test.

4.4 Goodness-of-fit tests

A *goodness-of-fit test* describes how well a statistical model fits the data. It determines whether the data come from a given probability distribution. We shall describe the two most popular such tests: the Pearson's χ^2 test (often called simply χ^2 test) and the Kolmogorov–Smirnov test (called KS test).

4.4.1 Pearson's χ^2 test

The null hypothesis for this test is that observed data come from a certain predetermined probability distribution. The test consists of grouping the data into bins and calculating the test statistic χ^2 is given by

$$\chi^2 := \sum_{i=1}^{n} \frac{(o_i - e_i)^2}{e_i}, \tag{4.47}$$

where o_i is the number of data in the ith bin, e_i is the expected number of data in the ith bin under the null hypothesis and n is the number of bins. The test statistic has approximately a χ^2 distribution with $n-1$ degrees of freedom when the number of data is large.

4.4.2 Kolmogorov–Smirnov test

The null hypothesis for the KS test is that the sample $X_1, \ldots, X_n$ is drawn from a certain cumulative distribution function F [so the random variables X_i $(i = 1, \ldots, n)$ are independent and identically distributed]. From n observations X_i $(i = 1, \ldots, n)$ one computes the *empirical* cumulative distribution function F_n,

$$F_n(x) := \frac{1}{n} \sum_{i=1}^{n} I_{(-\infty, x]}(X_i), \tag{4.48}$$

where $I_{(-\infty, x]}$ is the *indicator function* of the interval $(-\infty, x]$. For any set $A \subset \mathbb{R}$ the indicator function I_A is defined as

$$I_A(x) := \begin{cases} 1, & \text{if } x \in A, \\ 0, & \text{if } x \notin A. \end{cases} \tag{4.49}$$

The test statistic is given by

$$D_n := \sup_{x \in \mathbb{R}} |F_n(x) - F(x)|, \tag{4.50}$$

where $\sup S$ is the supremum of the set S. Thus the test statistic is the maximum distance between the cumulative distribution of the data and the cumulative distribution specified under the null hypothesis. It is known from *Glivenko–Cantelli* theorem that D_n converges to 0 as n tends to infinity.

Kolmogorov strengthened this result, by effectively providing the rate of the convergence $D_n \to 0$. He has shown that under the null hypothesis the quantity $\sqrt{n}\, D_n$ converges, as $n \to \infty$, to a random variable K defined by

$$K := \sup_{t \in [0,1]} |B(t)|, \tag{4.51}$$

where $B(t)$ is the so called Brownian bridge stochastic process (see Section 3.7.2 for its definition). The cumulative distribution function of the random variable K is given by

$$P(K \leq x) = 1 - 2 \sum_{i=1}^{\infty} (-1)^{i-1} \exp\left(-2i^2 x^2\right)$$

$$= \frac{\sqrt{2\pi}}{x} \sum_{i=1}^{\infty} \exp\left(-\frac{(2i-1)^2 \pi^2}{8x^2}\right). \tag{4.52}$$

The null hypothesis is rejected at a level α if

$$\sqrt{n}\, D_n > K_\alpha, \tag{4.53}$$

where K_α is determined by

$$P(K \leq K_\alpha) = 1 - \alpha. \tag{4.54}$$

The KS test applies only to continuous distributions and the distribution must be fully specified.

4.5 Higher-order spectra

If the stochastic process is Gaussian its first two moments completely specify its probability distribution. Studying higher-order moments of the process enables thus to detect its non-Gaussianity. Let us assume that x_n, $n \in \mathbb{Z}$ is a discrete-time zero-mean complex stationary stochastic process. Its third- and fourth-order cumulants are defined by

$$C_3(k, l) := \mathrm{E}\left\{x_n x_{n+k} x_{n+l}\right\}, \tag{4.55a}$$

$$
\begin{aligned}
C_4(k, l, m) := {} & \mathrm{E}\left\{x_n x_{n+k} x_{n+l} x_{n+m}\right\} \\
& - C(k)C(l-m) - C(l)C(k-m) - C(m)C(k-l),
\end{aligned} \tag{4.55b}
$$

where C is the covariance function [see Eq. (4.2)]. The zero-lag cumulants have special names: $C_3(0,0)$ and $C_4(0,0,0)$ are usually denoted by γ_3 and γ_4, the normalized quantities γ_3/σ^6 and γ_4/σ^8 are called *skewness* and *kurtosis*, respectively. The cumulants C_3 and C_4 have the following symmetry properties:

$$C_3(k, l) = C_3(l, k) = C_3(-k, l-k), \tag{4.56a}$$

$$C_4(k, l, m) = C_4(l, k, m) = C_4(k, m, l) = C_4(-k, l-k, m-k). \tag{4.56b}$$

We define the *bispectrum* $S_2(\omega_1, \omega_2)$ and the *trispectrum* $S_3(\omega_1, \omega_2, \omega_3)$ as the two- and three-dimensional Fourier transforms of $C_3(k, l)$ and

$C_4(k, l, m)$, respectively,

$$S_2(\omega_1, \omega_2) := \sum_{k=-N}^{N} \sum_{l=-N}^{N} C_3(k, l)\, e^{-i\omega_1 k} e^{-i\omega_2 l}, \tag{4.57a}$$

$$S_3(\omega_1, \omega_2, \omega_3) := \sum_{k=-N}^{N} \sum_{l=-N}^{N} \sum_{m=-N}^{N} C_4(k, l, m)\, e^{-i\omega_1 k} e^{-i\omega_2 l} e^{-i\omega_3 m}. \tag{4.57b}$$

Note that the bispectrum is a function of two frequencies, whereas the trispectrum is a function of three frequencies. In contrast with the power spectrum which is real valued and non-negative, bispectra and trispectra are complex-valued quantities. The symmetry properties of the bispectrum S_2 are given by

$$S_2(\omega_1, \omega_2) = S_2(\omega_2, \omega_1) = S_2(\omega_1, -\omega_1 - \omega_2)$$

$$= S_2(-\omega_1 - \omega_2, \omega_2) = S_2^*(-\omega_1, -\omega_2). \tag{4.58}$$

Symmetry properties of the trispectrum S_3 include:

$$S_3(\omega_1, \omega_2, \omega_3) = S_3(\omega_1, \omega_3, \omega_2) = S_3(\omega_2, \omega_1, \omega_3)$$

$$= S_3(-\omega_1, \omega_2 - \omega_1, \omega_3 - \omega_1) = S_3^*(-\omega_1, -\omega_2, -\omega_3). \tag{4.59}$$

Let x_k, $k \in \mathbb{Z}$ and u_l, $l \in \mathbb{Z}$ be two discrete-time stochastic processes and let u_l be independent and identically distributed random variables. We call the process x_k *linear* if it can be represented by

$$x_k = \sum_{l=0}^{N} a_l\, u_{k-l}, \tag{4.60}$$

where a_l are constant coefficients. If u_l is Gaussian (non-Gaussian), we say that x_l is linear Gaussian (non-Gaussian). In order to test for linearity and Gaussianity we examine the third-order cumulants of the data. The *bicoherence* is defined as

$$B(\omega_1, \omega_2) := \frac{S_2(\omega_1, \omega_2)}{S(\omega_1 + \omega_2) S(\omega_1) S(\omega_2)}, \tag{4.61}$$

where $S(\omega)$ is the spectral density of the process x_k and $S_2(\omega_1, \omega_2)$ is its bispectrum. If the process is Gaussian then its bispectrum and consequently its bicoherence is zero. One can easily show that if the process is linear then its bicoherence is constant. Thus if the bispectrum is not zero, then the process is non-Gaussian; if the bicoherence is not constant then the process is also non-linear. Consequently we have the following hypothesis testing problems:

- H_1: *the bispectrum of x_k is non-zero;*

- H_0: *the bispectrum of x_k is zero.*

If hypothesis H_1 holds, we can test for linearity, that is, we have a second hypothesis testing problem:

- H_1': *the bicoherence of x_k is not constant;*

- H_1'': *the bicoherence of x_k is a constant.*

If hypothesis H_1'' holds, the process is linear.

Using the above tests we can *detect* non-Gaussianity and, if the process is non-Gaussian, non-linearity of the process. The distribution of the test statistic $B(\omega_1, \omega_2)$, Eq. (4.61), can be calculated in terms of χ^2 distributions. For more details see [168].

5

Responses of detectors to gravitational waves

In this chapter we derive the responses of different detectors to a given gravitational wave described in a TT coordinate system related to the solar system barycenter by wave polarization functions h_+ and $h_\times$. We start in Section 5.1 by enumerating existing Earth-based gravitational-wave detectors, both laser interferometers and resonant bars. We give their geographical location and orientation with respect to local geographical directions.

In Section 5.2 we obtain a general response of a detector without assuming that the size of the detector is small compared to the wavelength of the gravitational wave. Such an approximation is considered in Section 5.3. In Section 5.4 we specialize our general formulae to the case of currently operating ground-based detectors and to the planned space-borne detector LISA.

5.1 Detectors of gravitational waves

There are two main methods of detecting gravitational waves that have been implemented in the currently working instruments. One method is to measure changes induced by gravitational waves on the distances between freely moving test masses using coherent trains of electromagnetic waves. The other method is to measure the deformation of large masses at their resonance frequencies induced by gravitational waves. The first idea is realized in laser interferometric detectors and Doppler tracking experiments [169, 170, 171, 172], whereas the second idea is implemented in resonant mass detectors [173, 174, 175].

114

Table 5.1. *Geographical positions and orientations of Earth-based laser interferometric gravitational-wave detectors. The angles ϕ and λ are the geodetic latitude (measured positively northwards) and longitude (measured positively westwards), respectively, of the detector's site. The angle γ determines the orientation of the detector's arms with respect to local geographical directions: γ is measured counter-clockwise from the East to the bisector of the interferometer arms. The angle ζ is the angle between the interferometer arms.*

Detector	ϕ	λ	γ	ζ
GEO600	$52°.25$	$-9°.81$	$68°.78$	$94°.33$
LIGO Hanford	$46°.46$	$119°.41$	$171°.00$	$90°.00$
LIGO Livingston	$30°.56$	$90°.77$	$242°.72$	$90°.00$
VIRGO	$43°.63$	$-10°.50$	$115°.57$	$90°.00$
TAMA300	$35°.68$	$-139°.54$	$225°.00$	$90°.00$

Table 5.2. *Geographical positions and orientations of Earth-based resonant gravitational-wave detectors. The angles ϕ and λ are the geodetic latitude (measured positively northwards) and longitude (measured positively westwards), respectively, of the detector's site. The angle γ determines the orientation of the detector with respect to local geographical directions: γ is measured counter-clockwise from the East to the bar's axis of symmetry.*

Detector	ϕ	λ	γ
ALLEGRO	$30°.41$	$-91°.18$	$130°.00$
AURIGA	$45°.35$	$11°.95$	$46°.00$
EXPLORER	$46°.45$	$-6°.20$	$51°.00$
NAUTILUS	$41°.82$	$-12°.67$	$46°.00$
NIOBE	$-31°.93$	$-115°.82$	$90°.00$

Currently networks of resonant detectors and laser interferometric detectors are working around the globe and collecting data. In Table 5.1 geographical positions and orientations of Earth-based laser interferometric gravitational-wave detectors are given whereas in Table 5.2 resonant detectors are listed.

5.2 Doppler shift between freely falling observers

A spacetime metric of a weak plane gravitational wave in the TT coordinate system (with coordinates $x^0 = ct$, $x^1 = x$, $x^2 = y$, $x^3 = z$) chosen such that the wave is traveling in the $+z$ direction, is described by the line element [see Eq. (1.68)]

$$ds^2 = -c^2\, dt^2 + \left(1 + h_+\left(t - \frac{z}{c}\right)\right) dx^2 + \left(1 - h_+\left(t - \frac{z}{c}\right)\right) dy^2$$

$$+ 2\, h_\times\left(t - \frac{z}{c}\right) dx\, dy + dz^2, \tag{5.1}$$

where h_+ and $h_\times$ are the two polarizations of the wave. The spacetime metric (5.1) admits three Killing vectors with components

$$K_1^\alpha = (0, 1, 0, 0), \tag{5.2a}$$

$$K_2^\alpha = (0, 0, 1, 0), \tag{5.2b}$$

$$K_3^\alpha = (1, 0, 0, 1). \tag{5.2c}$$

Let us consider a photon which travels along a null geodesic of the metric (5.1). The photon's world line is described by equations

$$x^\alpha = x^\alpha(\lambda), \tag{5.3}$$

where λ is some *affine* parameter along the world line. The 4-vector $dx^\alpha/d\lambda$ tangent to the geodesic (5.3) is null, i.e. it fulfills the equation

$$g_{\alpha\beta} \frac{dx^\alpha}{d\lambda} \frac{dx^\beta}{d\lambda} = 0. \tag{5.4}$$

Because with any Killing vector K^α one can relate a constant of motion along the null geodesic (5.3):

$$g_{\alpha\beta} K^\alpha \frac{dx^\beta}{d\lambda} = const, \tag{5.5}$$

with the three Killing vectors (5.2) we can relate three such constants:

$$g_{\alpha\beta} K_n^\alpha \frac{dx^\beta}{d\lambda} = c_n, \quad n = 1, 2, 3. \tag{5.6}$$

The three Eqs. (5.6) together with Eq. (5.4) form a system of four algebraic equations, which can be solved with respect to the four components $dx^\alpha/d\lambda$ of the tangent vector. This unique solution, accurate to the terms linear in h, reads

$$c\frac{dt}{d\lambda} = -\frac{c_1^2 + c_2^2 + c_3^2}{2c_3} + \frac{c_1^2 - c_2^2}{2c_3} h_+ + \frac{c_1 c_2}{c_3} h_\times + \mathcal{O}(h^2), \tag{5.7a}$$

$$\frac{dx}{d\lambda} = c_1 - c_1\, h_+ - c_2\, h_\times + \mathcal{O}(h^2), \tag{5.7b}$$

$$\frac{dy}{d\lambda} = c_2 + c_2\, h_+ - c_1\, h_\times + \mathcal{O}(h^2), \tag{5.7c}$$

$$\frac{dz}{d\lambda} = \frac{c_3^2 - c_1^2 - c_2^2}{2c_3} + \frac{c_1^2 - c_2^2}{2c_3} h_+ + \frac{c_1 c_2}{c_3} h_\times + \mathcal{O}(h^2). \tag{5.7d}$$

Let us now consider two observers which freely fall in the field (5.1) of the gravitational wave. Additionally let us also assume that their spatial coordinates remain constant [this condition indeed implies that the observers' world lines are geodesics of the metric (5.1)]. The observers' world lines are thus described by equations

$$t(\tau_a) = \tau_a, \quad x(\tau_a) = x_a, \quad y(\tau_a) = y_a, \quad z(\tau_a) = z_a, \quad a = 1, 2, \quad (5.8)$$

where (x_a, y_a, z_a) are constant spatial coordinates of the ath observer and τ_a, as it can be easily checked, is the proper time of this observer. The 4-velocity $U_a^\alpha := dx^\alpha/d\tau_a$ of the ath observer has thus components

$$U_a^\alpha = (1, 0, 0, 0), \quad a = 1, 2. \quad (5.9)$$

Let us imagine that our two observers measure, in their proper reference frames, the frequency of the *same* photon traveling along the world line (5.3). Let us call the events, at which the photon's frequency is measured, $\mathcal{E}_1$ and $\mathcal{E}_2$ (for observers 1 and 2, respectively). These events have the following spacetime coordinates:

$$\mathcal{E}_1 : (t_1, x_1, y_1, z_1), \quad \mathcal{E}_2 : (t_2, x_2, y_2, z_2). \quad (5.10)$$

We assume that $t_2 > t_1$. The photon's frequency ν is proportional to the energy of the photon measured by the observer, i.e.

$$\nu \propto -g_{\alpha\beta} U^\alpha \frac{dx^\beta}{d\lambda}. \quad (5.11)$$

One can rewrite Eq. (5.11), making use of Eqs. (5.1) and (5.9), in the form

$$\nu \propto \frac{dx^0}{d\lambda}. \quad (5.12)$$

Let us consider the ratio ν_2/ν_1 of the photon's frequencies measured by the two observers at the events $\mathcal{E}_2$ and $\mathcal{E}_1$. Let us denote

$$y_{12} := \frac{\nu_2}{\nu_1} - 1. \quad (5.13)$$

By virtue of Eqs. (5.12) and (5.7a) we obtain (keeping only terms linear in h):

$$y_{12} = c_+ \left(h_+(t_1 - z_1/c) - h_+(t_2 - z_2/c) \right)$$
$$+ c_\times \left(h_\times(t_1 - z_1/c) - h_\times(t_2 - z_2/c) \right) + \mathcal{O}(h^2), \quad (5.14)$$

where we have collected the three constants c_n into two new constant parameters c_+ and $c_\times$:

$$c_+ := \frac{c_1^2 - c_2^2}{c_1^2 + c_2^2 + c_3^2}, \quad c_\times := \frac{2 c_1 c_2}{c_1^2 + c_2^2 + c_3^2}. \quad (5.15)$$

It is convenient to express the constants c_+ and $c_\times$ [defined in Eq. (5.15)] in terms of angles describing the orientation of the photon's path with respect to the axes of the spatial coordinate system we use. From Eqs. (5.7) it follows that the components $dx^\alpha/d\lambda$ of the vector tangent to the photon's world line deviate from being constant by small quantities of the order h. Let us introduce the 3-vector $\mathbf{v}$ with components

$$\mathbf{v} := \left(\frac{dx}{d\lambda}, \frac{dy}{d\lambda}, \frac{dz}{d\lambda} \right), \tag{5.16}$$

together with its Euclidean length:

$$|\mathbf{v}| := \sqrt{\left(\frac{dx}{d\lambda}\right)^2 + \left(\frac{dy}{d\lambda}\right)^2 + \left(\frac{dz}{d\lambda}\right)^2}. \tag{5.17}$$

Equations (5.7) imply that

$$|\mathbf{v}| = \frac{c_1^2 + c_2^2 + c_3^2}{2\,|c_3|} + \mathcal{O}(h). \tag{5.18}$$

Let us also introduce three angles $\alpha, \beta, \gamma \in [0, \pi]$ such that their cosines are equal to

$$\cos\alpha = -\frac{2c_1 c_3}{c_1^2 + c_2^2 + c_3^2}, \tag{5.19a}$$

$$\cos\beta = -\frac{2c_2 c_3}{c_1^2 + c_2^2 + c_3^2}, \tag{5.19b}$$

$$\cos\gamma = \frac{c_1^2 + c_2^2 - c_3^2}{c_1^2 + c_2^2 + c_3^2}. \tag{5.19c}$$

It is easy to check that

$$\cos^2\alpha + \cos^2\beta + \cos^2\gamma = 1. \tag{5.20}$$

Moreover, making use of Eqs. (5.7), one can show that

$$\frac{1}{|\mathbf{v}|}\frac{dx}{d\lambda} = \cos\alpha + \mathcal{O}(h), \tag{5.21a}$$

$$\frac{1}{|\mathbf{v}|}\frac{dy}{d\lambda} = \cos\beta + \mathcal{O}(h), \tag{5.21b}$$

$$\frac{1}{|\mathbf{v}|}\frac{dz}{d\lambda} = \cos\gamma + \mathcal{O}(h). \tag{5.21c}$$

Equations (5.21) deliver a simple geometrical interpretation of the angles α, β, and γ: if one neglects the spacetime curvature caused by the gravitational wave, then α, β, and γ are the angles between the path of the photon in the 3-space and the coordinate axis x, y, or z, respectively. Making use of Eqs. (5.19), one can express the constants c_+ and $c_\times$ from Eq. (5.15) by cosines of the angles α, β, and γ:

$$c_+ = \frac{\cos^2\alpha - \cos^2\beta}{2(1-\cos\gamma)}, \quad c_\times = \frac{\cos\alpha\cos\beta}{1-\cos\gamma}. \tag{5.22}$$

Because the events $\mathcal{E}_1$ and $\mathcal{E}_2$ lie on the same null geodesic (5.3), the arguments $t_1 - z_1/c$ and $t_2 - z_2/c$ of the functions h_+ and $h_\times$ from Eq. (5.14) can be related to each other. Let us introduce the coordinate time duration t_{12} of the photon's trip from the event $\mathcal{E}_1$ to $\mathcal{E}_2$:

$$t_{12} := t_2 - t_1, \tag{5.23}$$

and the Euclidean coordinate distance L_{12} between the observers:

$$L_{12} := \sqrt{(x_2 - x_1)^2 + (y_2 - y_1)^2 + (z_2 - z_1)^2}. \tag{5.24}$$

Direct integration of Eqs. (5.7) gives (we assume that $\lambda = 0$ corresponds to the event $\mathcal{E}_1$):

$$ct(\lambda) = ct_1 - \frac{c_1^2 + c_2^2 + c_3^2}{2c_3}\lambda + \mathcal{O}(h), \tag{5.25a}$$

$$x(\lambda) = x_1 + c_1\lambda + \mathcal{O}(h), \tag{5.25b}$$

$$y(\lambda) = y_1 + c_2\lambda + \mathcal{O}(h), \tag{5.25c}$$

$$z(\lambda) = z_1 + \frac{c_3^2 - c_1^2 - c_2^2}{2c_3}\lambda + \mathcal{O}(h). \tag{5.25d}$$

From Eqs. (5.25) follows that

$$ct_{12} = L_{12} + \mathcal{O}(h). \tag{5.26}$$

Another useful relation one gets from Eqs. (5.25) and (5.19c). It reads

$$\frac{z_2 - z_1}{c} = t_{12}\cos\gamma + \mathcal{O}(h), \tag{5.27}$$

so we can write

$$t_2 - \frac{z_2}{c} = t_1 - \frac{z_1}{c} + \left(t_{12} - \frac{z_2 - z_1}{c}\right)$$

$$= t_1 - \frac{z_1}{c} - (1 - \cos\gamma)\frac{L_{12}}{c} + \mathcal{O}(h). \tag{5.28}$$

Making use of Eqs. (5.22) and (5.28), the ratio (5.14) of the photon's frequencies can be written in the form

$$
y_{12} = \frac{\cos^2 \alpha - \cos^2 \beta}{2(1 - \cos \gamma)} \left(h_+ \left(t_1 - \frac{z_1}{c} \right) - h_+ \left(t_1 - \frac{z_1}{c} + (1 - \cos \gamma) \frac{L_{12}}{c} \right) \right)
$$

$$
+ \frac{\cos \alpha \cos \beta}{1 - \cos \gamma} \left(h_\times \left(t_1 - \frac{z_1}{c} \right) - h_\times \left(t_1 - \frac{z_1}{c} + (1 - \cos \gamma) \frac{L_{12}}{c} \right) \right)
$$

$$
+ \mathcal{O} \left(h^2 \right). \tag{5.29}
$$

Let us introduce the three-dimensional matrix H of the spatial metric perturbation produced by the gravitational wave:

$$
\mathsf{H}(t) := \begin{pmatrix} h_+(t) & h_\times(t) & 0 \\ h_\times(t) & -h_+(t) & 0 \\ 0 & 0 & 0 \end{pmatrix}. \tag{5.30}
$$

Let us note that the form of the line element (5.1) implies that the functions $h_+ = h_+(t)$ and $h_\times = h_\times(t)$ describe the components of the wave-induced metric perturbation *at the origin* of the TT coordinate system (where $z = 0$). Let us also introduce the 3-vector $\mathbf{n}$ of unit Euclidean length directed along the line connecting the two observers. The components of this vector are arranged into the column matrix n [we thus distinguish here the 3-vector $\mathbf{n}$ from its components, which are the elements of the matrix n; we remember that the *same* 3-vector can be decomposed into components in *different* spatial coordinate systems]:

$$
\mathsf{n} := (\cos \alpha, \cos \beta, \cos \gamma)^\mathsf{T} = \begin{pmatrix} \cos \alpha \\ \cos \beta \\ \cos \gamma \end{pmatrix}, \tag{5.31}
$$

where the subscript T denotes *matrix transposition*. Then the frequency ratio y_{12} from Eq. (5.29) can be more compactly written as follows:

$$
y_{12} = \frac{1}{2(1 - \cos \gamma)}
$$

$$
\times \mathsf{n}^\mathsf{T} \cdot \left(\mathsf{H} \left(t_1 - \frac{z_1}{c} \right) - \mathsf{H} \left(t_1 - \frac{z_1}{c} + (1 - \cos \gamma) \frac{L_{12}}{c} \right) \right) \cdot \mathsf{n}
$$

$$
+ \mathcal{O}(h^2), \tag{5.32}
$$

where the dot means standard *matrix multiplication*. It is convenient to introduce the 3-vector $\mathbf{k}$ pointing to the gravitational-wave source. In our coordinate system the wave is traveling in the $+z$ direction, therefore the components of the 3-vector $\mathbf{k}$, arranged into the column matrix k, are

$$
\mathsf{k} = (0, 0, -1)^\mathsf{T}. \tag{5.33}
$$

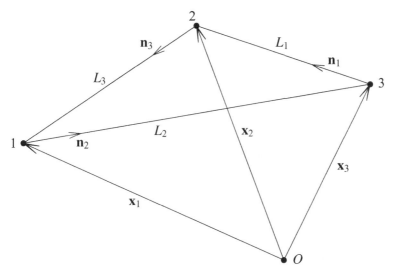

Fig. 5.1. Configuration of three freely falling particles as a detector of gravitational waves. The particles are labeled 1, 2, and 3. The Euclidean coordinate distances between the particles are denoted by L_a, where the index a corresponds to the opposite particle. The 3-vectors $\mathbf{n}_a$ of unit Euclidean lengths point between pairs of particles, with the orientation indicated.

Let us finally introduce the 3-vectors $\mathbf{x}_a$ $(a = 1, 2)$ describing the positions of the observers with respect to the origin of the coordinate system. Again the components of these 3-vectors we put into the column matrices x_a:

$$\mathsf{x}_a = (x_a, y_a, z_a)^\mathsf{T}, \quad a = 1, 2. \tag{5.34}$$

Making use of Eqs. (5.33)–(5.34) we rewrite the basic formula (5.32) in the following form [where we have also employed Eqs. (5.26) and (5.27)]

$$y_{12} = \frac{\mathsf{n}^\mathsf{T} \cdot \left(\mathsf{H}\left(t_1 + \dfrac{\mathsf{k}^\mathsf{T} \cdot \mathsf{x}_1}{c} \right) - \mathsf{H}\left(t_1 + \dfrac{L_{12}}{c} + \dfrac{\mathsf{k}^\mathsf{T} \cdot \mathsf{x}_2}{c} \right) \right) \cdot \mathsf{n}}{2(1 + \mathsf{k}^\mathsf{T} \cdot \mathsf{n})} + \mathcal{O}\left(h^2 \right). \tag{5.35}$$

Let us now complicate the situation by introducing the third observer, so we consider the configuration of three freely falling observers shown in Fig. 5.1. Let us also change our nomenclature by replacing the word "observer" by the word "particle" (to stress that in real detectors all measurements are done by appropriate devices without direct intervention of a human being). The configuration from Fig. 5.1 is general enough to obtain responses for all currently working and planned detectors. Two particles model a Doppler tracking experiment where one particle is the Earth and the other one is a distant spacecraft. Three particles model a ground-based laser interferometer where the particles (mirrors) are suspended from seismically isolated supports or a space-borne interferometer where the particles are shielded in satellites driven by drag-free control systems.

We denote by O the origin of the TT coordinate system. Let L_a ($a = 1, 2, 3$) be the Euclidean coordinate distance between the particles, where the index a corresponds to the opposite particle. Let $\mathbf{n}_a$ ($a = 1, 2, 3$) be the unit 3-vector along the lines joining the particles, and $\mathbf{x}_a$ ($a = 1, 2, 3$) be the 3-vector joining O and the ath particle.

Let us denote by ν_0 the frequency of the coherent beam used in the detector (laser light in the case of an interferometer and radio waves in the case of Doppler tracking). Let particle 1 emit the photon with frequency ν_0 at the moment t_0 towards particle 2, which registers the photon with frequency ν' at the moment $t' = t_0 + L_3/c + \mathcal{O}(h)$. The photon is immediately transponded (without change of frequency) back to particle 1, which registers the photon with frequency ν at the moment $t = t_0 + 2L_3/c + \mathcal{O}(h)$. We express the relative changes of the photon's frequency $y_{12} := (\nu' - \nu_0)/\nu_0$ and $y_{21} := (\nu - \nu')/\nu'$ as *functions of the instant of time t*. Making use of Eq. (5.35) we obtain

$$y_{12}(t) = \frac{1}{2(1 - \mathsf{k}^\mathsf{T} \cdot \mathsf{n}_3)}$$
$$\times \mathsf{n}_3^\mathsf{T} \cdot \left(\mathsf{H}\left(t - \frac{2L_3}{c} + \frac{\mathsf{k}^\mathsf{T} \cdot \mathsf{x}_1}{c} \right) - \mathsf{H}\left(t - \frac{L_3}{c} + \frac{\mathsf{k}^\mathsf{T} \cdot \mathsf{x}_2}{c} \right) \right) \cdot \mathsf{n}_3$$
$$+ \mathcal{O}(h^2), \tag{5.36a}$$

$$y_{21}(t) = \frac{1}{2(1 + \mathsf{k}^\mathsf{T} \cdot \mathsf{n}_3)}$$
$$\times \mathsf{n}_3^\mathsf{T} \cdot \left(\mathsf{H}\left(t - \frac{L_3}{c} + \frac{\mathsf{k}^\mathsf{T} \cdot \mathsf{x}_2}{c} \right) - \mathsf{H}\left(t + \frac{\mathsf{k}^\mathsf{T} \cdot \mathsf{x}_1}{c} \right) \right) \cdot \mathsf{n}_3$$
$$+ \mathcal{O}(h^2). \tag{5.36b}$$

The total frequency shift $y_{121} := (\nu - \nu_0)/\nu_0$ of the photon during its round trip can easily be computed from the one-way frequency shifts given above:

$$y_{121} = \frac{\nu}{\nu_0} - 1 = \frac{\nu}{\nu'}\frac{\nu'}{\nu_0} - 1$$
$$= (y_{21} + 1)(y_{12} + 1) - 1 = y_{12} + y_{21} + \mathcal{O}(h^2). \tag{5.37}$$

5.3 Long-wavelength approximation

In the long-wavelength approximation the reduced wavelength $\lambdabar = \lambda/(2\pi)$ of the gravitational wave is much larger than the size L of the detector.

The angular frequency of the wave equals $\omega = c/\lambda$. Time delays caused by the finite speed of the wave propagating across the detector are of the order $\Delta t \sim L/c$, but

$$\omega \Delta t \sim \frac{L}{\lambda} \ll 1, \tag{5.38}$$

so the time delays across the detector are much shorter than the period of the gravitational wave and can be neglected. It means that with a good accuracy the gravitational-wave field can be treated as being uniform (but time-dependent) in the space region that covers the entire detector.

Let us consider a wave with a dominant angular frequency ω. To detect such a wave one must collect data during a time interval longer (sometimes much longer) than the gravitational-wave period. It implies that in Eq. (5.32) the typical value of the quantity $\bar{t} := t_1 - z_1/c$ will be much larger than the retardation time $\Delta t := L_{12}/c$. Let us take any element h_{ij} of the matrix H, Eq. (5.30), and expand it with respect to Δt:

$$h_{ij}(\bar{t} + \Delta t) = h_{ij}(\bar{t}) + \dot{h}_{ij}(\bar{t})\,\Delta t + \frac{1}{2}\,\ddot{h}_{ij}(\bar{t})\,\Delta t^2 + \cdots$$

$$= h_{ij}(\bar{t})\left(1 + \frac{\dot{h}_{ij}(\bar{t})\,\Delta t}{h_{ij}(\bar{t})} + \frac{\ddot{h}_{ij}(\bar{t})\,\Delta t^2}{2\,h_{ij}(\bar{t})} + \cdots\right), \tag{5.39}$$

where overdot denotes differentiation with respect to time t. The time derivatives of h_{ij} can be estimated by $\dot{h}_{ij} \sim \omega h_{ij}$, $\ddot{h}_{ij} \sim \omega^2 h_{ij}$, and so on. It means that $\dot{h}_{ij}\Delta t/h_{ij} \sim \omega\Delta t$, $\ddot{h}_{ij}\Delta t^2/h_{ij} \sim (\omega\Delta t)^2$, and so on. We thus see that in the right-hand side of Eq. (5.39) the first term added to 1 is a small correction, and all the next terms are corrections of even higher order. We can therefore expand Eq. (5.32) with respect to Δt and keep terms only linear in Δt. After doing this one obtains the following formula for the relative frequency shift in the long-wavelength approximation:

$$y_{12}(\bar{t}) = -\frac{L_{12}}{2c}\,\mathsf{n}^\mathsf{T} \cdot \dot{\mathsf{H}}(\bar{t}) \cdot \mathsf{n} + \mathcal{O}(h^2). \tag{5.40}$$

For the configuration of particles shown in Fig. 5.1, the relative frequency shifts y_{12} and y_{21} given by Eqs. (5.36) can be written, by virtue of the formula (5.40), in the form

$$y_{12}(\bar{t}) = y_{21}(\bar{t}) = -\frac{L_3}{2c}\,\mathsf{n}_3^\mathsf{T} \cdot \dot{\mathsf{H}}(\bar{t}) \cdot \mathsf{n}_3 + \mathcal{O}(h^2), \tag{5.41}$$

so they are equal to each other up to terms $\mathcal{O}(h^2)$. The total round-trip frequency shift y_{121} [cf. Eq. (5.37)] is thus equal

$$y_{121}(\bar{t}) = -\frac{L_3}{c}\,\mathsf{n}_3^\mathsf{T} \cdot \dot{\mathsf{H}}(\bar{t}) \cdot \mathsf{n}_3 + \mathcal{O}(h^2). \tag{5.42}$$

There are important cases where the long-wavelength approximation is not valid. These include satellite Doppler tracking measurements and the space-borne LISA detector for gravitational-wave frequencies larger than a few mHz.

5.4 Responses of the solar-system-based detectors

Real gravitational-wave detectors do not move along geodesics of the spacetime metric related to the passing gravitational wave, as described in Section 5.2, because they move in the gravitational field of the solar system bodies, as in the case of the LISA spacecraft, or are fixed to the surface of Earth, as in the case of Earth-based laser interferometers or resonant bar detectors. The motion of the detector with respect to the *solar system barycenter* (SSB) will modulate the gravitational-wave signal registered by the detector.

The detector thus moves in spacetime with the metric which approximately can be written as

$$ds^2 = -(1 + 2\phi)\, c^2\, dt^2 + (1 - 2\phi)(dx^2 + dy^2 + dz^2) + h_{ij}^{\mathrm{TT}}\, dx^i\, dx^j,$$

(5.43)

where ϕ is the Newtonian potential (divided by c^2 to make it dimensionless) produced by different bodies of the solar system. The Newtonian gravitational field is weak, i.e. $|\phi| \ll 1$, therefore the disturbance e.g. of the photon's path introduced by this field is small, but it is still much larger than the disturbance caused by the passage of the gravitational wave, because we should expect that $|\phi| \gg |h|$. But both effects are small and can therefore be treated independently, i.e. all the effects of the order $\mathcal{O}(\phi h)$ can be neglected. The Newtonian-field-induced effects have to be "subtracted" before the gravitational-wave data analysis begins. This subtraction relies on accurate modeling of the motion of the detector with respect to the SSB using the solar system ephemeris. In the rest of this section we will assume that such subtraction was already done. Therefore we will put $\phi = 0$ into the line element (5.43) and reduce our analysis to consideration of the motion of particles and photons in the field of gravitational wave only.

After doing this it is possible to use in the computation of the responses of the real detectors the results we already derived in Section 5.2. We have to drop the assumption (made in Section 5.2) that the particles modeling the detector's parts are at rest with respect to the TT coordinate system – now their spatial coordinates can change in time, so the particles move along some world lines, which in general are not geodesics of the gravitational-wave-induced spacetime metric.

Let us choose the origin of the TT coordinate system of Section 5.2 to coincide with the SSB. The detailed computations show that as far as the velocities of the detector's parts (particles in the nomenclature of Section 5.2) with respect to the SSB are *non-relativistic*, which is the case for all existing or planned detectors, the formulae derived in Section 5.2 can still be used. The basic formulae (5.36) are valid, provided the column matrices n_a and x_a ($a = 1, 2, 3$) will be interpreted as made of the time-dependent components of the 3-vectors $\mathbf{n}_a$ and $\mathbf{x}_a$, respectively, computed in the SSB coordinate system. Equation (5.36b) takes now the form

$$
y_{21}(t) = \frac{1}{2(1 + \mathsf{k}^\mathsf{T} \cdot \mathsf{n}_3(t))}
$$

$$
\times \mathsf{n}_3^\mathsf{T}(t) \cdot \left(\mathsf{H} \left(t - \frac{L_3(t)}{c} + \frac{\mathsf{k}^\mathsf{T} \cdot \mathsf{x}_2(t)}{c} \right) - \mathsf{H} \left(t + \frac{\mathsf{k}^\mathsf{T} \cdot \mathsf{x}_1(t)}{c} \right) \right) \cdot \mathsf{n}_3(t)
$$

$$
+ \mathcal{O}(h^2). \tag{5.44}
$$

In this equation the spatial coordinates of the particles (which are components of the column matrices x_a, n_a and are needed to calculate the distances L_a, $a = 1, 2, 3$) are all computed at the same moment of time t. Corrections to these quantities coming from the retardation effects would lead to additional terms of the order of $\mathcal{O}(vh)$, but such terms we neglect.

Let us now introduce the *proper reference frame* of the detector with coordinates $(\hat{x}^\alpha)$. Because the motion of this frame with respect to the SSB is non-relativistic, we can assume that the transformation between the TT coordinates (x^α) and the proper-reference-frame coordinates $(\hat{x}^\alpha)$ has the form

$$
\hat{t} = t, \quad \hat{x}^i(t, x^k) = \hat{x}^i_{\hat{O}}(t) + O^i_j(t) \, x^j, \tag{5.45}
$$

where the functions $\hat{x}^i_{\hat{O}}(t)$ describe the motion of the origin $\hat{O}$ of the detector's proper reference frame with respect to the SSB, and the functions $O^i_j(t)$ account for the different and changing in time relative orientations of the spatial coordinate axes of the two reference frames. The transformation inverse to that of Eq. (5.45) reads

$$
t = \hat{t}, \quad x^i(t, \hat{x}^k) = (O^{-1})^i_j(t) \left(\hat{x}^j - \hat{x}^j_{\hat{O}}(t) \right). \tag{5.46}
$$

The functions O^i_j and $(O^{-1})^i_j$ are elements of two mutually inverse matrices, which we denote by O and O^{-1}, respectively. The following relations are thus fulfilled:

$$
O^i_j(t) \, (O^{-1})^j_k(t) = \delta^i_k, \quad (O^{-1})^i_j(t) \, O^j_k(t) = \delta^i_k, \tag{5.47}
$$

which in matrix notation read

$$
O(t) \cdot O(t)^{-1} = \mathsf{I}, \quad O(t)^{-1} \cdot O(t) = \mathsf{I}, \tag{5.48}
$$

where I is 3×3 identity matrix.

Let us consider the following scalar quantity, being the building block for responses of different gravitational-wave detectors [see Eq. (5.44)],

$$\Phi_a(t) := \frac{1}{2}\, n_a(t)^\mathsf{T} \cdot \mathsf{H}\big(t - t_\mathrm{r}(t)\big) \cdot n_a(t)$$

$$= \frac{1}{2}\, h_{ij}\big(t - t_\mathrm{r}(t)\big)\, n_a^i(t)\, n_a^j(t), \quad a = 1, 2, 3, \qquad (5.49)$$

where $t_\mathrm{r}(t)$ is some retardation [see Eq. (5.44)], h_{ij} are elements of the 3×3 matrix H, and n_a^i are elements of the 3×1 column matrix n_a. The value of Φ_a is invariant under the transformation (5.45), because n_a^i transform as components of a contravariant 3-vector and h_{ij} transform as components of a rank two tensor with two covariant indices [let us remember that the transformation (5.45) is non-relativistic]:

$$\hat{n}_a^i = O_j^i(t)\, n_a^j(t), \quad \hat{h}_{ij}(t) = (O^{-1})_i^k(t)\, (O^{-1})_j^l(t)\, h_{kl}(t), \qquad (5.50)$$

where we have indicated that the proper-reference-frame components $\hat{n}_a^i$ of the unit vector n_a usually can be treated as constant (at least to a good approximation). Let us introduce the 3×1 column matrix $\hat{n}_a$ with elements $\hat{n}_a^i$ and the 3×3 matrix $\hat{\mathsf{H}}$ with elements $\hat{h}_{ij}$. Then the transformation formulae from Eq. (5.50) in matrix notation read

$$\hat{n}_a = O(t) \cdot n_a(t), \quad \hat{\mathsf{H}}(t) = (O(t)^{-1})^\mathsf{T} \cdot \mathsf{H}(t) \cdot O(t)^{-1}. \qquad (5.51)$$

By virtue of Eqs. (5.51) the quantity Φ_a defined in Eq. (5.49) takes in the proper reference frame the following form:

$$\Phi_a(t) = \frac{1}{2}\, \hat{n}_a^\mathsf{T} \cdot \hat{\mathsf{H}}\big(t - t_\mathrm{r}(t)\big) \cdot \hat{n}_a. \qquad (5.52)$$

To get the above equality we have had to replace in some places $O(t - t_\mathrm{r}(t))$ by $O(t)$, which means that we again neglect all terms of the order of $\mathcal{O}(vh)$. If the transformation matrix O is *orthogonal*, then $O^{-1} = O^\mathsf{T}$, and by virtue of (5.51) the relation between the matrices H and $\hat{\mathsf{H}}$ reads

$$\hat{\mathsf{H}}(t) = O(t) \cdot \mathsf{H}(t) \cdot O(t)^\mathsf{T}. \qquad (5.53)$$

5.4.1 *LISA detector: time-delay interferometry*

The space-borne LISA detector [170] will have multiple readouts corresponding to the six laser Doppler shifts measured between different spacecraft and to the six intra-spacecraft Doppler shifts (measured between optical benches located at the same spacecraft). It is possible to combine, with suitable time delays, the time series of the different Doppler shifts to cancel both the frequency fluctuations of the lasers and the noise due to the mechanical vibrations of the optical benches. The technique used to devise such combinations is known as *time-delay interferometry* (TDI). The thorough review of the TDI technique can be found in Ref. [176].

The TDI responses were first derived under the assumption that the LISA spacecraft array is stationary (such responses are called sometimes *first-generation* TDI). But the rotational motion of the LISA array around the Sun and the relative motion of the spacecraft prevent the cancellation of the laser frequency fluctuations in the first-generation TDI responses. Therefore new combinations were devised that are capable of suppressing the laser frequency fluctuations for a rotating LISA array. They are called *second-generation* TDI.

Different second-generation TDI responses are described in detail e.g. in Refs. [177, 178]. They are rather complicated combinations of delayed time series of the relative laser frequency fluctuations measured between the spacecraft. We present here only the so called second-generation *Michelson observables* X_a, $a = 1, 2, 3$. We adopt the notation for the delay of the time series such that, for instance, $y_{12,3}(t) = y_{12}(t - L_3(t))$. Using this notation the Michelson observable X_1 reads (the observables X_2 and X_3 can be obtained by cyclical permutations of spacecraft labels):

$$X_1 = (y_{31} + y_{13,2}) + (y_{21} + y_{12,3})_{,22} - (y_{21} + y_{12,3}) - (y_{31} + y_{13,2})_{,33}$$

$$- \big((y_{31} + y_{13,2}) + (y_{21} + y_{12,3})_{,22}$$

$$- (y_{21} + y_{12,3}) - (y_{31} + y_{13,2})_{,33}\big)_{,2233}. \qquad (5.54)$$

The responses X_a can be constructed from expressions of the type (5.49), each of them can be written as a linear combination of the wave polarization functions h_+ and $h_\times$:

$$\Phi_a(t) = F_{a+}(t)\, h_+\big(t - t_r(t)\big) + F_{a\times}(t)\, h_\times\big(t - t_r(t)\big), \qquad (5.55)$$

where t_r is some retardation and the functions F_{a+} and $F_{a\times}$ are called the *beam-pattern* functions. In Appendix C.1 one can find derivation of the explicit formulae for the beam-patterns of the LISA detector.

5.4.2 Ground-based laser interferometric detector

For a ground-based interferometric detector the long-wavelength approximation can be applied. Let us consider a standard *Michelson* and *equal-arm* interferometric configuration. This configuration can be represented by Fig. 5.1, if we put $L_2 = L_3 = L$. The observed relative frequency shift $\Delta\nu(t)/\nu_0$ is equal to the difference of the round-trip frequency shifts in the two detector's arms:

$$\frac{\Delta\nu(t)}{\nu_0} = y_{131}(t) - y_{121}(t). \qquad (5.56)$$

This can be written, by virtue of Eq. (5.42), as

$$\frac{\Delta\nu(t)}{\nu_0} = \frac{L}{c}\left(\mathbf{n}_2^\mathsf{T}\cdot\dot{\mathsf{H}}\left(t-\frac{z_\mathrm{d}}{c}\right)\cdot\mathbf{n}_2 - \mathbf{n}_3^\mathsf{T}\cdot\dot{\mathsf{H}}\left(t-\frac{z_\mathrm{d}}{c}\right)\cdot\mathbf{n}_3\right), \qquad (5.57)$$

where z_d is the third spatial component of the 3-vector $\mathbf{r}_\mathrm{d} = (x_\mathrm{d}, y_\mathrm{d}, z_\mathrm{d})$ which describes the position of the characteristic point within the detector (represented in Fig. 5.1 by the particle number 1) with respect to the origin of the TT coordinates.

The difference $\Delta\phi(t)$ of the phase fluctuations measured, say, by a photo detector, is related to the corresponding relative frequency fluctuations $\Delta\nu(t)$ by

$$\frac{\Delta\nu(t)}{\nu_0} = \frac{1}{2\pi\nu_0}\frac{\mathrm{d}\Delta\phi(t)}{\mathrm{d}t}. \qquad (5.58)$$

Making use of (5.57) one can integrate Eq. (5.58), after this the phase change $\Delta\phi(t)$ can be written as

$$\Delta\phi(t) = 4\pi\,\nu_0\,L\,s(t), \qquad (5.59)$$

where we have introduced the dimensionless function s,

$$s(t) := \frac{1}{2}\left(\mathbf{n}_2^\mathsf{T}\cdot\mathsf{H}\left(t-\frac{z_\mathrm{d}}{c}\right)\cdot\mathbf{n}_2 - \mathbf{n}_3^\mathsf{T}\cdot\mathsf{H}\left(t-\frac{z_\mathrm{d}}{c}\right)\cdot\mathbf{n}_3\right), \qquad (5.60)$$

which we call the *response* of the interferometer to a gravitational wave.

Equations (5.59)–(5.60) we have derived in some TT reference frame assuming that the particles forming the triangle in Fig. 5.1 all have constant in time spatial coordinates with respect to this frame. We can choose the origin of the frame to coincide with the SSB. The reasoning presented at the beginning of the current section shows that the formula (5.60) can still be used in the case when the particles are moving with respect to the SSB with non-relativistic velocities, which is certainly true for the inteferometer parts fixed to the surface of Earth. Thus the response of the Earth-based inteferometric detector reads

$$s(t) = \frac{1}{2}\left(\mathbf{n}_2^\mathsf{T}(t)\cdot\mathsf{H}\left(t-\frac{z_\mathrm{d}(t)}{c}\right)\cdot\mathbf{n}_2(t) - \mathbf{n}_3^\mathsf{T}(t)\cdot\mathsf{H}\left(t-\frac{z_\mathrm{d}(t)}{c}\right)\cdot\mathbf{n}_3(t)\right),$$
$$(5.61)$$

where all quantities are computed here in the TT coordinates connected with the SSB. The same response (5.61) can be rewritten in terms of the proper-reference-frame quantities. It then reads

$$s(t) = \frac{1}{2}\left(\hat{\mathbf{n}}_2^\mathsf{T}\cdot\hat{\mathsf{H}}\left(t-\frac{z_\mathrm{d}(t)}{c}\right)\cdot\hat{\mathbf{n}}_2 - \hat{\mathbf{n}}_3^\mathsf{T}\cdot\hat{\mathsf{H}}\left(t-\frac{z_\mathrm{d}(t)}{c}\right)\cdot\hat{\mathbf{n}}_3\right), \qquad (5.62)$$

where the matrix $\hat{\mathsf{H}}$ is related to the matrix H by means of Eq. (5.53). In Eq. (5.62) the proper-reference-frame components $\hat{\mathbf{n}}_2$ and $\hat{\mathbf{n}}_3$ of the unit vectors along the inteferometer arms are treated as time-independent quantities. In fact they deviate from being constants by quantities of the order of $\mathcal{O}(h)$, therefore these deviations contribute to the response function only at the order of $\mathcal{O}(h^2)$ and as such they can be neglected.

The response function (5.62) can directly be derived in the proper reference frame of the interferometer by means of the equation of geodesic deviation. In this derivation the response s is defined as the relative change of the lengths of the two arms, i.e. $s(t) := \Delta L(t)/L$.

From Eq. (5.61) we see that the response function s depends linearly on the matrix H of gravitational-wave-induced perturbation of spatial metric. Taking into account the form (5.30) of the matrix H, it is obvious that the response function s is a linear combination of the two wave polarization functions h_+ and $h_\times$:

$$s(t) = F_+(t)\, h_+\left(t - \frac{z_\mathrm{d}(t)}{c}\right) + F_\times(t)\, h_\times\left(t - \frac{z_\mathrm{d}(t)}{c}\right), \qquad (5.63)$$

where F_+ and $F_\times$ are the *interferometric beam-pattern functions*. They depend on the location of the detector on Earth and on the position of the gravitational-wave source in the sky. In Appendix C.2.1 we derive the explicit formulae for the beam patterns F_+ and $F_\times$ of the interferometric detector.

Let us finally introduce the 3-vector $\mathbf{n}_0$ of unit Euclidean length directed from the SSB towards the gravitational-wave source. In the TT coordinate system considered here it has components $\mathbf{n}_0 = (0, 0, -1)$, so the z-component of the 3-vector $\mathbf{r}_\mathrm{d}$ connecting the SSB and the detector can be computed as

$$z_\mathrm{d}(t) = -\mathbf{n}_0 \cdot \mathbf{r}_\mathrm{d}(t). \qquad (5.64)$$

5.4.3 Ground-based resonant-bar detector

In the case of a resonant-bar detector the long-wavelength approximation is very accurate. If one additionally assumes that the frequency spectrum of the gravitational wave which hits the bar entirely lies within the sensitivity band of the detector, then the dimensionless response function s can be computed from the formula (see e.g. Section 9.5.2 in Ref. [16])

$$s(t) = \hat{\mathsf{n}}^\mathsf{T} \cdot \hat{\mathsf{H}}\left(t - \frac{z_\mathrm{d}(t)}{c}\right) \cdot \hat{\mathsf{n}}, \qquad (5.65)$$

where the column matrix $\hat{\mathsf{n}}$ is made of the proper-reference-frame components of the unit vector $\mathbf{n}$ directed along the symmetry axis of the bar. The response (5.65), as in the case of the interferometric detector considered in the previous subsection, can be written as a linear combination of the wave polarization functions h_+ and $h_\times$, i.e. it can be written in the form given in Eq. (5.63), but with beam-pattern functions F_+ and $F_\times$, which are different from the interferometric ones. In Appendix C.2.2 we derive the explicit formulae for the beam patterns F_+ and $F_\times$ of the resonant-bar detector.

The response (5.65) can be derived in the proper reference frame of the detector using the equation of geodesic deviation, the response s is then computed as $s(t) := \Delta L(t)/L$, where $\Delta L(t)$ is the wave-induced change of the proper length L of the bar.

6

Maximum-likelihood detection in Gaussian noise

In this chapter we describe in detail the detection of gravitational-wave signals and the estimation of their parameters using the maximum-likelihood (ML) principle. We assume that the noise in the detector is a *stationary* and *Gaussian* stochastic process. The detection of gravitational-wave signals in non-Gaussian noise is discussed in [2, 3].

We begin in Section 6.1 by discussing several tools needed for both theoretical evaluation of the ML data analysis method and for its practical implementation. They include splitting the parameters of the gravitational-wave signal into extrinsic and intrinsic ones and the concept of reduced likelihood function, called the $\mathcal{F}$-statistic, then optimal signal-to-noise ratio, Fisher matrix, false alarm and detection probabilities, number of templates in space spanned by the parameters of the signal, and suboptimal filtering.

In Sections 6.2 and 6.3 we study in detail the case of detection of several specific deterministic gravitational-wave signals by a single detector (in Section 6.2) and by a network of detectors (in Section 6.3). Finally the ML detection of stochastic gravitational-wave background is considered in Section 6.4.

6.1 Deterministic signals

Let us assume that the gravitational-wave signal s present in the data is a superposition of $\mathcal{L}$ component signals s_ℓ, $\ell = 1, \ldots, \mathcal{L}$,

$$s(t; \boldsymbol{\theta}) = \sum_{\ell=1}^{\mathcal{L}} s_\ell(t; \boldsymbol{\theta}_\ell), \qquad (6.1)$$

where the vector $\boldsymbol{\theta} = (\boldsymbol{\theta}_1, \ldots, \boldsymbol{\theta}_{\mathcal{L}})$ collects all the parameters of the signal s, and the vector $\boldsymbol{\theta}_\ell$, $\ell = 1, \ldots, \mathcal{L}$, comprises all the parameters of the ℓth component signal s_ℓ. We restrict ourselves to signals, s, for which each component s_ℓ has the form

$$s_\ell(t; \boldsymbol{\theta}_\ell) = \sum_{i=1}^{n} a_{\ell i}\, h_i(t; \boldsymbol{\xi}_\ell), \qquad (6.2)$$

i.e. the signal s_ℓ is a linear combination of the n basic time-dependent *waveforms* h_i with constant coefficients $a_{\ell i}$ (which can be interpreted as *amplitudes* of the waveforms h_i), the vector $\boldsymbol{\xi}_\ell$ collects the remaining parameters of the ℓth signal. We thus assume that all the component signals s_ℓ are of the same form, i.e. all are the linear combinations of the *same* waveforms h_i.

Let us collect the amplitude parameters $a_{\ell i}$ and the waveforms h_i into column vectors (i.e. $n \times 1$ matrices),

$$\mathsf{a}_\ell := \begin{pmatrix} a_{\ell 1} \\ \vdots \\ a_{\ell n} \end{pmatrix}, \quad \mathsf{h}(t; \boldsymbol{\xi}_\ell) := \begin{pmatrix} h_1(t; \boldsymbol{\xi}_\ell) \\ \vdots \\ h_n(t; \boldsymbol{\xi}_\ell) \end{pmatrix}. \qquad (6.3)$$

With this notation the signal (6.2) can compactly be written in the following form:

$$s_\ell(t; \boldsymbol{\theta}_\ell) = \mathsf{a}_\ell^\mathsf{T} \cdot \mathsf{h}(t; \boldsymbol{\xi}_\ell), \qquad (6.4)$$

where T stands for the matrix transposition and $\cdot$ denotes matrix multiplication.

6.1.1 The $\mathcal{F}$-statistic

Let x be the data stream and let the noise in the data be an *additive*, *zero-mean*, *Gaussian*, and *stationary* stochastic process. Then the log likelihood function for the signal (6.1) is given by

$$\ln \Lambda[x; \boldsymbol{\theta}] = \sum_{\ell=1}^{\mathcal{L}} (x(t)|s_\ell(t; \boldsymbol{\theta}_\ell)) - \frac{1}{2} \sum_{\ell=1}^{\mathcal{L}} \sum_{\ell'=1}^{\mathcal{L}} (s_\ell(t; \boldsymbol{\theta}_\ell)|s_{\ell'}(t; \boldsymbol{\theta}_{\ell'})), \qquad (6.5)$$

where the scalar product $(\cdot|\cdot)$ is defined in Eq. (3.99). It is convenient to relate with each component signal s_ℓ an $n \times 1$ column matrix N_ℓ which depends linearly on the data x and has components

$$N_{\ell i}[x; \boldsymbol{\xi}_\ell] := (x|h_i(t; \boldsymbol{\xi}_\ell)), \quad \ell = 1, \ldots, \mathcal{L}, \quad i = 1, \ldots, n, \qquad (6.6)$$

and with any two component signals s_ℓ and $s_{\ell'}$ an $n \times n$ square matrix $M_{\ell\ell'}$ which does not depend on the data x and has components

$$M_{\ell\ell'\,ij}(\boldsymbol{\xi}_\ell, \boldsymbol{\xi}_{\ell'}) := (h_i(t; \boldsymbol{\xi}_\ell)|h_j(t; \boldsymbol{\xi}_{\ell'})), \quad \ell, \ell' = 1, \ldots, \mathcal{L}, \quad i, j = 1, \ldots, n.$$
(6.7)

With this notation the log likelihood function (6.5) can be put in the form

$$\ln \Lambda[x; \boldsymbol{\theta}] = \sum_{\ell=1}^{\mathcal{L}} \mathsf{a}_\ell^\mathsf{T} \cdot \mathsf{N}_\ell[x; \boldsymbol{\xi}_\ell] - \frac{1}{2} \sum_{\ell=1}^{\mathcal{L}} \sum_{\ell'=1}^{\mathcal{L}} \mathsf{a}_\ell^\mathsf{T} \cdot \mathsf{M}_{\ell\ell'}(\boldsymbol{\xi}_\ell, \boldsymbol{\xi}_{\ell'}) \cdot \mathsf{a}_{\ell'}. \quad (6.8)$$

We can make our notation even more compact. Let us define the $(n\mathcal{L}) \times 1$ column matrices a and N made of all $n \times 1$ column matrices a_ℓ and N_ℓ, respectively:

$$\mathsf{a} := \begin{pmatrix} \mathsf{a}_1 \\ \vdots \\ \mathsf{a}_\mathcal{L} \end{pmatrix}, \quad \mathsf{N}[x; \boldsymbol{\xi}] := \begin{pmatrix} \mathsf{N}_1[x; \boldsymbol{\xi}_1] \\ \vdots \\ \mathsf{N}_\mathcal{L}[x; \boldsymbol{\xi}_\mathcal{L}] \end{pmatrix}, \quad (6.9)$$

where $\boldsymbol{\xi} = (\boldsymbol{\xi}_1, \ldots, \boldsymbol{\xi}_\mathcal{L})$. Let us also arrange all matrices $\mathsf{M}_{\ell\ell'}$ into the single $(n\mathcal{L}) \times (n\mathcal{L})$ matrix M:

$$\mathsf{M}(\boldsymbol{\xi}) := \begin{pmatrix} \mathsf{M}_{11}(\boldsymbol{\xi}_1, \boldsymbol{\xi}_1) & \cdots & \mathsf{M}_{1\mathcal{L}}(\boldsymbol{\xi}_1, \boldsymbol{\xi}_\mathcal{L}) \\ \cdots\cdots\cdots\cdots\cdots\cdots\cdots\cdots\cdots\cdots\cdots \\ \mathsf{M}_{\mathcal{L}1}(\boldsymbol{\xi}_\mathcal{L}, \boldsymbol{\xi}_1) & \cdots & \mathsf{M}_{\mathcal{L}\mathcal{L}}(\boldsymbol{\xi}_\mathcal{L}, \boldsymbol{\xi}_\mathcal{L}) \end{pmatrix}. \quad (6.10)$$

The matrix M is symmetric, which follows from the equality $M_{\ell\ell'\,ij} = M_{\ell'\ell\,ji}$ [see the definition (6.7)]. Making use of the matrices a, M, and N, the log likelihood ratio (6.8) can be written concisely as

$$\ln \Lambda[x; \mathsf{a}, \boldsymbol{\xi}] = \mathsf{a}^\mathsf{T} \cdot \mathsf{N}[x; \boldsymbol{\xi}] - \frac{1}{2} \mathsf{a}^\mathsf{T} \cdot \mathsf{M}(\boldsymbol{\xi}) \cdot \mathsf{a}, \quad (6.11)$$

where we have taken into account that $\boldsymbol{\theta} = (\mathsf{a}, \boldsymbol{\xi})$.

The ML equations for the amplitude parameters a can be solved explicitly. From (6.11) we obtain

$$\frac{\partial \ln \Lambda}{\partial \mathsf{a}} = \mathsf{N} - \mathsf{M} \cdot \mathsf{a}, \quad (6.12)$$

so the ML estimators $\hat{\mathsf{a}}$ of the amplitudes a are given by

$$\hat{\mathsf{a}}[x; \boldsymbol{\xi}] = \mathsf{M}(\boldsymbol{\xi})^{-1} \cdot \mathsf{N}[x; \boldsymbol{\xi}]. \quad (6.13)$$

Replacing in Eq. (6.11) the amplitudes a by their ML estimators $\hat{\mathsf{a}}$, we obtain a function that we call the $\mathcal{F}$-*statistic*:

$$\mathcal{F}[x; \boldsymbol{\xi}] := \ln \Lambda[x; \hat{\mathsf{a}}[x; \boldsymbol{\xi}], \boldsymbol{\xi}] = \frac{1}{2} \mathsf{N}[x; \boldsymbol{\xi}]^\mathsf{T} \cdot \mathsf{M}(\boldsymbol{\xi})^{-1} \cdot \mathsf{N}[x; \boldsymbol{\xi}]. \quad (6.14)$$

We shall call the amplitudes a *extrinsic* parameters and the remaining parameters $\boldsymbol{\xi}$ *intrinsic* parameters. We have just demonstrated that for extrinsic parameters we can obtain their ML estimators in a closed analytic form and eliminate them from the likelihood function. In this way we obtain a reduced likelihood function, the $\mathcal{F}$-statistic, that depends (non-linearly) only on the intrinsic parameters $\boldsymbol{\xi}$.

The procedure to estimate all the parameters of the gravitational-wave signal (6.1) consists of two parts. The first part is to find the (local) maxima of the $\mathcal{F}$-statistic in the intrinsic parameters $\boldsymbol{\xi}$-space. The ML estimators of the intrinsic parameters $\boldsymbol{\xi}$ are those for which the $\mathcal{F}$-statistic attains a maximum. The second part is to calculate the estimators of the extrinsic parameters a from the analytic formula (6.13), where the matrix M and the correlations N are calculated for the intrinsic parameters equal to their ML estimators obtained from the first part of the analysis. We call this procedure the *maximum-likelihood estimation*. In Section 7.3 we discuss the efficient algorithms needed to find the (local) maxima of the $\mathcal{F}$-statistic.

Let us mention the case of *non-interfering* signals, i.e. the case when all the scalar products $(h_i(t; \boldsymbol{\xi}_\ell)|h_j(t; \boldsymbol{\xi}_{\ell'}))$ are equal to zero when $\ell \neq \ell'$, i.e. when waveforms h_i and h_j belong to different component signals in the set of $\mathcal{L}$ such signals present in the data. As a result the matrix M becomes block diagonal and the likelihood ratio splits into the sum of $\mathcal{L}$ terms, each term corresponding to a different component signal. Consequently, we can consider each component signal separately and for each such signal we can obtain ML estimators of the amplitude parameters and reduce the likelihood ratio to obtain the $\mathcal{F}$-statistic.

In the rest of this section we shall study the $\mathcal{F}$-statistic and the parameter estimation for a one-component signal s, therefore we shall drop the index ℓ in our notation.

6.1.2 Signal-to-noise ratio and the Fisher matrix

Let us consider the gravitational-wave signal s which is a linear combination of the n waveforms h_i, $i = 1, \dots, n$,

$$s(t; \boldsymbol{\theta}) = \sum_{i=1}^{n} a_i \, h_i(t; \boldsymbol{\xi}) = \mathsf{a}^\mathsf{T} \cdot \mathsf{h}(t; \boldsymbol{\xi}), \tag{6.15}$$

where the vector $\boldsymbol{\theta} = (\mathsf{a}, \boldsymbol{\xi})$ comprises all the signal's parameters, which split into amplitude or extrinsic parameters a and intrinsic parameters (on which the waveforms h_i depend) $\boldsymbol{\xi} = (\xi_1, \dots \xi_m)$, so m is the number of the intrinsic parameters. As in the more general case considered in the previous section, we arrange the extrinsic parameters a_i and the waveforms h_i into $n \times 1$ column matrices a and h,

$$
\mathsf{a} := \begin{pmatrix} a_1 \\ \vdots \\ a_n \end{pmatrix}, \quad \mathsf{h}(t; \boldsymbol{\xi}) := \begin{pmatrix} h_1(t; \boldsymbol{\xi}) \\ \vdots \\ h_n(t; \boldsymbol{\xi}) \end{pmatrix}. \tag{6.16}
$$

We also introduce the data-dependent $n \times 1$ column matrix N and the square $n \times n$ matrix M with components

$$
N_i[x; \boldsymbol{\xi}] := (x|h_i(t; \boldsymbol{\xi})), \quad M_{ij}(\boldsymbol{\xi}) := (h_i(t; \boldsymbol{\xi})|h_j(t; \boldsymbol{\xi})), \quad i, j = 1, \ldots, n. \tag{6.17}
$$

Then the ML estimators â of the amplitude parameters a are given by Eq. (6.13) and the $\mathcal{F}$-statistic for the signal (6.15) has the form displayed in Eq. (6.14).

The detectability of the signal s is determined by the signal-to-noise ratio ρ. It depends on all the signal's parameters $\boldsymbol{\theta}$ and is given by the equation [see Eq. (3.100)]

$$
\rho(\boldsymbol{\theta}) = \sqrt{(s(t; \boldsymbol{\theta})|s(t; \boldsymbol{\theta}))}. \tag{6.18}
$$

This equation for the signal (6.15) leads to the following formula:

$$
\rho(\mathsf{a}, \boldsymbol{\xi}) = \sqrt{\mathsf{a}^{\mathsf{T}} \cdot \mathsf{M}(\boldsymbol{\xi}) \cdot \mathsf{a}}. \tag{6.19}
$$

The accuracy of estimation of the signal's parameters $\boldsymbol{\theta}$ is determined by Fisher information matrix Γ. From the general definition of Γ given by Eq. (3.134) and the expression for the likelihood ratio Λ for Gaussian noise given in Eq. (3.98), one easily finds that the components of Γ (in the case of Gaussian noise) can be computed from the formula

$$
\Gamma_{ij} = \left(\frac{\partial s}{\partial \theta_i} \middle| \frac{\partial s}{\partial \theta_j} \right). \tag{6.20}
$$

For the signal (6.15) the parameter vector $\boldsymbol{\theta} = (a_1, \ldots, a_n, \xi_1, \ldots, \xi_m)$, so the matrix Γ has dimensions $(n + m) \times (n + m)$. To distinguish between extrinsic and intrinsic parameter indices, let us introduce calligraphic lettering to denote the intrinsic parameter indices: $\xi_{\mathcal{A}}$, $\mathcal{A} = 1, \ldots, m$. Let us also introduce two families of the auxiliary $n \times n$ square matrices which depend on the intrinsic parameters $\boldsymbol{\xi}$ only. First we define m matrices $\mathsf{F}_{(\mathcal{A})}$ with components (we put the index $\mathcal{A}$ within parentheses to stress that it serves here as the label of the matrix)

$$
F_{(\mathcal{A})ij}(\boldsymbol{\xi}) := \left(h_i(t; \boldsymbol{\xi}) \middle| \frac{\partial h_j(t; \boldsymbol{\xi})}{\partial \xi_{\mathcal{A}}} \right), \quad \mathcal{A} = 1, \ldots, m, \quad i, j = 1, \ldots, n, \tag{6.21a}
$$

and then we define m^2 matrices $\mathsf{S}_{(AB)}$ with components

$$S_{(AB)ij}(\boldsymbol{\xi}) := \left(\frac{\partial h_i(t;\boldsymbol{\xi})}{\partial \xi_A} \middle| \frac{\partial h_j(t;\boldsymbol{\xi})}{\partial \xi_B} \right), \quad A, B = 1, \ldots, m, \quad i, j = 1, \ldots, n.$$

(6.21b)

The Fisher matrix Γ [with components given by Eq. (6.20)] for the signal (6.15) can be written in terms of block matrices for the two sets of extrinsic a and intrinsic $\boldsymbol{\xi}$ parameters in the following form:

$$\Gamma(\mathsf{a}, \boldsymbol{\xi}) = \begin{pmatrix} \Gamma_{\mathsf{aa}}(\boldsymbol{\xi}) & \Gamma_{\mathsf{a}\boldsymbol{\xi}}(\mathsf{a}, \boldsymbol{\xi}) \\ \Gamma_{\mathsf{a}\boldsymbol{\xi}}(\mathsf{a}, \boldsymbol{\xi})^{\mathsf{T}} & \Gamma_{\boldsymbol{\xi}\boldsymbol{\xi}}(\mathsf{a}, \boldsymbol{\xi}) \end{pmatrix},$$

(6.22)

where Γ_{aa} is an $n \times n$ matrix with components $(\partial s/\partial a_i|\partial s/\partial a_j)$ $(i, j = 1, \ldots, n)$, $\Gamma_{\mathsf{a}\boldsymbol{\xi}}$ is an $n \times m$ matrix with components $(\partial s/\partial a_i|\partial s/\partial \xi_A)$ $(i = 1, \ldots, n, \mathcal{A} = 1, \ldots, m)$, and finally $\Gamma_{\boldsymbol{\xi}\boldsymbol{\xi}}$ is an $m \times m$ matrix with components $(\partial s/\partial \xi_A|\partial s/\partial \xi_B)$ $(\mathcal{A}, \mathcal{B} = 1, \ldots, m)$. Making use of the definitions (6.21) one can write the explicit form of these matrices as follows:

$$\Gamma_{\mathsf{aa}}(\boldsymbol{\xi}) = \mathsf{M}(\boldsymbol{\xi}),$$

(6.23a)

$$\Gamma_{\mathsf{a}\boldsymbol{\xi}}(\mathsf{a}, \boldsymbol{\xi}) = \left(\mathsf{F}_{(1)}(\boldsymbol{\xi}) \cdot \mathsf{a} \cdots \mathsf{F}_{(m)}(\boldsymbol{\xi}) \cdot \mathsf{a} \right),$$

(6.23b)

$$\Gamma_{\boldsymbol{\xi}\boldsymbol{\xi}}(\mathsf{a}, \boldsymbol{\xi}) = \begin{pmatrix} \mathsf{a}^{\mathsf{T}} \cdot \mathsf{S}_{(11)}(\boldsymbol{\xi}) \cdot \mathsf{a} & \cdots & \mathsf{a}^{\mathsf{T}} \cdot \mathsf{S}_{(1m)}(\boldsymbol{\xi}) \cdot \mathsf{a} \\ \cdots\cdots\cdots\cdots\cdots\cdots\cdots\cdots\cdots\cdots\cdots\cdots \\ \mathsf{a}^{\mathsf{T}} \cdot \mathsf{S}_{(m1)}(\boldsymbol{\xi}) \cdot \mathsf{a} & \cdots & \mathsf{a}^{\mathsf{T}} \cdot \mathsf{S}_{(mm)}(\boldsymbol{\xi}) \cdot \mathsf{a} \end{pmatrix}.$$

(6.23c)

The notation introduced above means that the matrix $\Gamma_{\mathsf{a}\boldsymbol{\xi}}$ can be thought of as a $1 \times m$ row matrix made of $n \times 1$ column matrices $\mathsf{F}_{(A)} \cdot \mathsf{a}$ $(\mathcal{A} = 1, \ldots, m)$, thus the general formula for the component of this matrix is

$$\left(\Gamma_{\mathsf{a}\boldsymbol{\xi}}\right)_{i\mathcal{A}} = \left(\mathsf{F}_{(A)} \cdot \mathsf{a}\right)_i = \sum_{j=1}^{n} F_{(A)ij}\, a_j, \quad \mathcal{A} = 1, \ldots, m, \quad i = 1, \ldots, n;$$

(6.24a)

and the general component of the matrix $\Gamma_{\boldsymbol{\xi}\boldsymbol{\xi}}$ is given by

$$\left(\Gamma_{\boldsymbol{\xi}\boldsymbol{\xi}}\right)_{\mathcal{A}\mathcal{B}} = \mathsf{a}^{\mathsf{T}} \cdot \mathsf{S}_{(AB)} \cdot \mathsf{a} = \sum_{i=1}^{n}\sum_{j=1}^{n} S_{(AB)ij}\, a_i a_j, \quad \mathcal{A}, \mathcal{B} = 1, \ldots, m.$$

(6.24b)

The *covariance matrix* C, which approximates the expected covariances of the ML estimators of the signal's parameters, is defined as Γ^{-1}. Let us write this matrix in terms of block matrices corresponding to extrinsic and intrinsic parameters,

$$C(a, \xi) = \begin{pmatrix} C_{aa}(a, \xi) & C_{a\xi}(a, \xi) \\ C_{a\xi}(a, \xi)^T & C_{\xi\xi}(a, \xi) \end{pmatrix}. \tag{6.25}$$

Applying to Eq. (6.22) the standard formula for the inverse of a block matrix [179], we can express the matrices C_{aa}, $C_{a\xi}$, and $C_{\xi\xi}$ in terms of the matrices $\Gamma_{aa} = M$, $\Gamma_{a\xi}$, and $\Gamma_{\xi\xi}$. We obtain

$$C_{aa}(a, \xi) = M(\xi)^{-1} + M(\xi)^{-1} \cdot \Gamma_{a\xi}(a, \xi) \cdot \overline{\Gamma}(a, \xi)^{-1} \cdot \Gamma_{a\xi}(a, \xi)^T \cdot M(\xi)^{-1}, \tag{6.26a}$$

$$C_{a\xi}(a, \xi) = -M(\xi)^{-1} \cdot \Gamma_{a\xi}(a, \xi) \cdot \overline{\Gamma}(a, \xi)^{-1}, \tag{6.26b}$$

$$C_{\xi\xi}(a, \xi) = \overline{\Gamma}(a, \xi)^{-1}, \tag{6.26c}$$

where we have introduced the new $m \times m$ matrix:

$$\overline{\Gamma}(a, \xi) := \Gamma_{\xi\xi}(a, \xi) - \Gamma_{a\xi}(a, \xi)^T \cdot M(\xi)^{-1} \cdot \Gamma_{a\xi}(a, \xi). \tag{6.27}$$

We shall call the matrix $\overline{\Gamma}$ (which is the *Schur complement* of the matrix M) the *projected Fisher matrix* (onto the space of intrinsic parameters). Because the matrix $\overline{\Gamma}$ is the inverse of the intrinsic-parameter submatrix $C_{\xi\xi}$ of the covariance matrix C, it expresses the information available about the intrinsic parameters that takes into account the correlations with the extrinsic parameters. Note that $\overline{\Gamma}$ is still a function of the extrinsic parameters.

We next define the *normalized projected Fisher matrix* $\overline{\overline{\Gamma}}$

$$\overline{\overline{\Gamma}}(a, \xi) := \frac{\overline{\Gamma}(a, \xi)}{\rho(a, \xi)^2}, \tag{6.28}$$

which is the $m \times m$ square matrix. Making use of the definition (6.27) and Eqs. (6.24) we can show that the $\mathcal{AB}$ component of this matrix can be written in the form

$$\overline{\overline{\Gamma}}_{\mathcal{AB}}(a, \xi) = \frac{a^T \cdot A_{(\mathcal{AB})}(\xi) \cdot a}{a^T \cdot M(\xi) \cdot a}, \quad \mathcal{A}, \mathcal{B} = 1, \dots, m, \tag{6.29}$$

where $A_{(\mathcal{AB})}$ is the $n \times n$ matrix defined as

$$A_{(\mathcal{AB})}(\xi) := S_{(\mathcal{AB})}(\xi) - F_{(\mathcal{A})}(\xi)^T \cdot M(\xi)^{-1} \cdot F_{(\mathcal{B})}(\xi), \quad \mathcal{A}, \mathcal{B} = 1, \dots, m. \tag{6.30}$$

From the Rayleigh principle [179] it follows that the minimum value of the component $\overline{\overline{\Gamma}}_{\mathcal{AB}}$ is given by the smallest eigenvalue of the matrix $M^{-1} \cdot A_{(\mathcal{AB})}$. Similarly, the maximum value of the component $\overline{\overline{\Gamma}}_{\mathcal{AB}}$ is given by the largest eigenvalue of that matrix.

Because the trace of a matrix is equal to the sum of its eigenvalues, the $m \times m$ square matrix $\widetilde{\Gamma}$ with components

$$\widetilde{\Gamma}_{\mathcal{AB}}(\boldsymbol{\xi}) := \frac{1}{n}\mathrm{Tr}\left(\mathsf{M}(\boldsymbol{\xi})^{-1}\cdot\mathsf{A}_{(\mathcal{AB})}(\boldsymbol{\xi})\right), \quad \mathcal{A},\mathcal{B} = 1,\ldots,m, \quad (6.31)$$

expresses the information available about the intrinsic parameters, averaged over the possible values of the extrinsic parameters. Note that the factor $1/n$ is specific to the case of n extrinsic parameters. We shall call $\widetilde{\Gamma}$ the *reduced Fisher matrix*. This matrix is a function of the intrinsic parameters alone. We shall see that the reduced Fisher matrix plays a key role in the signal processing theory that we present here. It is used in the calculation of the threshold for statistically significant detection and in the formula for the number of templates needed to do a given search.

6.1.3 False alarm and detection probabilities

Statistical properties of the $\mathcal{F}$-statistic for known intrinsic parameters. We shall first obtain the false alarm and detection probability density functions (pdfs) in the case when the intrinsic parameters $\boldsymbol{\xi}$ of the signal are known. In this case the $\mathcal{F}$-statistic is a quadratic function of random variables that are linear correlations of the data with filters, see Eq. (6.14). As we assume that the noise in data is Gaussian, $\mathcal{F}$-statistic is a quadratic form of Gaussian random variables. Consequently the $\mathcal{F}$-statistic has a distribution related to the χ^2 distribution. One can show that $2\mathcal{F}$ has a χ^2 distribution with n degrees of freedom (where n is the number of the extrinsic parameters a) when the signal is absent and non-central χ^2 distribution with n degrees of freedom and non-centrality parameter equal to square of the signal-to-noise ratio ρ when the signal is present [31]. As a result the pdfs p_0 and p_1 of $\mathcal{F}$ when the intrinsic parameters are known and when respectively the signal is absent and present are given by[1]

$$p_0(\mathcal{F}) = \frac{\mathcal{F}^{n/2-1}}{(n/2-1)!}\exp(-\mathcal{F}), \tag{6.32a}$$

$$p_1(\rho,\mathcal{F}) = \frac{(2\mathcal{F})^{(n/2-1)/2}}{\rho^{n/2-1}}I_{n/2-1}\left(\rho\sqrt{2\mathcal{F}}\right)\exp\left(-\mathcal{F}-\frac{1}{2}\rho^2\right), \tag{6.32b}$$

where n is the number of degrees of freedom of the χ^2 distribution and $I_{n/2-1}$ is the modified Bessel function of the first kind and order $n/2-1$.

---

[1] In the current chapter we use the factorial symbol $n!$ also in the cases when n is not a positive integer, i.e. we are employing the definition $n! := \Gamma(n+1)$, where Γ denotes the gamma function. In the present chapter we reserve the Greek letter Γ for denoting different kinds of Fisher matrices.

The expectation values and the variances of the $\mathcal{F}$-statistic when signal is absent and present read

$$E_0\{\mathcal{F}\} = \frac{n}{2}, \qquad\qquad \mathrm{Var}_0\{\mathcal{F}\} = \frac{n}{2}, \qquad\qquad (6.33a)$$

$$E_1\{\mathcal{F}\} = \frac{n}{2} + \frac{\rho^2}{2}, \qquad \mathrm{Var}_1\{\mathcal{F}\} = \frac{n}{2} + \rho^2, \qquad (6.33b)$$

where the subscript "0" means that signal is absent whereas the subscript "1" means that signal is present.

The *false alarm probability* P_F is the probability that $\mathcal{F}$ exceeds a certain threshold $\mathcal{F}_0$ when there is no signal. In our case we have

$$P_F(\mathcal{F}_0) := \int_{\mathcal{F}_0}^{\infty} p_0(\mathcal{F})\,\mathrm{d}\mathcal{F} = \exp(-\mathcal{F}_0) \sum_{k=0}^{n/2-1} \frac{\mathcal{F}_0^k}{k!}. \qquad (6.34)$$

The *probability of detection* P_D is the probability that $\mathcal{F}$ exceeds the threshold $\mathcal{F}_0$ when the signal-to-noise ratio is equal to ρ:

$$P_D(\rho, \mathcal{F}_0) := \int_{\mathcal{F}_0}^{\infty} p_1(\rho, \mathcal{F})\,\mathrm{d}\mathcal{F}. \qquad (6.35)$$

The integral in the above formula can be expressed in terms of the generalized Marcum Q-function [130, 180, 181], $P_D(\rho, \mathcal{F}_0) = Q(\rho, \sqrt{2\mathcal{F}_0})$. We thus see that when the noise in the detector is Gaussian and the intrinsic parameters are known the probability of detection of the signal depends on a single quantity: the optimal signal-to-noise ratio ρ.

False alarm probability. We return now to the case when the intrinsic parameters $\boldsymbol{\xi}$ are not known. Then the statistic $\mathcal{F}[x; \boldsymbol{\xi}]$ given by Eq. (6.14) is a certain generalized multiparameter stochastic process called the *random field* (see monographs [182, 183] for a comprehensive discussion of random fields). If vector $\boldsymbol{\xi}$ has one component the random field is simply a stochastic process. For random fields we can define the *mean* m and the *autocovariance function* C just in the same way as we define such quantities for a stochastic process:

$$m(\boldsymbol{\xi}) := E\{\mathcal{F}[x; \boldsymbol{\xi}]\}, \qquad (6.36a)$$

$$C(\boldsymbol{\xi}, \boldsymbol{\xi}') := E\{(\mathcal{F}[x; \boldsymbol{\xi}] - m(\boldsymbol{\xi}))(\mathcal{F}[x; \boldsymbol{\xi}'] - m(\boldsymbol{\xi}'))\}, \qquad (6.36b)$$

where $\boldsymbol{\xi}$ and $\boldsymbol{\xi}'$ are two values of the intrinsic parameter set and E denotes the expectation value. As usual the autocovariance function can be rewritten in a more convenient way as

$$C(\boldsymbol{\xi}, \boldsymbol{\xi}') = E\{\mathcal{F}[x; \boldsymbol{\xi}]\mathcal{F}[x; \boldsymbol{\xi}']\} - m(\boldsymbol{\xi})\,m(\boldsymbol{\xi}'). \qquad (6.37)$$

Let us compute the mean and the autocovariance function of the $\mathcal{F}$-statistic in the case when the signal is absent in the data, i.e. when $x(t) = n(t)$. We will employ the following useful formulae, which are valid for any deterministic waveforms h_i, $1 = 1, \ldots, 4$:

$$E\left\{(n|h_1)(n|h_2)\right\} = (h_1|h_2), \tag{6.38a}$$

$$E\left\{(n|h_1)(n|h_2)(n|h_3)(n|h_4)\right\} = (h_1|h_2)(h_3|h_4) + (h_1|h_3)(h_2|h_4)$$
$$+ (h_1|h_4)(h_2|h_3). \tag{6.38b}$$

Making use of Eqs. (6.36)–(6.38) we obtain that the mean m of the $\mathcal{F}$-statistic does not depend on the intrinsic parameters $\boldsymbol{\xi}$ and is equal

$$m(\boldsymbol{\xi}) = \frac{1}{2}n, \tag{6.39}$$

and its autocovariance function $\mathcal{C}$ is given by

$$\mathcal{C}(\boldsymbol{\xi}, \boldsymbol{\xi}') = \frac{1}{2}\mathrm{Tr}\left(\mathsf{Q}(\boldsymbol{\xi}, \boldsymbol{\xi}') \cdot \mathsf{M}(\boldsymbol{\xi}')^{-1} \cdot \mathsf{Q}(\boldsymbol{\xi}, \boldsymbol{\xi}')^\mathsf{T} \cdot \mathsf{M}(\boldsymbol{\xi})^{-1}\right), \tag{6.40}$$

where we have introduced a new $n \times n$ matrix Q with components defined as

$$Q_{ij}(\boldsymbol{\xi}, \boldsymbol{\xi}') := \left(h_i(t; \boldsymbol{\xi})|h_j(t; \boldsymbol{\xi}')\right), \quad i, j = 1, \ldots, n. \tag{6.41}$$

Because $\mathsf{Q}(\boldsymbol{\xi}, \boldsymbol{\xi}) = \mathsf{M}(\boldsymbol{\xi})$, we can easily check that $\mathcal{C}(\boldsymbol{\xi}, \boldsymbol{\xi}) = n/2$.

One can estimate the false alarm probability in the following way [31]: the autocovariance function $\mathcal{C}$ tends to zero as the displacement $\Delta\boldsymbol{\xi} := \boldsymbol{\xi}' - \boldsymbol{\xi}$ increases ($\mathcal{C}$ is maximal and equal to $n/2$ for $\Delta\boldsymbol{\xi} = \mathbf{0}$). We can thus divide the parameter space into *elementary cells* such that in each cell the autocovariance function $\mathcal{C}$ among points in the cell is appreciably different from zero. The realizations of the random field within a cell will be correlated (dependent) whereas realizations of the random field within each cell and outside the cell are almost uncorrelated (independent). Thus the number of cells covering the parameter space gives an estimate of the number of independent realizations of the random field. We choose the elementary cell to coincide with the *correlation hypervolume*, i.e. with a compact region with boundary defined by the requirement that at the boundary the correlation $\mathcal{C}$ equals half of its maximum value. The elementary cell with its origin at the point $\boldsymbol{\xi}$ is defined as the set of the points $\boldsymbol{\xi}'$ in the parameter space which fulfill the inequality

$$\mathcal{C}(\boldsymbol{\xi}, \boldsymbol{\xi}') \leq \frac{1}{2}\mathcal{C}(\boldsymbol{\xi}, \boldsymbol{\xi}) = \frac{1}{4}n. \tag{6.42}$$

The equality in the above formula holds when $\boldsymbol{\xi}'$ is on the cell's boundary.

To estimate the number of cells we perform the Taylor expansion of the autocorrelation function (6.40) with respect to $\boldsymbol{\xi}'$ around $\boldsymbol{\xi}' = \boldsymbol{\xi}$ up to the second-order terms:

$$
\mathcal{C}(\boldsymbol{\xi}, \boldsymbol{\xi}') \cong \frac{1}{2}n + \sum_{A=1}^{m} \frac{\partial \mathcal{C}(\boldsymbol{\xi}, \boldsymbol{\xi}')}{\partial \xi'_A}\bigg|_{\boldsymbol{\xi}'=\boldsymbol{\xi}} \Delta \xi_A
$$

$$
+ \frac{1}{2} \sum_{A=1}^{m} \sum_{B=1}^{m} \frac{\partial^2 \mathcal{C}(\boldsymbol{\xi}, \boldsymbol{\xi}')}{\partial \xi'_A \partial \xi'_B}\bigg|_{\boldsymbol{\xi}'=\boldsymbol{\xi}} \Delta \xi_A \Delta \xi_B. \tag{6.43}
$$

As $\mathcal{C}$ attains its maximum value when $\boldsymbol{\xi}' = \boldsymbol{\xi}$, we have

$$
\frac{\partial \mathcal{C}(\boldsymbol{\xi}, \boldsymbol{\xi}')}{\partial \xi'_A}\bigg|_{\boldsymbol{\xi}'=\boldsymbol{\xi}} = 0, \quad A = 1, \ldots, m. \tag{6.44}
$$

Let us introduce the symmetric $m \times m$ matrix G which depends only on the intrinsic parameters $\boldsymbol{\xi}$ and has components

$$
G_{AB}(\boldsymbol{\xi}) := -\frac{1}{2}\frac{1}{\mathcal{C}(\boldsymbol{\xi}, \boldsymbol{\xi})} \frac{\partial^2 \mathcal{C}(\boldsymbol{\xi}, \boldsymbol{\xi}')}{\partial \xi'_A \partial \xi'_B}\bigg|_{\boldsymbol{\xi}'=\boldsymbol{\xi}}, \quad A, B = 1, \ldots, m. \tag{6.45}
$$

Making use of the definition (6.45) and Eqs. (6.43) and (6.44), the inequality (6.42) which determines the elementary cell can approximately be rewritten as

$$
\sum_{A=1}^{m} \sum_{B=1}^{m} G_{AB}(\boldsymbol{\xi}) \Delta \xi_A \Delta \xi_B \leq \frac{1}{2}. \tag{6.46}
$$

After some lengthy algebra one can show that G is precisely equal to the reduced Fisher matrix $\tilde{\Gamma}$ given by Eq. (6.31) (see Appendix B in Ref. [32]),

$$
\mathsf{G} = \tilde{\Gamma}. \tag{6.47}
$$

If the components of the matrix G are constant (i.e. independent of the values of the intrinsic parameters $\boldsymbol{\xi}$ of the signal), the inequality (6.46) does not depend on where the origin $\boldsymbol{\xi}$ of the elementary cell is located. In this case the correlation hypervolume is approximated by a *correlation hyperellipsoid*. The m-dimensional Euclidean volume V_c of the elementary cell defined by Eq. (6.46) equals

$$
V_c = \frac{(\pi/2)^{m/2}}{(m/2)! \sqrt{\det \mathsf{G}}}. \tag{6.48}
$$

We estimate the number N_c of elementary cells by dividing the total Euclidean volume V of the m-dimensional parameter space by the volume

V_c of the elementary cell, i.e. we have

$$N_c = \frac{V}{V_c}. \tag{6.49}$$

To estimate the number of cells in the case when the components of the matrix G are not constant, i.e. when they depend on the values of the intrinsic parameters $\boldsymbol{\xi}$, instead of Eq. (6.49) we write the following formula

$$N_c = \frac{(m/2)!}{(\pi/2)^{m/2}} \int_V \sqrt{\det \mathsf{G}(\boldsymbol{\xi})} \, dV. \tag{6.50}$$

This procedure can be thought of as interpreting the matrix G as the metric on the parameter space. This interpretation appeared for the first time in the context of gravitational-wave data analysis in the work by Owen [29], where an analogous integral formula was proposed for the number of templates (see Section 6.1.4) needed to perform a search for gravitational-wave signals from coalescing binaries. The concept of the number of cells was introduced in [31] and it is a generalization of the idea of an effective number of samples introduced in [184] for the case of a coalescing binary signal.

    We approximate the pdf of the $\mathcal{F}$-statistic in each cell by the pdf $p_0(\mathcal{F})$ when the parameters are known [in our case it is given by Eq. (6.32a)]. The values of the $\mathcal{F}$-statistic in each cell can be considered as independent random variables. The probability that $\mathcal{F}$ does not exceed the threshold $\mathcal{F}_0$ in a given cell is $1 - P_F(\mathcal{F}_0)$, where $P_F(\mathcal{F}_0)$ is given by Eq. (6.34). Consequently the probability that $\mathcal{F}$ does not exceed the threshold $\mathcal{F}_0$ in *all* the N_c cells is $\left[1 - P_F(\mathcal{F}_0)\right]^{N_c}$. The probability P_F^T that $\mathcal{F}$ exceeds $\mathcal{F}_0$ in *one or more* cells is thus given by

$$P_F^T(\mathcal{F}_0) = 1 - \left[1 - P_F(\mathcal{F}_0)\right]^{N_c}. \tag{6.51}$$

This *by definition* is the false alarm probability when the phase parameters are unknown. The expected number of false alarms N_F is given by

$$N_F = N_c P_F(\mathcal{F}_0). \tag{6.52}$$

Using Eq. (6.52) we can express the false alarm probability P_F^T from Eq. (6.51) in terms of the expected number of false alarms. Using $\lim_{n\to\infty}(1 + x/n)^n = \exp(x)$ we have that for a large number of cells

$$P_F^T(\mathcal{F}_0) \cong 1 - \exp(-N_F). \tag{6.53}$$

It follows from the above formula that asymptotically the number of false alarms is approximately Poisson distributed [Eq. (3.21)] with expected number of events equal to N_F. When the expected number of false alarms is small (much less than 1), Eq. (6.53) leads to the equality $P_F^T \cong N_F$.

Another approach to calculate the false alarm probability can be found in the monograph [130]. Namely one can use the theory of *level crossing* by stochastic processes. A classic exposition of this theory for the case of a stochastic process, i.e. for a one-dimensional random field, can be found in monograph [185]. The case of m-dimensional random fields is treated in [182] and important recent contributions are contained in Ref. [186].

For a stochastic process $n(t)$ it is clear how to define an *upcrossing* of the level u. We say that n has an upcrossing of u at t_0 if there exists $\epsilon > 0$, such that $n(t) \leq u$ in the interval $(t_0 - \epsilon, t_0)$, and $n(t) \geq u$ in $(t_0, t_0 + \epsilon)$. Then under suitable regularity conditions of the stochastic process (involving the differentiability of the process and the existence of its appropriate moments), one can calculate the mean number of upcrossings per unit parameter interval (in the one-dimensional case the parameter is usually the time t).

For the case of an m-dimensional random field the situation is more complicated. Somehow we need to count the number of times a random field crosses a fixed hypersurface. Let $\mathcal{F}(\boldsymbol{\xi})$ be an m-dimensional real-valued random field where parameters $\boldsymbol{\xi} = (\xi_1, \ldots, \xi_m)$ belong to m-dimensional Euclidean space $\mathbb{R}^m$. We say that the random field $\mathcal{F}$ is *homogeneous* if its mean is constant and its autocovariance function $\mathcal{C}(\boldsymbol{\xi}, \boldsymbol{\xi}')$ depends only on the difference $\boldsymbol{\xi} - \boldsymbol{\xi}'$. Such defined homogeneous random fields are also called *second-order* or *wide-sense homogeneous* fields.

Let C be a compact subset of $\mathbb{R}^m$. We define the *excursion set* of the random field $\mathcal{F}(\boldsymbol{\xi})$ inside C above the level $\mathcal{F}_0$ as

$$A_{\mathcal{F}}(\mathcal{F}_0, C) := \{\boldsymbol{\xi} \in C : \mathcal{F}(\boldsymbol{\xi}) \geq \mathcal{F}_0\}. \tag{6.54}$$

It was found [182] that when the excursion set does not intersect the boundary of the set C, then a suitable analogue of the mean number of level crossings is the expectation value of the *Euler characteristic* χ of the set $A_{\mathcal{F}}$. For simplicity we shall denote $\chi[A_{\mathcal{F}}(\mathcal{F}_0, C)]$ by $\chi_{\mathcal{F}_0}$. It turns out that using Morse theory the expectation value of the Euler characteristic of $A_{\mathcal{F}}$ can be given in terms of certain multidimensional integrals (see Ref. [182], Theorem 5.2.1). Closed-form formulae were obtained for homogeneous m-dimensional Gaussian fields ([182], Theorem 5.3.1). Worsley ([186], Theorem 3.5) obtained explicit formulae for m-dimensional homogeneous χ^2 fields. We quote here the most general results and give a few special cases.

We say that the random field $U(\boldsymbol{\xi})$, $\boldsymbol{\xi} \in \mathbb{R}^m$, is a χ^2 random field if $U(\boldsymbol{\xi}) = \sum_{l=1}^{n} X_l(\boldsymbol{\xi})^2$, where $X_1(\boldsymbol{\xi}), \ldots, X_n(\boldsymbol{\xi})$ are real-valued, independent, identically distributed, and homogeneous Gaussian random fields with zero mean and unit variance. We say that $U(\boldsymbol{\xi})$ is a *generalized* χ^2

field if the Gaussian fields $X_l(\boldsymbol{\xi})$ are not necessarily independent. Let $2\mathcal{F}(\boldsymbol{\xi})$ be a χ^2 random field and let $X_l(\boldsymbol{\xi})$, $l = 1, \ldots, n$, be the component Gaussian fields. Then under suitable regularity conditions (differentiability of the random fields and the existence of appropriate moments of their distributions) the expectation value of the Euler characteristic of the excursion set reads ([186], Theorem 3.5)

$$\mathrm{E}\{\chi_{\mathcal{F}_0}\} = \frac{V\sqrt{\det \mathsf{L}}}{\pi^{m/2}(n/2-1)!}\,\mathcal{F}_0^{(n-m)/2}\,W_{m,n}(\mathcal{F}_0)\exp(-\mathcal{F}_0). \quad (6.55)$$

In Eq. (6.55), V is the volume of the set $\boldsymbol{C}$ and the square $m \times m$ matrix L is defined by

$$L_{AB} := -\frac{\partial^2 \mathcal{S}(\boldsymbol{\xi})}{\partial \xi_A \partial \xi_B}\bigg|_{\boldsymbol{\xi}=0}, \quad A, B = 1, \ldots, m, \quad (6.56)$$

where $\mathcal{S}$ is the autocorrelation function of each Gaussian random field $X_l(\boldsymbol{\xi})$, the function $W_{m,n}(\mathcal{F}_0)$ is a polynomial of degree $m-1$ in $\mathcal{F}_0$ given by

$$W_{m,n}(\mathcal{F}_0) := \frac{(m-1)!}{(-2)^{m-1}}\sum_{j=0}^{[(m-1)/2]}\sum_{k=0}^{m-1-2j}\binom{n-1}{m-1-2j-k}2^k\frac{(-\mathcal{F}_0)^{j+k}}{j!\,k!},$$

$$(6.57)$$

where $[N]$ denotes the greatest integer $\leq N$. We have the following special cases:

$$W_{1,n}(\mathcal{F}_0) = 1, \quad (6.58\text{a})$$

$$W_{2,n}(\mathcal{F}_0) = \mathcal{F}_0 - \frac{1}{2}(n-1), \quad (6.58\text{b})$$

$$W_{3,n}(\mathcal{F}_0) = \mathcal{F}_0^2 - \left(n - \frac{1}{2}\right)\mathcal{F}_0 + \frac{1}{4}(n-1)(n-2), \quad (6.58\text{c})$$

$$W_{4,n}(\mathcal{F}_0) = \mathcal{F}_0^3 - \frac{3}{2}n\,\mathcal{F}_0^2 + \frac{3}{4}(n-1)^2\mathcal{F}_0 - \frac{1}{8}(n-1)(n-2)(n-3). \quad (6.58\text{d})$$

It has rigorously been shown that the probability distribution of the Euler characteristic of the excursion set of a homogeneous Gaussian random field asymptotically approaches a Poisson distribution (see Ref. [182], Theorem 6.9.3). It has been argued that the same holds for χ^2 fields. It has also been shown for m-dimensional homogeneous χ^2 fields that asymptotically the level surfaces of the local maxima of the field are m-dimensional ellipsoids. Thus for a large threshold the excursion set consists of disjoint and simply connected (i.e. without holes) sets. Remembering that we assume that the excursion set does not intersect the boundary of the parameter space, the Euler characteristic of the excursion set is simply the number of connected components of the excursion set. Thus we can expect that for a χ^2 random field the expected number of level crossings by the field, i.e. in the language of signal detection theory the expected number of false alarms, has a Poisson distribution. Consequently the probability that $\mathcal{F}$ does not cross a threshold $\mathcal{F}_0$ is given by $\exp(-\mathrm{E}\{\chi_{\mathcal{F}_0}\})$ and the probability that there is at least one level crossing (i.e. for our signal detection problem the false alarm probability P_F^T) is given by

$$P_F^T(\mathcal{F}_0) = P(\mathcal{F} \geq \mathcal{F}_0) \cong 1 - \exp\left(-\mathrm{E}\{\chi_{\mathcal{F}_0}\}\right). \qquad (6.59)$$

From Eqs. (6.53) and (6.59) we see that to compare the two approaches presented above it is enough to compare the expected number of false alarms N_F with $\mathrm{E}\{\chi_{\mathcal{F}_0}\}$. For χ^2 fields we have $\mathsf{G} = 2\mathsf{L}$. Thus asymptotically (i.e. for large thresholds $\mathcal{F}_0$) we obtain

$$\frac{N_F}{\mathrm{E}\{\chi_{\mathcal{F}_0}\}} \to 2^m \, (m/2)! \, \mathcal{F}_0^{-m/2} \quad \text{as} \quad \mathcal{F}_0 \to \infty. \qquad (6.60)$$

In the two approaches the numbers of false alarms decrease exponentially with the threshold $\mathcal{F}_0$. Equation (6.60) implies that they asymptotically differ by a multiplicative term which is a power of the threshold.

Detection probability. When the gravitational-wave signal is present in the data, x, a precise calculation of the pdf of the $\mathcal{F}$-statistic is very difficult because the presence of the signal makes the data stochastic process x non-stationary. As a first approximation we can estimate the probability of detection of the signal when the parameters are unknown by the probability of detection when the parameters of the signal are known [given by Eq. (6.35)]. This approximation assumes that when the signal is present the true values of the phase parameters fall within the cell where $\mathcal{F}$-statistic has a maximum. This approximation will be the better the higher the signal-to-noise ratio ρ.

6.1.4 Number of templates

To search for gravitational-wave signals we evaluate the $\mathcal{F}$-statistic on a *grid* in the intrinsic-parameter space. The grid has to be sufficiently fine in order that the loss of signals is minimized. In order to estimate the number of points of the grid, or in other words the number of *templates* that we need to search for a signal, the natural quantity to study is the expectation value of the $\mathcal{F}$-statistic when the signal is present.

Let us thus assume that the data x contains the gravitational-wave signal $s(t; \boldsymbol{\theta})$ of the form given in Eq. (6.15), so the parameters $\boldsymbol{\theta}$ of the signal consist of amplitude (or extrinsic) parameters a and intrinsic parameters ξ, $\theta = (a, \xi)$. The data x will be correlated with the filters $h_i(t; \xi')$ $(i = 1, \ldots, n)$ parametrized by the values ξ' of the intrinsic parameters. The $\mathcal{F}$-statistic can thus be written in the form [see Eq. (6.14)]

$$\mathcal{F}[x(t; a, \xi); \xi'] = \frac{1}{2} N[x(t; a, \xi); \xi']^\mathsf{T} \cdot M(\xi')^{-1} \cdot N[x(t; a, \xi); \xi'], \quad (6.61)$$

where the components of the $n \times 1$ matrix N and $n \times n$ matrix M are defined in Eqs. (6.17). The expectation value of the $\mathcal{F}$-statistic (6.61) is equal

$$\mathrm{E}\left\{\mathcal{F}[x(t; a, \xi); \xi']\right\} = \frac{1}{2}\left(n + a^\mathsf{T} \cdot Q(\xi, \xi') \cdot M(\xi')^{-1} \cdot Q(\xi, \xi')^\mathsf{T} \cdot a\right),$$

$$(6.62)$$

where the components of the $n \times n$ matrix Q are defined in Eq. (6.41). Let us rewrite this expectation value in the following form

$$\mathrm{E}\left\{\mathcal{F}[x(t; a, \xi); \xi']\right\}$$
$$= \frac{1}{2}\left(n + \rho(a, \xi)^2 \frac{a^\mathsf{T} \cdot Q(\xi, \xi') \cdot M(\xi')^{-1} \cdot Q(\xi, \xi')^\mathsf{T} \cdot a}{a^\mathsf{T} \cdot M(\xi) \cdot a}\right), \quad (6.63)$$

where ρ is the signal-to-noise ratio [it is given by Eq. (6.19)].

Let us also define the *normalized correlation function* $\bar{C}$,

$$\bar{C}(a, \xi, \xi') := \frac{a^\mathsf{T} \cdot Q(\xi, \xi') \cdot M(\xi')^{-1} \cdot Q(\xi, \xi')^\mathsf{T} \cdot a}{a^\mathsf{T} \cdot M(\xi) \cdot a}. \quad (6.64)$$

From the Rayleigh principle [179] it follows that the minimum of the normalized correlation function $\bar{C}$ is equal to the smallest eigenvalue of the matrix $M(\xi)^{-1} \cdot Q(\xi, \xi') \cdot M(\xi')^{-1} \cdot Q(\xi, \xi')^\mathsf{T}$ whereas the maximum is given by its largest eigenvalue. We define the *reduced correlation function* $\tilde{C}$ as

$$\tilde{C}(\xi, \xi') := \frac{1}{2} \mathrm{Tr}\left(M(\xi)^{-1} \cdot Q(\xi, \xi') \cdot M(\xi')^{-1} \cdot Q(\xi, \xi')^\mathsf{T}\right). \quad (6.65)$$

The advantage of the reduced correlation function is that it depends only on the intrinsic parameters and thus it is suitable for studying the number of grid points on which the $\mathcal{F}$-statistic needs to be evaluated. We also note that the reduced correlation function $\tilde{C}$ precisely coincides with the autocovariance function $\mathcal{C}$ of the $\mathcal{F}$-statistic given by Eq. (6.40).

Like in the calculation of the number of cells in order to estimate the number of templates we perform Taylor expansion of $\mathcal{C}$ up to second-order terms around the true values of the parameters and we obtain an equation analogous to Eq. (6.46),

$$\sum_{\mathcal{A}=1}^{m}\sum_{\mathcal{B}=1}^{m} \mathsf{G}_{\mathcal{A}\mathcal{B}}(\boldsymbol{\xi})\,\Delta\xi_{\mathcal{A}}\,\Delta\xi_{\mathcal{B}} = 1 - \mathcal{C}_0, \qquad (6.66)$$

where G is given by Eq. (6.45) and

$$\mathcal{C}_0 = \frac{\tilde{C}(\boldsymbol{\xi},\boldsymbol{\xi}')}{\tilde{C}(\boldsymbol{\xi},\boldsymbol{\xi})} = 2\frac{\tilde{C}(\boldsymbol{\xi},\boldsymbol{\xi}')}{n}. \qquad (6.67)$$

By arguments identical to those employed in deriving a formula for the number of cells we arrive at the following formula for the number of templates:

$$N_{\mathrm{t}} = \frac{1}{(1-\mathcal{C}_0)^{m/2}}\frac{(m/2)!}{\pi^{m/2}}\int_V \sqrt{\det \mathsf{G}(\boldsymbol{\xi})}\,\mathrm{d}V. \qquad (6.68)$$

When $\mathcal{C}_0 = 1/2$ the above formula coincides with the formula for the number N_{c} of cells, Eq. (6.50).

We would like to place the templates sufficiently closely to each other so that the loss of signals is minimized. Thus $1 - \mathcal{C}_0$ needs to be chosen to be sufficiently small. The formula (6.68) for the number of templates assumes that the templates are placed in the centers of hyperspheres and that the hyperspheres fill the parameter space without holes. This latter assumption can not be satisfied in practice and the formula (6.68) is a lower bound for the number of templates. The true number of templates is given by

$$N_{\mathrm{t}} = \frac{\Theta(m)}{(1-\mathcal{C}_0)^{m/2}}\frac{(m/2)!}{\pi^{m/2}}\int_V \sqrt{\det \mathsf{G}(\boldsymbol{\xi})}\,\mathrm{d}V, \qquad (6.69)$$

where Θ is the so called *thickness* of the grid. Thickness is always greater than one (and it is equal to one only in the trivial case of a one-dimensional parameter space). The definition of Θ and the problem of finding a grid with the smallest thickness are introduced in Section 7.2.1. The formula (6.69) is equivalent to the original formula derived by Owen for the case of a gravitational-wave signal from an inspiraling binary [29]. Owen [29] also introduced a geometric approach to the problem of template placement by identifying the Fisher matrix with a metric on the parameter space.

6.1.5 Suboptimal filtering

Very often to extract signals from the noise we shall use filters that are not optimal. We may have to choose approximate, *suboptimal* filters because we do not know the exact form of the signal (this is almost always the case in practice) or in order to reduce the computational cost and simplify the analysis of the data.

Let us assume that we are looking for the gravitational-wave signal of the form given in Eq. (6.15). Then the most natural and simplest way to proceed is to use as our statistic the $\mathcal{F}$-statistic (6.14) in which the filters $h'_i(t;\boldsymbol{\zeta})$ $(i = 1,\ldots,n)$ are the approximate ones instead of the optimal ones $h_i(t;\boldsymbol{\xi})$ $(i = 1,\ldots,n)$ matched to the signal. In general the functions $h'_i(t;\boldsymbol{\zeta})$ will be different from the functions $h_i(t;\boldsymbol{\xi})$ used in optimal filtering and also the set of parameters $\boldsymbol{\zeta}$ will be different from the set of parameters $\boldsymbol{\xi}$ in optimal filters. We shall call this procedure the *suboptimal filtering* and we shall denote the suboptimal statistic by $\mathcal{F}_{\mathrm{s}}$. It is given by the relation [see Eqs. (6.14) and (6.17)]

$$\mathcal{F}_{\mathrm{s}}[x;\boldsymbol{\zeta}] := \frac{1}{2}\,\mathsf{N}_{\mathrm{s}}[x;\boldsymbol{\zeta}]^{\mathsf{T}}\cdot\mathsf{M}_{\mathrm{s}}(\boldsymbol{\zeta})^{-1}\cdot\mathsf{N}_{\mathrm{s}}[x;\boldsymbol{\zeta}], \qquad (6.70)$$

where we have introduced the data-dependent $n \times 1$ column matrix N_{s} and the square $n \times n$ matrix M_{s} with components

$$N_{\mathrm{s}\,i}[x;\boldsymbol{\zeta}] := (x|h'_i(t;\boldsymbol{\zeta})), \quad M_{\mathrm{s}\,ij}(\boldsymbol{\zeta}) := (h'_i(t;\boldsymbol{\zeta})|h'_j(t;\boldsymbol{\zeta})), \quad i,j = 1,\ldots,n. \tag{6.71}$$

We need a measure of how well a given suboptimal filter performs. To find such a measure we calculate the expectation value of the suboptimal statistic in the case when the data contains the gravitational-wave signal $s = s(t;\mathbf{a},\boldsymbol{\xi})$ given in Eq. (6.15). We get [see Eq. (6.62)]

$$\mathrm{E}\,\{\mathcal{F}_{\mathrm{s}}[x(t;\mathbf{a},\boldsymbol{\xi});\boldsymbol{\zeta}]\} = \frac{1}{2}\left(n + \mathbf{a}^{\mathsf{T}}\cdot\mathsf{Q}_{\mathrm{s}}(\boldsymbol{\xi},\boldsymbol{\zeta})\cdot\mathsf{M}_{\mathrm{s}}(\boldsymbol{\zeta})^{-1}\cdot\mathsf{Q}_{\mathrm{s}}(\boldsymbol{\xi},\boldsymbol{\zeta})^{\mathsf{T}}\cdot\mathbf{a}\right), \tag{6.72}$$

where the matrix Q_{s} has components

$$Q_{\mathrm{s}\,ij}(\boldsymbol{\xi},\boldsymbol{\zeta}) := (h_i(t;\boldsymbol{\xi})|h'_j(t;\boldsymbol{\zeta})), \quad i,j = 1,\ldots,n. \tag{6.73}$$

Let us rewrite the expectation value (6.72) in the following form

$$\mathrm{E}\,\{\mathcal{F}_{\mathrm{s}}[x(t;\mathbf{a},\boldsymbol{\xi});\boldsymbol{\zeta}]\}$$
$$= \frac{1}{2}\left(n + \rho(\mathbf{a},\boldsymbol{\xi})^2\,\frac{\mathbf{a}^{\mathsf{T}}\cdot\mathsf{Q}_{\mathrm{s}}(\boldsymbol{\xi},\boldsymbol{\zeta})\cdot\mathsf{M}_{\mathrm{s}}(\boldsymbol{\zeta})^{-1}\cdot\mathsf{Q}_{\mathrm{s}}(\boldsymbol{\xi},\boldsymbol{\zeta})^{\mathsf{T}}\cdot\mathbf{a}}{\mathbf{a}^{\mathsf{T}}\cdot\mathsf{M}(\boldsymbol{\xi})\cdot\mathbf{a}}\right), \tag{6.74}$$

where ρ is the optimal signal-to-noise ratio given in Eq. (6.19). The expectation value (6.74) reaches its maximum equal to $n/2 + \rho^2/2$ when the filter is perfectly matched to the signal. A natural measure of the performance of a suboptimal filter is thus the quantity FF defined by

$$\mathrm{FF}(\boldsymbol{\xi}) := \max_{(\mathbf{a},\boldsymbol{\zeta})}\sqrt{\frac{\mathbf{a}^{\mathsf{T}}\cdot\mathsf{Q}_{\mathrm{s}}(\boldsymbol{\xi},\boldsymbol{\zeta})\cdot\mathsf{M}_{\mathrm{s}}(\boldsymbol{\zeta})^{-1}\cdot\mathsf{Q}_{\mathrm{s}}(\boldsymbol{\xi},\boldsymbol{\zeta})^{\mathsf{T}}\cdot\mathbf{a}}{\mathbf{a}^{\mathsf{T}}\cdot\mathsf{M}(\boldsymbol{\xi})\cdot\mathbf{a}}}. \tag{6.75}$$

We shall call the quantity FF the *generalized fitting factor*. From the Rayleigh principle follows that the generalized fitting factor is the maximum of the largest eigenvalue of the matrix $\mathsf{M}(\boldsymbol{\xi})^{-1} \cdot \mathsf{Q}_\mathrm{s}(\boldsymbol{\xi}, \boldsymbol{\zeta}) \cdot \mathsf{M}_\mathrm{s}(\boldsymbol{\zeta})^{-1} \cdot \mathsf{Q}_\mathrm{s}(\boldsymbol{\xi}, \boldsymbol{\zeta})^\mathsf{T}$ over the intrinsic parameters of the signal.

In the case of a one-amplitude signal given by

$$s(t; A_0, \boldsymbol{\xi}) = A_0\, h(t; \boldsymbol{\xi}), \tag{6.76}$$

the generalized fitting factor defined in Eq. (6.75) reduces to the fitting factor introduced by Apostolatos [28]:

$$\mathrm{FF}(\boldsymbol{\xi}) = \max_{\boldsymbol{\zeta}} \frac{|(h(t; \boldsymbol{\xi})|h'(t; \boldsymbol{\zeta}))|}{\sqrt{(h(t; \boldsymbol{\xi})|h(t; \boldsymbol{\xi}))}\sqrt{(h'(t; \boldsymbol{\zeta})|h'(t; \boldsymbol{\zeta}))}}. \tag{6.77}$$

Fitting factor is thus the ratio of the maximal signal-to-noise ratio that can be achieved with suboptimal filtering to the signal-to-noise ratio obtained when we use a perfectly matched, optimal filter. We note that for the signal given by Eq. (6.76), FF is independent of the value of the amplitude A_0.

For the case of the signal of the form

$$s(t; A_0, \phi_0, \boldsymbol{\xi}) = A_0 \cos\left(\phi(t; \boldsymbol{\xi}) + \phi_0\right), \tag{6.78}$$

where ϕ_0 is a constant initial phase, the maximum over the angle ϕ_0 in Eq. (6.77) can be obtained analytically. We assume that over the frequency bandwidth of the signal the spectral density of the noise is nearly constant and that over the observation time $\cos\left(\phi(t; \boldsymbol{\xi})\right)$ oscillates rapidly. Then the fitting factor is approximately given by

$$\mathrm{FF}(\boldsymbol{\xi}) \cong \max_{\boldsymbol{\zeta}} \left\{ \left(\int_0^{T_\mathrm{o}} \cos\left(\phi(t; \boldsymbol{\xi}) - \phi'(t; \boldsymbol{\zeta})\right) \mathrm{d}t\right)^2 \right. $$
$$\left. + \left(\int_0^{T_\mathrm{o}} \sin\left(\phi(t; \boldsymbol{\xi}) - \phi'(t; \boldsymbol{\zeta})\right) \mathrm{d}t\right)^2 \right\}^{1/2}. \tag{6.79}$$

In designing suboptimal filters one faces the issue of how small a fitting factor one can accept. A popular rule of thumb is accepting FF = 0.97. Assuming that the amplitude of the signal and consequently the signal-to-noise ratio decreases inversely proportionally to the distance from the source, this corresponds to 10% loss of the signals that would be detected by a matched filter.

Proposals for good suboptimal search templates for the case of coalescing binaries are given in [187, 188] and for the case of spinning neutron stars in [189, 190].

6.2 Case studies: deterministic signals

In this section we shall consider the maximum-likelihood detection of several deterministic gravitational-wave signals. We shall study signals produced by gravitational waves coming from spinning neutron stars, inspiraling compact binaries, and supernova explosions.

6.2.1 Periodic signal from a spinning neutron star

Let us assume that the neutron star can be modeled as a triaxial ellipsoid rotating about its principal axis. Gravitational waves emitted by such source were studied in Section 2.5 and their polarization functions h_+ and $h_\times$ are given by Eqs. (2.65) and (2.66). We rewrite here the form of these functions valid at the SSB [i.e. we put $\mathbf{x} = \mathbf{0}$ in Eqs. (2.65) and (2.66)],

$$h_+(t) = -h_0 \frac{1 + \cos^2 \iota}{2} \cos(\phi(t) + \phi_0), \tag{6.80a}$$

$$h_\times(t) = -h_0 \cos \iota \sin(\phi(t) + \phi_0), \tag{6.80b}$$

where the dimensionless amplitude h_0 is given in Eq. (2.67), ι $(0 \le \iota \le \pi)$ is the angle between the angular momentum vector of the star and the line of sight, ϕ_0 is some constant initial phase [which is twice the initial phase of Eq. (2.65)], and the time-dependent part of the waveforms' phase is equal to

$$\phi(t) = 2\pi \sum_{k=0}^{s} f_k \frac{t^{k+1}}{(k+1)!}. \tag{6.81}$$

Here f_0 is the instantaneous gravitational-wave frequency (equal to twice the instantaneous rotational frequency of the neutron star) and f_k $(k = 1, \ldots, s)$ is the kth time derivative of the gravitational-wave frequency evaluated at time $t = 0$ at the SSB.

Let us consider the detection of these waves by means of an Earth-based detector. The response function of such detector, both the laser interferometer or the resonant bar, is of the form (see Sections 5.4.2 and 5.4.3)

$$s(t) = F_+(t)\, h_+\left(t + \frac{\mathbf{n}_0 \cdot \mathbf{r}_\mathrm{d}(t)}{c}\right) + F_\times(t)\, h_\times\left(t + \frac{\mathbf{n}_0 \cdot \mathbf{r}_\mathrm{d}(t)}{c}\right), \tag{6.82}$$

where $\mathbf{n}_0$ is the constant unit vector in the direction of the star in the SSB reference frame, $\mathbf{r}_\mathrm{d}$ is the position vector of the detector in that frame.

The detector's beam-pattern functions F_+ and $F_\times$ are equal

$$F_+(t) = \sin\zeta\left(a(t)\cos 2\psi + b(t)\sin 2\psi\right), \tag{6.83a}$$

$$F_\times(t) = \sin\zeta\left(b(t)\cos 2\psi - a(t)\sin 2\psi\right), \tag{6.83b}$$

where ζ is the angle between the interferometer arms or it has to be replaced by $90°$ for bar detectors, ψ is the polarization angle of the wave, and the functions a and b are the amplitude modulation functions. They are periodic functions with the period equal to 1 sidereal day and they depend on the position of the detector on Earth. Their explicit form for the interferometers is given in Appendix C.2.1 and for the resonant detectors in Appendix C.2.2.

Making use of Eqs. (6.80), (6.82) and (6.83), the detector's response function s can be written as a superposition of the four time-dependent functions h_k, $k = 1, \ldots, 4$,

$$s(t;\boldsymbol{\theta}) = \sum_{k=1}^{4} A_k\, h_k(t;\boldsymbol{\xi}), \tag{6.84}$$

where the vector $\boldsymbol{\theta} = (\mathbf{A}, \boldsymbol{\xi})$ collects all parameters of the signal, which split into extrinsic (or amplitude) parameters $\mathbf{A} = (A_1, \ldots, A_4)$ and intrinsic parameters $\boldsymbol{\xi}$. The extrinsic parameters are the four constant amplitudes A_k given by

$$A_1 := h_{0+}\cos 2\psi\cos\phi_0 - h_{0\times}\sin 2\psi\sin\phi_0, \tag{6.85a}$$

$$A_2 := h_{0+}\sin 2\psi\cos\phi_0 + h_{0\times}\cos 2\psi\sin\phi_0, \tag{6.85b}$$

$$A_3 := -h_{0+}\cos 2\psi\sin\phi_0 - h_{0\times}\sin 2\psi\cos\phi_0, \tag{6.85c}$$

$$A_4 := -h_{0+}\sin 2\psi\sin\phi_0 + h_{0\times}\cos 2\psi\cos\phi_0, \tag{6.85d}$$

where we have introduced the amplitudes h_{0+} and $h_{0\times}$ of the individual wave polarizations:

$$h_{0+} := \frac{1}{2}h_0(1 + \cos^2\iota), \tag{6.86a}$$

$$h_{0\times} := h_0\cos\iota. \tag{6.86b}$$

The four functions of time h_k depend on the intrinsic parameters $\boldsymbol{\xi}$ and are defined as follows:

$$h_1(t;\boldsymbol{\xi}) := a(t;\delta,\alpha)\cos\phi(t;\mathbf{f},\delta,\alpha), \tag{6.87a}$$

$$h_2(t;\boldsymbol{\xi}) := b(t;\delta,\alpha)\cos\phi(t;\mathbf{f},\delta,\alpha), \tag{6.87b}$$

$$h_3(t;\boldsymbol{\xi}) := a(t;\delta,\alpha)\sin\phi(t;\mathbf{f},\delta,\alpha), \tag{6.87c}$$

$$h_4(t;\boldsymbol{\xi}) := b(t;\delta,\alpha)\sin\phi(t;\mathbf{f},\delta,\alpha). \tag{6.87d}$$

Here $\boldsymbol{\xi} = (\mathbf{f}, \delta, \alpha)$, where the angles δ (declination) and α (right ascension) are equatorial coordinates determining the position of the source in the sky and the "frequency vector" $\mathbf{f} := (f_0, f_1, \ldots, f_s)$ collects the frequency parameter f_0 and the s spindown parameters. The phase ϕ is given by

$$\phi(t; \boldsymbol{\xi}) = 2\pi \sum_{k=0}^{s} \frac{f_k}{(k+1)!} \left(t + \frac{\mathbf{n}_0(\delta, \alpha) \cdot \mathbf{r}_\mathrm{d}(t)}{c} \right)^{k+1}. \tag{6.88}$$

The dominant term in the phase (6.88) is $2\pi f_0 t$. Typical gravitational-wave frequencies f_0 are contained in the range from a few Hz to a few kHz. The gravitational-wave signal from a rotating neutron star is a nearly periodic signal that is weakly amplitude and phase modulated due to the intrinsic variation of a star's rotation frequency and the motion of the detector with respect to the star. Moreover the amplitude of this signal is expected to be very small. Consequently, the detection of a signal requires observation time T_o that is very long compared to the gravitational-wave period $P_0 = 1/f_0$. This allows us to expand the phase given in Eq. (6.88) and keep only the terms with the highest powers of time t. Consequently, we approximate the phase $\phi(t)$ by

$$\phi(t; \boldsymbol{\xi}) \cong 2\pi \sum_{k=0}^{s} f_k \frac{t^{k+1}}{(k+1)!} + \frac{2\pi}{c} \mathbf{n}_0(\delta, \alpha) \cdot \mathbf{r}_\mathrm{d}(t) \sum_{k=0}^{s} f_k \frac{t^k}{k!}. \tag{6.89}$$

A detailed discussion of the validity of this approximated model of the phase is contained in Appendix A of Ref. [31].

We further assume that over the frequency bandwidth of the signal the spectral density of the detector's noise is almost constant and can be approximated by $S_0 := S(f_0)$. Then for any function h, with the Fourier transform concentrated around the frequency f_0, the scalar product of h with some other function h' can be approximated by

$$(h|h') \cong 2\frac{T_\mathrm{o}}{S_0} \langle hh' \rangle, \tag{6.90}$$

where we have assumed that the observational interval is $[0, T_\mathrm{o}]$, so the time-averaging operator $\langle \cdot \rangle$ is defined as

$$\langle g \rangle := \frac{1}{T_\mathrm{o}} \int_0^{T_\mathrm{o}} g(t) \, \mathrm{d}t. \tag{6.91}$$

We also expect that the period P_0 of the gravitational wave is much smaller than one day and much smaller than the observation time T_o. Making use of Eq. (6.90) we can approximate the log likelihood function by

$$\ln \Lambda[x(t); \boldsymbol{\theta}] = (x(t)|s(t; \boldsymbol{\theta})) - \frac{1}{2}(s(t; \boldsymbol{\theta})|s(t; \boldsymbol{\theta}))$$

$$\cong 2\frac{T_o}{S_0} \left(\langle x(t)s(t; \boldsymbol{\theta}) \rangle - \frac{1}{2}\langle s(t; \boldsymbol{\theta})^2 \rangle \right). \qquad (6.92)$$

The ML equations for the amplitudes A_k are given by

$$\frac{\partial \ln \Lambda}{\partial A_k} = 0, \quad k = 1, \ldots, 4. \qquad (6.93)$$

One easily finds that for the signal (6.84) the above set of equations is equivalent to the following set of linear algebraic equations

$$\sum_{j=1}^{4} \mathcal{M}_{ij} A_j = \mathcal{N}_i, \quad i = 1, \ldots, 4, \qquad (6.94)$$

where the components of the matrix $\mathcal{M}$ and the vector $\mathcal{N}$ are given by

$$\mathcal{M}_{ij} := \langle h_i \, h_j \rangle, \quad \mathcal{N}_i := \langle x \, h_i \rangle, \quad i, j = 1, \ldots, 4. \qquad (6.95)$$

Since over a typical observation time T_o the phase ϕ will have very many oscillations, then to a very good accuracy we have

$$\langle h_1 \, h_3 \rangle \cong 0, \quad \langle h_1 \, h_4 \rangle \cong 0, \quad \langle h_2 \, h_3 \rangle \cong 0, \quad \langle h_2 \, h_4 \rangle \cong 0, \qquad (6.96)$$

and also

$$\langle h_1 \, h_1 \rangle \cong \langle h_3 \, h_3 \rangle \cong \frac{1}{2}A,$$

$$\langle h_2 \, h_2 \rangle \cong \langle h_4 \, h_4 \rangle \cong \frac{1}{2}B, \qquad (6.97)$$

$$\langle h_1 \, h_2 \rangle \cong \langle h_3 \, h_4 \rangle \cong \frac{1}{2}C,$$

where we have introduced

$$A := \langle a^2 \rangle, \quad B := \langle b^2 \rangle, \quad C := \langle ab \rangle. \qquad (6.98)$$

With these approximations the matrix $\mathcal{M}$ is given by

$$\mathcal{M} \cong \frac{1}{2} \begin{pmatrix} A & C & 0 & 0 \\ C & B & 0 & 0 \\ 0 & 0 & A & C \\ 0 & 0 & C & B \end{pmatrix}. \qquad (6.99)$$

Assuming that $a \neq b$, $A \neq 0$, and $B \neq 0$, the explicit expressions for the ML estimators $\hat{A}_k$ of the amplitudes A_k are given by

$$\hat{A}_1 \cong \frac{B\langle xh_1\rangle - C\langle xh_2\rangle}{D}, \tag{6.100a}$$

$$\hat{A}_2 \cong \frac{A\langle xh_2\rangle - C\langle xh_1\rangle}{D}, \tag{6.100b}$$

$$\hat{A}_3 \cong \frac{B\langle xh_3\rangle - C\langle xh_4\rangle}{D}, \tag{6.100c}$$

$$\hat{A}_4 \cong \frac{A\langle xh_4\rangle - C\langle xh_3\rangle}{D}, \tag{6.100d}$$

where D is defined as

$$D := AB - C^2. \tag{6.101}$$

The second-order partial derivatives of the log likelihood function with respect to A_k are given by

$$\frac{\partial^2 \ln \Lambda}{\partial A_i \partial A_j} = -\mathcal{M}_{ij}. \tag{6.102}$$

Since $a \neq b$ it follows from the Cauchy–Schwarz inequality that $D > 0$. Thus as also $A > 0$ and $B > 0$, the matrix $\mathcal{M}$ is positive-definite. Therefore the extrema of the log likelihood function with respect to A_k are the local maxima.

The ML estimators (6.100) of the amplitudes A_i are substituted for the amplitudes A_i in the likelihood function (6.92) giving the following reduced likelihood function that we call the $\mathcal{F}$-*statistic*:

$$\mathcal{F}[x; \xi] := \ln \Lambda[x; \hat{\mathbf{A}}, \xi]$$

$$\cong \frac{2}{S_0 T_o} \frac{B|F_a|^2 + A|F_b|^2 - 2C\Re(F_a F_b^*)}{D}. \tag{6.103}$$

In the expression above F_a and F_b are the integrals given by:

$$F_a := \int_0^{T_o} x(t)a(t)\exp[-i\phi(t)]\,dt, \tag{6.104a}$$

$$F_b := \int_0^{T_o} x(t)b(t)\exp[-i\phi(t)]\,dt. \tag{6.104b}$$

It can be shown [31] that the $\mathcal{F}$-statistic given by Eq. (6.103) has a χ^2 distribution with four degrees of freedom when signal is absent. This follows from the fact that $\mathcal{F}$ is a function of two complex quantities F_a and F_b and by a suitable transformation it can be converted into the sum of

4 squares of real quantities that are linear functions of the data x which we assume is Gaussian.

We shall now give compact analytic formulae for the signal-to-noise ratio and the Fisher matrix of the signal considered here. The signal-to-noise ratio ρ we compute using Eq. (6.19). We first observe that the matrix M from Eq. (6.17) is equal to [see Eqs. (6.90) and (6.95)]

$$\mathsf{M} \cong \frac{2T_\mathrm{o}}{S_0}\mathcal{M},\tag{6.105}$$

then we make use of Eqs. (6.85), (6.86) and (6.99) to express the signal-to-noise ratio in the following form (which was found in Ref. [191]):

$$\rho^2 \cong h_\mathrm{o}^2 \frac{T_\mathrm{o}}{S_0}\left(\alpha_1\,A + \alpha_2\,B + 2\alpha_3\,C\right),\tag{6.106}$$

where we have introduced (defining $z := \cos\iota$):

$$\alpha_1(z,\psi) := \frac{1}{4}(1+z^2)^2\cos^2 2\psi + z^2\sin^2 2\psi,\tag{6.107a}$$

$$\alpha_2(z,\psi) := \frac{1}{4}(1+z^2)^2\sin^2 2\psi + z^2\cos^2 2\psi,\tag{6.107b}$$

$$\alpha_3(z,\psi) := \frac{1}{4}(1-z^2)^2\sin 2\psi\,\cos 2\psi.\tag{6.107c}$$

The signal-to-noise ratio ρ does not depend on the initial phase ϕ_0. When the signal is present in the detector's data, the $\mathcal{F}$-statistic has a non-central χ^2 distribution of four degrees of freedom with non-centrality parameter equal to ρ^2.

The signal-to-noise ratio ρ is a complicated function of the angles α, δ, ψ, and ι. It is sometimes useful to consider an average value of ρ independent of these angles. Such averaging is performed according to the definition:

$$\langle\cdots\rangle_{\alpha,\delta,\psi,\iota} := \frac{1}{2\pi}\int_0^{2\pi} d\alpha\times\frac{1}{2}\int_{-1}^{1} d\sin\delta\times\frac{1}{2\pi}\int_0^{2\pi} d\psi\times\frac{1}{2}\int_{-1}^{1} d\cos\iota\,(\cdots).$$

$$\tag{6.108}$$

Note that because $\delta \in [-\pi/2, \pi/2]$ integration over $\sin\delta$ rather than $\cos\delta$ is involved in Eq. (6.108). The above averaging procedure means that there is no preferred orientation of the star's rotational axis with respect to the line of sight and no preferred position of the star in the sky. Averaging over the angles α, δ, ψ, and ι yields the result which does not depend also on the position of the detector on the Earth and on the orientation of its arms. The averaging procedure (6.108) applied to the signal-to-noise ratio

(6.106) gives

$$\sqrt{\langle \rho^2 \rangle_{\alpha,\delta,\psi,\iota}} \cong \frac{2}{5} h_0 \sqrt{\frac{T_o}{S_0}}. \tag{6.109}$$

In order to calculate the reduced Fisher matrix $\tilde{\Gamma}$ for the signal s we first need to compute the matrices $\mathsf{F}_{(A)}$ and $\mathsf{S}_{(AB)}$ introduced in Eqs. (6.21). The 4×4 matrices $\mathsf{F}_{(A)}$ can be expressed in terms of the two 2×2 matrices $\mathsf{F}_{(A)1}$ and $\mathsf{F}_{(A)2}$,

$$\mathsf{F}_{(A)} \cong \frac{T_o}{S_0} \begin{pmatrix} \mathsf{F}_{(A)1} & \mathsf{F}_{(A)2} \\ -\mathsf{F}_{(A)2} & \mathsf{F}_{(A)1} \end{pmatrix}, \tag{6.110}$$

where [we denote here partial derivatives $\partial(\cdots)/\partial_A$ by $\partial_A \cdots$]

$$\mathsf{F}_{(A)1} \cong \begin{pmatrix} \langle a\,\partial_A a \rangle & \langle a\,\partial_A b \rangle \\ \langle b\,\partial_A a \rangle & \langle b\,\partial_A b \rangle \end{pmatrix}, \quad \mathsf{F}_{(A)2} \cong \begin{pmatrix} \langle a^2\,\partial_A \phi \rangle & \langle ab\,\partial_A \phi \rangle \\ \langle ab\,\partial_A \phi \rangle & \langle b^2\,\partial_A \phi \rangle \end{pmatrix}. \tag{6.111}$$

The 4×4 matrices $\mathsf{S}_{(AB)}$ can also be expressed in terms of the 2×2 matrices $\mathsf{S}_{(AB)1}$ and $\mathsf{S}_{(AB)2}$,

$$\mathsf{S}_{(AB)} \cong \frac{T_o}{S_0} \begin{pmatrix} \mathsf{S}_{(AB)1} & \mathsf{S}_{(AB)2} \\ -\mathsf{S}_{(AB)2} & \mathsf{S}_{(AB)1} \end{pmatrix}, \tag{6.112}$$

where

$$\mathsf{S}_{(AB)1} \cong \begin{pmatrix} \langle (\partial_A a)(\partial_B a) + a^2 (\partial_A \phi)(\partial_B \phi) \rangle \\ \langle (\partial_A b)(\partial_B a) + ab\,(\partial_A \phi)(\partial_B \phi) \rangle \end{pmatrix}$$
$$\begin{pmatrix} \langle (\partial_A a)(\partial_B b) + ab\,(\partial_A \phi)(\partial_B \phi) \rangle \\ \langle (\partial_A b)(\partial_B b) + b^2 (\partial_A \phi)(\partial_B \phi) \rangle \end{pmatrix}, \tag{6.113a}$$

$$\mathsf{S}_{(AB)2} \cong \begin{pmatrix} \langle a\,\big((\partial_A a)(\partial_B \phi) - (\partial_B a)(\partial_A \phi)\big) \rangle \\ \langle a\,(\partial_A b)(\partial_B \phi) - b\,(\partial_B a)(\partial_A \phi) \rangle \end{pmatrix}$$
$$\begin{pmatrix} \langle b\,(\partial_A a)(\partial_B \phi) - a\,(\partial_B b)(\partial_A \phi) \rangle \\ \langle b\,\big((\partial_A b)(\partial_B \phi) - (\partial_B b)(\partial_A \phi)\big) \rangle \end{pmatrix}. \tag{6.113b}$$

Making use of Eqs. (6.110)–(6.113) one can compute the projected Fisher matrix $\bar{\Gamma}$, the normalized Fisher matrix $\bar{\bar{\Gamma}}$, and finally the reduced Fisher matrix $\tilde{\Gamma}$ by means of Eqs. (6.27), (6.28), and (6.31), respectively.

Earth-based detectors look for gravitational waves with frequencies $\gtrsim 1$ Hz. For such high frequencies and observational times of the order of few hours or longer, one can neglect the derivatives of the amplitude modulation functions a and b compared to the derivatives of the phase ϕ, i.e.

$$|\partial_{\mathcal{A}}\phi| \gg |\partial_{\mathcal{A}}a|, \; |\partial_{\mathcal{A}}b| \qquad (6.114)$$

(see Section IIID of Ref. [191] for justification). Making use of the assumption (6.114) we can give more explicit forms for the matrices $\overline{\Gamma}$, $\overline{\overline{\Gamma}}$, and $\widetilde{\Gamma}$ [191]. First of all the matrices $\mathsf{F}_{(\mathcal{A})}$ and $S_{(\mathcal{AB})}$ reduce in this case to

$$\mathsf{F}_{(\mathcal{A})} \cong \frac{T_o}{S_0} \begin{pmatrix} 0 & 0 & \mathsf{F}_{(\mathcal{A})13} & \mathsf{F}_{(\mathcal{A})14} \\ 0 & 0 & \mathsf{F}_{(\mathcal{A})14} & \mathsf{F}_{(\mathcal{A})24} \\ -\mathsf{F}_{(\mathcal{A})13} & -\mathsf{F}_{(\mathcal{A})14} & 0 & 0 \\ -\mathsf{F}_{(\mathcal{A})14} & -\mathsf{F}_{(\mathcal{A})24} & 0 & 0 \end{pmatrix}, \qquad (6.115a)$$

$$S_{(\mathcal{AB})} \cong \frac{T_o}{S_0} \begin{pmatrix} P^1_{\mathcal{AB}} & P^3_{\mathcal{AB}} & 0 & 0 \\ P^3_{\mathcal{AB}} & P^2_{\mathcal{AB}} & 0 & 0 \\ 0 & 0 & P^1_{\mathcal{AB}} & P^3_{\mathcal{AB}} \\ 0 & 0 & P^3_{\mathcal{AB}} & P^2_{\mathcal{AB}} \end{pmatrix}, \qquad (6.115b)$$

where the non-zero components of the matrices are equal to:

$$\mathsf{F}_{(\mathcal{A})13} = \langle a^2\, \partial_{\mathcal{A}}\phi \rangle, \quad \mathsf{F}_{(\mathcal{A})24} = \langle b^2\, \partial_{\mathcal{A}}\phi \rangle, \quad \mathsf{F}_{(\mathcal{A})14} = \langle ab\, \partial_{\mathcal{A}}\phi \rangle, \quad (6.116a)$$

$$P^1_{\mathcal{AB}} = \langle a^2\, (\partial_{\mathcal{A}}\phi)\, (\partial_{\mathcal{B}}\phi) \rangle, \quad P^2_{\mathcal{AB}} = \langle b^2\, (\partial_{\mathcal{A}}\phi)\, (\partial_{\mathcal{B}}\phi) \rangle,$$
$$P^3_{\mathcal{AB}} = \langle ab\, (\partial_{\mathcal{A}}\phi)\, (\partial_{\mathcal{B}}\phi) \rangle. \qquad (6.116b)$$

The projected Fisher matrix $\overline{\Gamma}$ defined in Eq. (6.27) is given by

$$\overline{\Gamma}_{\mathcal{AB}} \cong h_0^2\, \frac{T_o}{S_0} \left(\alpha_1\, m^1_{\mathcal{AB}} + \alpha_2\, m^2_{\mathcal{AB}} + 2\alpha_3\, m^3_{\mathcal{AB}} \right), \qquad (6.117)$$

where the quantities α_r $(r=1,2,3)$ are defined in (6.107) and where

$$m^r_{\mathcal{AB}} := P^r_{\mathcal{AB}} - Q^r_{\mathcal{AB}}, \quad r = 1,2,3, \qquad (6.118)$$

with $P^r_{\mathcal{AB}}$ defined in Eq. (6.116b) and $Q^r_{\mathcal{AB}}$ given by

$$D\, Q^1_{\mathcal{AB}} = A \langle ab\, \partial_{\mathcal{A}}\phi \rangle \langle ab\, \partial_{\mathcal{B}}\phi \rangle + B \langle a^2 \partial_{\mathcal{A}}\phi \rangle \langle a^2 \partial_{\mathcal{B}}\phi \rangle$$
$$- 2C \langle a^2 \partial_{\mathcal{A}}\phi \rangle \langle ab\, \partial_{\mathcal{B}}\phi \rangle, \qquad (6.119a)$$

$$D\, Q^2_{\mathcal{AB}} = A \langle b^2 \partial_{\mathcal{A}}\phi \rangle \langle b^2 \partial_{\mathcal{B}}\phi \rangle + B \langle ab\, \partial_{\mathcal{A}}\phi \rangle \langle ab\, \partial_{\mathcal{B}}\phi \rangle$$
$$- 2C \langle ab\, \partial_{\mathcal{A}}\phi \rangle \langle b^2 \partial_{\mathcal{B}}\phi \rangle, \qquad (6.119b)$$

$$D\, Q^3_{\mathcal{AB}} = A \langle ab\, \partial_{\mathcal{A}}\phi \rangle \langle b^2 \partial_{\mathcal{B}}\phi \rangle + B \langle ab\, \partial_{\mathcal{A}}\phi \rangle \langle a^2 \partial_{\mathcal{B}}\phi \rangle$$
$$- C \left[\langle b^2 \partial_{\mathcal{A}}\phi \rangle \langle a^2 \partial_{\mathcal{B}}\phi \rangle + \langle ab\, \partial_{\mathcal{A}}\phi \rangle \langle ab\, \partial_{\mathcal{B}}\phi \rangle \right]. \qquad (6.119c)$$

The projected Fisher matrix $\overline{\Gamma}$ (similarly like the signal-to-noise ratio ρ) does not depend on the initial phase parameter ϕ_0. The normalized Fisher matrix $\overline{\overline{\Gamma}}$ and the reduced Fisher matrix $\widetilde{\Gamma}$ are defined by Eq. (6.28) and

Eq. (6.31), respectively. They are explicitly given by

$$\overline{\overline{\Gamma}}_{AB} \cong \frac{\alpha_1 \, m^1_{AB} + \alpha_2 \, m^2_{AB} + 2\alpha_3 \, m^3_{AB}}{\alpha_1 \, A + \alpha_2 \, B + 2\alpha_3 \, C}, \tag{6.120a}$$

$$\widetilde{\Gamma}_{AB} \cong \frac{B \, m^1_{AB} + A \, m^2_{AB} - 2C \, m^3_{AB}}{2D}. \tag{6.120b}$$

The normalized matrix $\overline{\overline{\Gamma}}$ does not depend on the amplitude h_0 and the initial phase ϕ_0, whereas the reduced Fisher matrix $\widetilde{\Gamma}$ does not depend on any of the extrinsic parameters h_0, ϕ_0, ψ, and ι, it only depends on the intrinsic parameters $\boldsymbol{\xi}$.

6.2.2 Chirp signal from an inspiraling compact binary

In this section we consider the detection of the gravitational-wave *chirp* signal, i.e. the signal being the response of the detector to gravitational waves coming from an inspiraling binary system made of compact objects. We assume that this signal is to be detected by some Earth-based laser interferometric detector. We also restrict ourselves to a binary system made of bodies moving along *circular* orbits. The wave polarization functions h_+ and $h_\times$ for such gravitational-wave source were computed in Section 2.4 and they are given in Eqs. (2.35) there, which written at the SSB [so we put $\mathbf{x} = \mathbf{0}$ in Eqs. (2.35)] read

$$h_+(t) = -\frac{4(G\mathcal{M})^{5/3}}{c^4 R} \frac{1 + \cos^2 \iota}{2} \omega\big(t_{\mathrm{r}}(t)\big)^{2/3} \cos 2\phi\big(t_{\mathrm{r}}(t)\big), \tag{6.121a}$$

$$h_\times(t) = -\frac{4(G\mathcal{M})^{5/3}}{c^4 R} \cos \iota \, \omega\big(t_{\mathrm{r}}(t)\big)^{2/3} \sin 2\phi\big(t_{\mathrm{r}}(t)\big), \tag{6.121b}$$

where $\mathcal{M}$ is the chirp mass of the binary [defined in Eq. (2.34)]; R, the distance from the SSB to the binary's center-of-mass; ι ($0 \leq \iota \leq \pi$), the angle between the orbital angular momentum vector and the line of sight; ϕ, the instantaneous orbital phase; $\omega := \dot{\phi}$, the instantaneous orbital angular frequency; and $t_{\mathrm{r}}(t) := t - R/c$, the retarded time.

The dimensionless response function s of the interferometric detector to gravitational waves with polarization functions (6.121) can be written in the form of Eq. (6.82) from Section 6.2.1. But here we assume that the duration of the whole gravitational-wave signal is short enough compared to the time scales related both to the rotational and the orbital motion of the Earth, that the modulation effects related to the motion of the detector with respect to the SSB can be neglected. It implies that in Eq. (6.82) the beam-patterns F_+ and $F_\times$ and the term $\mathbf{n}_0 \cdot \mathbf{r}_{\mathrm{d}}$ can be treated as essentially time-independent quantities during the time interval in which the chirp signal is observed. Moreover, due to the intrinsic instrumental noise, each interferometric detector can see any signals only in a certain frequency window. It means that one can introduce the *time of arrival* t_i of the signal, i.e. the instant of time at which the instantaneous waveform's angular frequency $\omega_{gw}(t)$ *registered by the detector* is equal to some fixed value ω_i, which represents the low-frequency cut-off of the detector's frequency window,

$$\omega_{gw}(t_i) = \omega_i. \tag{6.122}$$

All this means that the detector's response function s can be written in the following form:

$$s(t) = \Theta(t - t_i)\left(F_+ h_+\left(t + \frac{\mathbf{n}_0 \cdot \mathbf{r}_d}{c}\right) + F_\times h_\times\left(t + \frac{\mathbf{n}_0 \cdot \mathbf{r}_d}{c}\right)\right), \tag{6.123}$$

where Θ is the step function defined as

$$\Theta(t) := \begin{cases} 0, & t < 0, \\ 1, & t \geq 0, \end{cases} \tag{6.124}$$

and where the quantities F_+, $F_\times$, and $\mathbf{r}_d$ are treated as (approximately) independent of time constants. Collecting Eqs. (6.121) and (6.123) together we get

$$s(t) = -\frac{4(G\mathcal{M})^{5/3}}{c^4 R}\, \Theta(t - t_i)\, \omega(t + \Delta)^{2/3}$$

$$\times \left(\frac{1 + \cos^2 \iota}{2} F_+ \cos 2\phi(t + \Delta) + \cos \iota\, F_\times \sin 2\phi(t + \Delta)\right), \tag{6.125}$$

where we have introduced $\Delta := \mathbf{n}_0 \cdot \mathbf{r}_d/c - R/c$.

Now we will connect the initial phase ϕ_0 and the initial angular frequency ω_0 of the *orbital motion* with the initial phase ϕ_i and the initial angular frequency ω_i of the *waveform* observed by the detector. The time evolution of the orbital angular frequency ω is, in the case of circular orbits, described by a first-order ordinary differential equation of the form

$$\frac{d\omega}{dt} = \Omega(\omega) \quad \Longrightarrow \quad \int_{\omega_0}^{\omega} \frac{d\omega'}{\Omega(\omega')} = t - t_0, \tag{6.126}$$

where Ω is a known function of ω. Let us denote by $\omega(t; t_0, \omega_0)$ the general solution to this equation which fulfills the initial condition $\omega(t_0; t_0, \omega_0) = \omega_0$, so ω_0 is the value of the angular frequency at $t = t_0$. The time evolution of the orbital phase ϕ is then obtained from the equation $d\phi/dt = \omega$. Its general solution can be written as

$$\phi(t; t_0, \omega_0, \phi_0) = \phi_0 + \int_{t_0}^{t} \omega(t'; t_0, \omega_0) \, dt', \qquad (6.127)$$

where ϕ_0 is the value of the phase at $t = t_0$, i.e. $\phi(t_0; t_0, \omega_0, \phi_0) = \phi_0$.

By virtue of Eq. (6.125) the instantaneous phase ϕ_{gw} of the waveform equals

$$\phi_{gw}(t; t_i, \omega_i, \phi_i) := 2\,\phi(t + \Delta; t_0, \omega_0, \phi_0). \qquad (6.128)$$

The notation introduced here means that we want to parametrize the phase ϕ_{gw} by the time of arrival t_i and the initial values ϕ_i and ω_i of the waveform's phase and angular frequency, respectively, both evaluated at $t = t_i$. The instantaneous angular frequency ω_{gw} of the waveform is defined as the time derivative of the instantaneous phase,

$$\omega_{gw}(t; t_i, \omega_i) := \dot{\phi}_{gw}(t; t_i, \omega_i, \phi_i) = 2\,\omega(t + \Delta; t_0, \omega_0)\,(1 + \dot{\Delta}). \qquad (6.129)$$

Here $\dot{\Delta} = \mathbf{n}_0 \cdot \mathbf{v}_d/c$, where $\mathbf{v}_d := \dot{\mathbf{r}}_d$ is the (approximately constant) velocity of the detector with respect to the SSB. The relation between the instantaneous phase and angular frequency of the waveform is thus

$$\phi_{gw}(t; t_i, \omega_i, \phi_i) = \phi_i + \int_{t_i}^{t} \omega_{gw}(t'; t_i, \omega_i) \, dt'. \qquad (6.130)$$

The time of arrival t_i of the signal is defined by the requirement (6.122), which rewritten in our new notation reads

$$\omega_{gw}(t_i; t_i, \omega_i) = 2\,\omega(t_i + \Delta; t_0, \omega_0)\,(1 + \dot{\Delta}) = \omega_i. \qquad (6.131)$$

We can use this equation to express the instant of time t_0 as a function of the instant of time $t_i + \Delta$ and the angular frequencies ω_0, $\bar{\omega}_i$, where

$$\bar{\omega}_i := \frac{\omega_i}{2\,(1 + \dot{\Delta})}. \qquad (6.132)$$

Then we plug the result into $\omega(t; t_0, \omega_0)$ to get

$$\omega\big(t; t_0(t_i + \Delta, \omega_0, \bar{\omega}_i), \omega_0\big) = \omega\big(t; t_i + \Delta, \bar{\omega}_i\big), \qquad (6.133)$$

which follows from the uniqueness of the solution of the ordinary differential equation for the fixed initial conditions. By virtue of Eqs. (6.129) and (6.133) the instantaneous angular frequency ω_{gw} of the waveform is equal to

$$\omega_{gw}(t; t_i, \omega_i) = 2\,(1 + \dot{\Delta})\,\omega\big(t + \Delta; t_i + \Delta, \bar{\omega}_i\big)$$
$$= 2\,(1 + \dot{\Delta})\,\omega\big(t; t_i, \bar{\omega}_i\big), \qquad (6.134)$$

where the last equality follows from the fact that the general solution $w(t; t_0, w_0)$ of Eq. (6.126) depends on t and t_0 only through the difference $t - t_0$ [therefore $w(t + \tau; t_0 + \tau, w_0) = w(t; t_0, w_0)$ for any constant time shift τ].

Because the velocity of the detector with respect to the SSB is non-relativistic, $|\mathbf{v_d}|/c \ll 1$, in the rest of this section we will use $\dot{\Delta} \cong 0$, i.e. we will employ

$$\omega_{gw}(t; t_i, \omega_i) \cong 2\,w\left(t; t_i, \bar{\omega}_i\right), \quad \bar{\omega}_i \cong \frac{\omega_i}{2}. \tag{6.135}$$

Making use of Eqs. (6.127)–(6.128), (6.133), and (6.135), we can rewrite the phase of the wavefrom in the following way

$$\phi_{gw}(t; t_i, \omega_i, \phi_i) = 2\,\phi_0 + 2\int_{t_0}^{t+\Delta} w(t'; t_0, w_0)\,dt' \tag{6.136a}$$

$$= 2\,\phi_0 + 2\int_{t_0}^{t+\Delta} w\left(t' - \Delta; t_i, \bar{\omega}_i\right)dt' \tag{6.136b}$$

$$= 2\,\phi_0 + 2\int_{t_0-\Delta}^{t} w\left(t'; t_i, \bar{\omega}_i\right)dt' \tag{6.136c}$$

$$= 2\,\phi_0 + \int_{t_0-\Delta}^{t} \omega_{gw}\left(t'; t_i, \omega_i\right)dt'. \tag{6.136d}$$

Comparing Eqs. (6.130) and (6.136d) we get the following link between the initial phases ϕ_i and ϕ_0:

$$\phi_i = 2\,\phi_0 + 2\int_{t_0-\Delta}^{t_i} w\left(t'; t_i, \bar{\omega}_i\right)dt'. \tag{6.137}$$

One can show that the numerical value of the right-hand side of Eq. (6.137) does not (approximately) depend on the location of the detector on Earth. By virtue of Eqs. (6.130), (6.135), and (6.127), the relation between the instantaneous phase ϕ_{gw} of the waveform "seen" by the detector and the instantaneous phase ϕ of the orbital motion of the binary system is the following:

$$\phi_{gw}(t; t_i, \omega_i, \phi_i) = 2\,\phi\left(t; t_i, \bar{\omega}_i, \frac{\phi_i}{2}\right). \tag{6.138}$$

The response function (6.125) written in terms of the functions ϕ_{gw} and $\omega_{gw} = \dot{\phi}_{gw}$ reads

$$s(t) = -h_0\,\Theta(t - t_i)\,\omega_{gw}(t)^{2/3}$$

$$\times \left(\frac{1 + \cos^2\iota}{2}\,F_+\cos\phi_{gw}(t) + \cos\iota\,F_\times\sin\phi_{gw}(t)\right), \tag{6.139}$$

where we have introduced the constant h_0 defined as

$$h_0 := \frac{2^{4/3} (G\mathcal{M})^{5/3}}{c^4 R}. \tag{6.140}$$

Let us now introduce the vector $\boldsymbol{\theta}$ consisting of all the parameters on which the gravitational-wave signal (6.139) depends, so we can write $s = s(t; \boldsymbol{\theta})$. Below we introduce two independent amplitude parameters A_c and A_s, so $\boldsymbol{\theta} = (A_c, A_s, \boldsymbol{\xi})$, where $\boldsymbol{\xi}$ comprises all the parameters on which the gravitational-wave angular frequency depends: $\omega_{\mathrm{gw}} = \omega_{\mathrm{gw}}(t; \boldsymbol{\xi})$. One of these parameters is the time of arrival t_{i} of the signal. Let us also single out the initial phase parameter ϕ_{i} by writing

$$\phi_{\mathrm{gw}}(t; \phi_{\mathrm{i}}, \boldsymbol{\xi}) = \phi_{\mathrm{i}} + \bar{\phi}_{\mathrm{gw}}(t; \boldsymbol{\xi}), \qquad \bar{\phi}_{\mathrm{gw}}(t; \boldsymbol{\xi}) := \int_{t_{\mathrm{i}}}^{t} \omega_{\mathrm{gw}}(t'; \boldsymbol{\xi}) \, \mathrm{d}t'. \tag{6.141}$$

Making use of all these conventions the response function s of Eq. (6.139) can be put into the following form:

$$s(t; A_c, A_s, \boldsymbol{\xi}) = A_c \, h_c(t; \boldsymbol{\xi}) + A_s \, h_s(t; \boldsymbol{\xi}), \tag{6.142}$$

where the time dependent functions h_c and h_s are given by

$$h_c(t; \boldsymbol{\xi}) := \Theta(t - t_{\mathrm{i}}) \, \omega_{\mathrm{gw}}(t; \boldsymbol{\xi})^{2/3} \cos \bar{\phi}_{\mathrm{gw}}(t; \boldsymbol{\xi}), \tag{6.143a}$$

$$h_s(t; \boldsymbol{\xi}) := \Theta(t - t_{\mathrm{i}}) \, \omega_{\mathrm{gw}}(t; \boldsymbol{\xi})^{2/3} \sin \bar{\phi}_{\mathrm{gw}}(t; \boldsymbol{\xi}), \tag{6.143b}$$

and the amplitude parameters A_c and A_s are defined as follows

$$A_c := -h_0 \left(F_+ \frac{1 + \cos^2 \iota}{2} \cos \phi_{\mathrm{i}} + F_\times \cos \iota \sin \phi_{\mathrm{i}} \right), \tag{6.144a}$$

$$A_s := -h_0 \left(-F_+ \frac{1 + \cos^2 \iota}{2} \sin \phi_{\mathrm{i}} + F_\times \cos \iota \cos \phi_{\mathrm{i}} \right). \tag{6.144b}$$

The beam-pattern functions F_+ and $F_\times$ depend on the position of the gravitational-wave source in the sky (which can be represented by the right ascension α and declination δ of the binary) and the polarization angle ψ of the wave. Their numerical values can be computed from the general formulae given in Appendix C.2.1.

The logarithm of the likelihood function,

$$\ln \Lambda[x(t); \boldsymbol{\theta}] = (x(t)|s(t; \boldsymbol{\theta})) - \frac{1}{2}(s(t; \boldsymbol{\theta})|s(t; \boldsymbol{\theta})), \tag{6.145}$$

for the chirp signal (6.142) can be written as

$$\ln \Lambda = A_c(x|h_c) + A_s(x|h_s) - \frac{1}{2}A_c^2(h_c|h_c) - A_c A_s(h_c|h_s) - \frac{1}{2}A_s^2(h_s|h_s). \tag{6.146}$$

The ML estimators of the amplitudes A_c and A_s are obtained by solving the equations

$$\frac{\partial \ln \Lambda}{\partial A_c} = 0, \quad \frac{\partial \ln \Lambda}{\partial A_s} = 0, \tag{6.147}$$

which, by virtue of Eq. (6.146), form the set of two linear equations for A_c and A_s,

$$A_c(h_c|h_c) + A_s(h_c|h_s) = (x|h_c), \tag{6.148a}$$
$$A_c(h_c|h_s) + A_s(h_s|h_s) = (x|h_s). \tag{6.148b}$$

Their unique solution determines the explicit analytic expressions for the ML estimators of the amplitudes,

$$\hat{A}_c = \frac{S(x|h_c) - M(x|h_s)}{D}, \tag{6.149a}$$

$$\hat{A}_s = \frac{C(x|h_s) - M(x|h_c)}{D}, \tag{6.149b}$$

where we have used the definitions

$$C := (h_c|h_c), \quad S := (h_s|h_s), \quad M := (h_c|h_s), \quad D := SC - M^2. \tag{6.150}$$

Substituting the estimators (6.149) into the likelihood function (6.146) we obtain the $\mathcal{F}$-statistic for the chirp signal:

$$\mathcal{F}[x; \boldsymbol{\xi}] := \ln \Lambda[x; \hat{A}_c, \hat{A}_s, \boldsymbol{\xi}]$$
$$= \frac{S(x|h_c)^2 + C(x|h_s)^2 - 2M(x|h_c)(x|h_s)}{2D}. \tag{6.151}$$

The scalar products C, S, and M appearing in Eq. (6.151) can be analytically calculated making use of the *stationary phase approximation* (see Appendix A). Within this approximation it holds

$$N := (h_c|h_c) \cong (h_s|h_s), \quad (h_c|h_s) \cong 0, \tag{6.152}$$

so the $\mathcal{F}$-statistic (6.151) simplifies then to

$$\mathcal{F} \cong \frac{1}{2N}\left((x|h_c)^2 + (x|h_s)^2\right). \tag{6.153}$$

Sometimes it is useful to represent the gravitational-wave chirp signal (6.142) in the following form:

$$s(t; A_0, \xi_0, \boldsymbol{\xi}) = A_0\, h(t; \xi_0, \boldsymbol{\xi}), \tag{6.154}$$

where the time-dependent function h is equal

$$h(t; \xi_0, \boldsymbol{\xi}) := \Theta(t - t_i)\, \omega_{\text{gw}}(t; \boldsymbol{\xi})^{2/3} \cos\left(\bar{\phi}_{\text{gw}}(t; \boldsymbol{\xi}) - \xi_0\right)$$
$$= h_c(t; \boldsymbol{\xi}) \cos \xi_0 + h_s(t; \boldsymbol{\xi}) \sin \xi_0. \tag{6.155}$$

The amplitude parameter A_0 is defined as

$$A_0 := h_0 \sqrt{\left(F_+ \frac{1 + \cos^2 \iota}{2}\right)^2 + (F_\times \cos \iota)^2}, \tag{6.156}$$

whereas the new initial phase parameter ξ_0 is determined by equations

$$\sin(\xi_0 + \phi_i) = -\frac{F_+ \dfrac{1 + \cos^2 \iota}{2}}{\sqrt{\left(F_+ \dfrac{1 + \cos^2 \iota}{2}\right)^2 + (F_\times \cos \iota)^2}}, \tag{6.157a}$$

$$\cos(\xi_0 + \phi_i) = -\frac{F_\times \cos \iota}{\sqrt{\left(F_+ \dfrac{1 + \cos^2 \iota}{2}\right)^2 + (F_\times \cos \iota)^2}}. \tag{6.157b}$$

Equations (6.157) lead to the relation

$$\xi_0 = -\phi_i + \operatorname{atan}\left(\frac{F_+(1 + \cos^2 \iota)}{2 F_\times \cos \iota}\right). \tag{6.158}$$

The logarithm of the likelihood ratio (6.145) in terms of the parameters $(A_0, \xi_0, \boldsymbol{\xi})$ reads

$$\ln \Lambda[x(t); A_0, \xi_0, \boldsymbol{\xi}] = A_0(x(t)|h(t; \xi_0, \boldsymbol{\xi})) - \frac{1}{2}A_0^2(h(t; \xi_0, \boldsymbol{\xi})|h(t; \xi_0, \boldsymbol{\xi})). \tag{6.159}$$

Maximization of (6.159) with respect to A_0 leads to

$$\hat{A}_0 = \frac{(x|h)}{(h|h)}. \tag{6.160}$$

We replace in (6.159) A_0 by $\hat{A}_0$ and express the result in terms of $\tan \xi_0$:

$$\ln \Lambda[x(t); \hat{A}_0, \xi_0, \boldsymbol{\xi}] = \frac{(x|h)^2}{2(h|h)} = \frac{\left((x|h_c) + (x|h_s)\tan \xi_0\right)^2}{2\left(C + 2M \tan \xi_0 + S \tan^2 \xi_0\right)}, \tag{6.161}$$

where we have used the notation introduced in (6.150). The maximum of (6.161) with respect to ξ_0 is attained for

$$\hat{\xi}_0 = \operatorname{atan}\left(\frac{C(x|h_s) - M(x|h_c)}{S(x|h_c) - M(x|h_s)}\right). \tag{6.162}$$

Replacing in (6.161) the initial phase ξ_0 by its ML estimator $\hat{\xi}_0$, we obtain the $\mathcal{F}$-statistic again in the form given by Eq. (6.151).

6.2.3 Signal from a supernova explosion

The gravitational-wave signal s from a supernova explosion can be expressed in the form of the general equation (6.4) with a single constant amplitude parameter A_0 and a single time-dependent function h,

$$s(t; \boldsymbol{\theta}) = A_0\, h(t; \boldsymbol{\xi}). \tag{6.163}$$

The function h is an *impulse* that can be expressed as

$$h(t; \boldsymbol{\xi}) = \frac{1}{\sqrt{\sqrt{2\pi}\tau}} \exp\left(-\frac{(t - t_0)^2}{4\tau^2}\right). \tag{6.164}$$

Here the parameter vector $\boldsymbol{\xi} = (t_0, \tau)$, where t_0 is related to the *time-of-arrival* of the signal and τ is related to its *time duration*. The gravitational-wave signal from a supernova explosion depends thus on the three parameters: $\boldsymbol{\theta} = (A_0, t_0, \tau)$.

The *time duration* Δ_t of any gravitational-wave signal s which has the form of an impulse and its *frequency bandwidth* Δ_f can be defined through equations

$$\Delta_t^2 = \frac{\int_{-\infty}^{\infty} (t - t_0)^2 s(t)^2 \, \mathrm{d}t}{\int_{-\infty}^{\infty} s(t)^2 \, \mathrm{d}t}, \qquad \Delta_f^2 = \frac{\int_{-\infty}^{\infty} f^2 |\tilde{s}(f)|^2 \, \mathrm{d}f}{\int_{-\infty}^{\infty} |\tilde{s}(f)|^2 \, \mathrm{d}f}. \tag{6.165}$$

The product $\mathrm{TB} = \Delta_t \Delta_f$ is called the *time-bandwidth* product. We have the following "uncertainty relation" for the TB product:

$$\Delta_t \Delta_f \geq \frac{1}{4\pi}. \tag{6.166}$$

It follows from the above relation that signal cannot be both arbitrarily time-limited and band-limited.

The Fourier transform $\tilde{s}$ of the signal s defined in Eqs. (6.163) and (6.164) can be explicitly computed. It is equal to

$$\tilde{s}(f) = A_0 \sqrt{2\sqrt{2\pi}\,\tau}\, \exp(-2\pi \mathrm{i} f t_0 - 4\pi^2 f^2 \tau^2). \tag{6.167}$$

We can thus easily compute time duration and frequency bandwidth of this signal. Taking into account that

$$\int_{-\infty}^{\infty} s(t)^2 \, \mathrm{d}t = \int_{-\infty}^{\infty} |\tilde{s}(f)|^2 \, \mathrm{d}f = A_0^2, \tag{6.168}$$

we obtain

$$\Delta_t = \tau, \quad \Delta_f = \frac{1}{4\pi\tau}. \tag{6.169}$$

We see that the time-bandwidth product for the gravitational-wave signal given in Eqs. (6.163) and (6.164) achieves its minimum value equal to $1/(4\pi)$.

For simplicity let us now assume that the noise in the detector is *white* with *two-sided* spectral density $N \cong 1$. Then the matrices M, $\mathsf{F}_{(A)}$, and $\mathsf{S}_{(AB)}$ introduced in Section 6.1.2 (where $\mathcal{A} = 1, 2$ with $\xi_1 \equiv t_0$ and $\xi_2 \equiv \tau$) can easily be calculated. Because the signal (6.163) depends on only one amplitude parameter A_0, all the matrices M, $\mathsf{F}_{(A)}$, and $\mathsf{S}_{(AB)}$ reduce to numbers (which can be treated as 1×1 matrices). The matrix (number) M defined in Eq. (6.17) equals

$$\mathsf{M}(\boldsymbol{\xi}) = (h(t; \boldsymbol{\xi})|h(t; \boldsymbol{\xi})) \cong 1. \tag{6.170}$$

The matrices (numbers) $\mathsf{F}_{(A)}$ introduced in Eq. (6.21a) read

$$\mathsf{F}_{(1)} = \mathsf{F}_{(1)} \cong 0. \tag{6.171}$$

Finally, the matrices (numbers) $\mathsf{S}_{(AB)}$ defined in Eqs. (6.21b) are equal to

$$\mathsf{S}_{(11)} \cong \frac{1}{4\tau^2}, \quad \mathsf{S}_{(12)} = \mathsf{S}_{(21)} \cong 0, \quad \mathsf{S}_{(22)} \cong \frac{1}{2\tau^2}. \tag{6.172}$$

Making use of Eqs. (6.170)–(6.172) we easily compute the signal-to-noise ratio ρ and the Fisher matrix Γ for the signal (6.163). The signal-to-noise ratio is given by [see Eq. (6.19)]

$$\rho \cong A_0, \tag{6.173}$$

and the Fisher matrix is equal to [see Eqs. (6.23)]

$$\Gamma \cong \begin{pmatrix} 1 & 0 & 0 \\ 0 & \dfrac{1}{4}\dfrac{A_0^2}{\tau^2} & 0 \\ 0 & 0 & \dfrac{1}{2}\dfrac{A_0^2}{\tau^2} \end{pmatrix}. \tag{6.174}$$

The lower bounds on the rms errors of the ML estimators of the parameters A_0, t_0, and τ are thus given by

$$\sigma(A_0) \cong 1, \quad \sigma(t_0) \cong 2\frac{\tau}{A_0}, \quad \sigma(\tau) \cong \sqrt{2}\frac{\tau}{A_0}. \tag{6.175}$$

Let us also note that for the signal we consider here the normalized projected Fisher matrix $\overline{\overline{\Gamma}}$ [defined in Eq. (6.28)] and the reduced Fisher matrix $\widetilde{\Gamma}$ [defined in Eq. (6.31)] coincide and they are simply related with the projected Fisher matrix $\bar{\bar{\Gamma}}$ [defined in Eq. (6.27)]. All these three matrices are equal

$$\frac{\bar{\Gamma}}{A_0^2} = \bar{\bar{\Gamma}} = \tilde{\Gamma} \cong \begin{pmatrix} \dfrac{1}{4}\dfrac{1}{\tau^2} & 0 \\ 0 & \dfrac{1}{2}\dfrac{1}{\tau^2} \end{pmatrix}. \tag{6.176}$$

The logarithm of the likelihood ratio for the signal (6.163) reads [see Eqs. (6.6) and (6.11)]

$$\ln \Lambda[x; A_0, \boldsymbol{\xi}] = A_0 \left(x(t)|h(t;\boldsymbol{\xi}) \right) - \frac{1}{2} A_0^2 \left(h(t;\boldsymbol{\xi})|h(t;\boldsymbol{\xi}) \right)$$

$$\cong A_0 \left(x(t)|h(t;\boldsymbol{\xi}) \right) - \frac{1}{2} A_0^2, \tag{6.177}$$

where we have used that $(h|h) \cong 1$ [see Eq. (6.170)]. The ML estimator $\hat{A}_0$ of the amplitude A_0 is thus equal to

$$\hat{A}_0 \cong \left(x(t)|h(t;\boldsymbol{\xi}) \right). \tag{6.178}$$

Substituting this into Eq. (6.177) we get the $\mathcal{F}$-statistic:

$$\mathcal{F}[x; \boldsymbol{\xi}] = \ln \Lambda[x; \hat{A}_0, \boldsymbol{\xi}] \cong \frac{1}{2} \left(x(t)|h(t;\boldsymbol{\xi}) \right)^2. \tag{6.179}$$

6.3 Network of detectors

Several gravitational-wave detectors can observe gravitational waves coming from the same source. For example a network of bar detectors can observe a gravitational-wave burst from the same supernova explosion or a network of laser interferometers can detect the inspiral of the same compact binary system. The space-borne LISA detector can be considered as a network of three detectors, which can make three independent measurements of the same gravitational-wave signal. Simultaneous observations are also possible among different types of detectors, for example a search for supernova bursts can be performed simultaneously by bar and interferometric detectors [192].

We shall consider the general case of a network of N detectors. Let thus x_I $(I = 1, \ldots, N)$ be the data collected by the Ith detector. The data x_I is a sum of the noise n_I and eventually a gravitational-wave signal s_I registered in the Ith detector (so we assume that the noise in all detectors is additive),

$$x_I(t) = n_I(t) + s_I(t), \quad I = 1, \ldots, N. \tag{6.180}$$

We collect all the data streams, all the noises, and all the gravitational-wave signals into column vectors (each with N components) denoted
respectively by $\mathbf{x}$, $\mathbf{n}$, and $\mathbf{s}$,

$$\mathbf{x}(t) := \begin{pmatrix} x_1(t) \\ x_2(t) \\ \vdots \\ x_N(t) \end{pmatrix}, \quad \mathbf{n}(t) := \begin{pmatrix} n_1(t) \\ n_2(t) \\ \vdots \\ n_N(t) \end{pmatrix}, \quad \mathbf{s}(t) := \begin{pmatrix} s_1(t) \\ s_2(t) \\ \vdots \\ s_N(t) \end{pmatrix}. \tag{6.181}$$

We can thus concisely write

$$\mathbf{x}(t) = \mathbf{n}(t) + \mathbf{s}(t). \tag{6.182}$$

Let as also assume that each noise n_I is a stationary, Gaussian, and continuous stochastic process with zero mean. Then the log likelihood function
for the whole network is given by

$$\ln \Lambda[\mathbf{x}] = (\mathbf{x}|\mathbf{s}) - \frac{1}{2}(\mathbf{s}|\mathbf{s}), \tag{6.183}$$

where the scalar product $(\cdot | \cdot)$ is defined as

$$(\mathbf{x}|\mathbf{y}) := 4 \Re \int_0^\infty \tilde{\mathbf{x}}^\mathsf{T} \cdot \mathbf{S}^{-1} \cdot \tilde{\mathbf{y}}^* \, df. \tag{6.184}$$

Here $\mathbf{S}$ is the square $N \times N$ matrix which is the *one-sided* cross spectral
density matrix of the noises in the detectors. It is defined through the
relation (here E as usual denotes the expectation value)

$$\mathrm{E}\big\{\tilde{\mathbf{n}}(f) \cdot \tilde{\mathbf{n}}(f')^{*\mathsf{T}}\big\} = \frac{1}{2}\delta(f - f')\,\mathbf{S}(f). \tag{6.185}$$

The analysis is greatly simplified if the cross spectrum matrix $\mathbf{S}$ is
diagonal. This means that the noises in various detectors are uncorrelated. This is the case when the detectors of the network are in widely
separated locations like for example the two LIGO detectors. However
this assumption is not always satisfied. An important case is LISA detector, where the noises of the three independent responses are correlated.
Nevertheless for the case of LISA one can find a set of three combinations
for which the noises are uncorrelated [193, 194]. When the cross spectrum
matrix is diagonal, the optimum $\mathcal{F}$-statistic is just the sum of $\mathcal{F}$-statistics
in each detector.

Derivation of the likelihood function for an arbitrary network of detectors can be found in [195]. Applications of optimal filtering for the special
cases of observations of coalescing binaries by networks of ground-based
detectors are given in [22, 196, 197], and for the case of stellar-mass binaries observed by LISA detector in [32, 33]. A least-square fit solution for
the estimation of sky location of a source of gravitational waves by a

network of detectors for the case of a broadband burst was obtained in [198].

There is also another important method for analyzing the data from a network of detectors – search for coincidences of events among detectors. This analysis is particularly important when we search for supernovae bursts the waveforms of which are not very well known. Such signals can be easily mimicked by non-Gaussian behavior of the detector noise. The idea is to filter the data optimally in each of the detectors and obtain candidate events. Then one compares parameters of candidate events, like for example times of arrivals of the bursts, among the detectors in the network. This method is widely used in the search for supernovae by networks of bar detectors [199].

6.3.1 Chirp signal from an inspiraling compact binary

In this section we consider detection of the chirp signal by a network of N Earth-based detectors. We assume that the noises in the detectors are independent, so the cross spectrum matrix [defined through Eq. (6.185)] is diagonal and the likelihood ratio for the network is equal to the product of likelihood ratios for the individual detectors.

With a single detector we can estimate the following four parameters of the chirp waveform: its time-of-arrival, the chirp mass, amplitude of the waveform, and its initial phase. The waveform's amplitude and the initial phase depend on several "astrophysical" parameters: position of the binary in the sky, inclination and orientation of its orbital plane, and distance to the binary. In order to determine all these astrophysical parameters we need a network of detectors and we need to solve the *inverse problem*, i.e. the problem to estimate parameters of the binary by processing the data from all detectors. It was shown [17] that we need at least *three* detectors to solve the inverse problem for inspiraling binaries.

There are two ways to process the data from the network in order to solve the inverse problem. We can process the data separately in each detector, obtain the estimators of the parameters of each waveform, and then implement a least-squares procedure to obtain estimators of the astrophysical parameters. Alternatively we can process simultaneously data streams from all the detectors to obtain directly estimators of all the astrophysical parameters.

We first need to choose a common coordinate system (x_n, y_n, z_n) for the whole network. We adapt here the coordinate system introduced in Ref. [198] (see also [196, 200]), which distinguishes some three out of all N detectors of the network. Let us label these distinguished detectors by 1, 2, and 3. The (x_n, y_n) plane of the coordinate system coincides with the plane defined by the positions of the detectors 1, 2, and 3. The origin is at the position of the detector 1. The positive x_n semi-axis passes through the positions of the detectors 1 and 2, the direction of the y_n axis is chosen in such a way that the y_n component of the vector joining the detector 1 with the detector 3 is *positive*, and the direction of the z_n axis is chosen to form a right-handed coordinate system.

Let $\mathbf{r}_{12}, \ldots, \mathbf{r}_{1N}$ be the vectors connecting the detector 1 with the detector 2, ..., N, respectively. In the network coordinate system just introduced their components (expressed in time units) read

$$\mathbf{r}_{12}/c = (a, 0, 0), \tag{6.186a}$$

$$\mathbf{r}_{13}/c = (b_1, b_2, 0), \tag{6.186b}$$

$$\mathbf{r}_{1J}/c = (c_{1J}, c_{2J}, c_{3J}), \quad J = 4, \ldots, N, \tag{6.186c}$$

where always $a > 0$ and $b_2 > 0$. We also have

$$\mathbf{r}_{1J} = \mathbf{r}_J - \mathbf{r}_1, \quad J = 2, \ldots, N, \tag{6.187}$$

where $\mathbf{r}_I$ is the vector connecting the center of the Earth with the Ith detector. If $\mathbf{d}_I$ denotes the unit vector in the direction of the vector $\mathbf{r}_I$, then

$$\mathbf{r}_I = R_\oplus \mathbf{d}_I, \tag{6.188}$$

where $R_\oplus$ stands for the radius of the Earth. Combining Eqs. (6.186)–(6.188) together we obtain the following formulae for the coefficients a, b_1, and b_2:

$$a = \frac{R_\oplus}{c} |\mathbf{d}_2 - \mathbf{d}_1|, \tag{6.189a}$$

$$b_1 = \left(\frac{R_\oplus}{c}\right)^2 \frac{1}{a} (\mathbf{d}_2 - \mathbf{d}_1) \cdot (\mathbf{d}_3 - \mathbf{d}_1), \tag{6.189b}$$

$$b_2 = \sqrt{\left(\frac{R_\oplus}{c}\right)^2 |\mathbf{d}_3 - \mathbf{d}_1|^2 - b_1^2}. \tag{6.189c}$$

Let us now introduce the Euler angles θ, ϕ, and ψ (where θ and ϕ give the incoming direction of the wave and ψ is the wave's polarization angle), which determine the orientation of the network coordinate system with respect to the wave TT coordinate system. Then the detectors' beam-pattern functions F_{I+} and $F_{I\times}$ are of the form

$$\begin{aligned} F_{I+}(\theta, \phi, \psi) &= a_I(\theta, \phi) \cos 2\psi + b_I(\theta, \phi) \sin 2\psi, \\ F_{I\times}(\theta, \phi, \psi) &= b_I(\theta, \phi) \cos 2\psi - a_I(\theta, \phi) \sin 2\psi, \end{aligned} \quad I = 1, \ldots, N, \tag{6.190}$$

where a_I and b_I also depend on the position of the detector on Earth.

Let $\mathbf{n}_0$ be the unit vector pointing to the position of the gravitational-wave source in the sky. In the network coordinate system $(x_\mathrm{n}, y_\mathrm{n}, z_\mathrm{n})$ it has components

$$\mathbf{n}_0 = (-\sin\theta\sin\phi, \sin\theta\cos\phi, -\cos\theta). \tag{6.191}$$

Let us finally introduce the $N-1$ independent time delays $\tau_{12}, \ldots, \tau_{1N}$ between the detectors $(1,2), \ldots, (1,N)$, respectively. Then we have

$$\tau_{1J} := t_{\mathrm{i}1} - t_{\mathrm{i}J} = \frac{\mathbf{n}_0 \cdot \mathbf{r}_{1J}}{c}, \quad J = 2, \ldots, N, \tag{6.192}$$

where $t_{\mathrm{i}J}$ is the time of arrival of some gravitational-wave signal into the Jth detector. Making use of Eqs. (6.186) and (6.191)–(6.192) one can express the time delays τ_{1J} $(J = 2, \ldots, N)$ as functions of the angles θ and ϕ:

$$\tau_{12}(\theta, \phi) = -a\sin\theta\sin\phi, \tag{6.193a}$$

$$\tau_{13}(\theta, \phi) = \sin\theta(b_2\cos\phi - b_1\sin\phi), \tag{6.193b}$$

$$\tau_{1J}(\theta, \phi) = -c_{1J}\sin\theta\sin\phi + c_{2J}\sin\theta\cos\phi - c_{3J}\cos\theta, \quad J = 4, \ldots, N. \tag{6.193c}$$

The response functions s_I $(I = 1, \ldots, N)$ of the detectors in the network to the same gravitational wave described in the TT coordinate system related to the SSB by some polarization functions h_+ and $h_\times$, can be written as

$$s_1(t) = \Theta(t - t_{\mathrm{i}1})\big(F_{+1}h_+(t) + F_{\times 1}h_\times(t)\big), \tag{6.194a}$$

$$s_J(t) = \Theta(t - t_{\mathrm{i}J})\big(F_{+J}h_+(t + \tau_{1J}) + F_{\times J}h_\times(t + \tau_{1J})\big), \quad J = 2, \ldots, N, \tag{6.194b}$$

where the step function Θ is defined in Eq. (6.124) and where we have used the following definitions of the polarization functions h_+ and $h_\times$:

$$h_+(t) := -h_0\frac{1 + \cos^2\iota}{2}\omega_\mathrm{gw}(t)^{2/3}\cos\phi_\mathrm{gw}(t), \tag{6.195a}$$

$$h_\times(t) := -h_0\cos\iota\,\omega_\mathrm{gw}(t)^{2/3}\sin\phi_\mathrm{gw}(t). \tag{6.195b}$$

These definitions together with Eqs. (6.194) mean that the above defined functions h_+ and $h_\times$ describe the gravitational-wave induced metric perturbation computed at the position of the detector number 1 [more precisely: the values of h_+ and $h_\times$ at a time t are equal to the values of the functions h_+ and $h_\times$ given by Eqs. (6.121) and evaluated at time $t + \mathbf{n}_0 \cdot \mathbf{r}_{\mathrm{d}1}/c$, where $\mathbf{r}_{\mathrm{d}1}$ is the vector connecting the SSB and the position of the detector 1, see Eq. (6.123)]. Making use of Eqs. (6.194) and

(6.195), we rewrite the response functions s_I in the following form:

$$s_1(t) = \Theta(t - t_{i1})A_{01}\,\omega_{gw}(t)^{2/3}\cos\left(\bar{\phi}_{gw}(t) - \xi_{01}\right), \tag{6.196a}$$

$$s_J(t) = \Theta(t - t_{iJ})A_{0J}\,\omega_{gw}(t + \tau_{1J})^{2/3}\cos\left(\bar{\phi}_{gw}(t + \tau_{1J}) - \xi_{0J}\right),$$
$$J = 2, \ldots, N, \tag{6.196b}$$

where $\bar{\phi}_{gw}(t) = \phi_{gw}(t) - \phi_i$, $\phi_i = \phi_{gw}(t_{i1})$ is the initial value of the gravi-tational-wave phase. The amplitude parameters A_{0I} are given by [compare Eqs. (6.156)]

$$A_{0I}(R, \mathcal{M}, \theta, \phi, \psi, \iota) = \frac{2^{4/3}(G\mathcal{M})^{5/3}}{c^4 R}$$
$$\times \left(\left(F_{+I}(\theta, \phi, \psi)\frac{1 + \cos^2\iota}{2}\right)^2 + (F_{\times I}(\theta, \phi, \psi)\cos\iota)^2\right)^{1/2}, \tag{6.197a}$$

and the new initial phase parameters ξ_{0I} are determined by the relation [compare Eqs. (6.158)]

$$\xi_{0I}(\theta, \phi, \psi, \iota, \phi_i) = -\phi_i + \text{atan}\left(\frac{F_{+I}(\theta, \phi, \psi)(1 + \cos^2\iota)}{2F_{\times I}(\theta, \phi, \psi)\cos\iota}\right). \tag{6.197b}$$

Still another useful representation of the responses s_I reads

$$s_I(t) = A_{0I}\left(h_{cI}(t)\cos\xi_{0I} + h_{sI}(t)\sin\xi_{0I}\right), \tag{6.198}$$

where

$$h_{cI}(t) := \Theta(t - t_{iI})\omega_{gw}(t + \tau_{1I})^{2/3}\cos\left(\bar{\phi}_{gw}(t + \tau_{1I})\right), \tag{6.199a}$$

$$h_{sI}(t) := \Theta(t - t_{iI})\omega_{gw}(t + \tau_{1I})^{2/3}\sin\left(\bar{\phi}_{gw}(t + \tau_{1I})\right). \tag{6.199b}$$

We restrict now to the simplest case of time evolution of the gravita-tional-wave frequency ω_{gw} describing adiabatic inspiral of the circularized binary. Moreover, we take into account only the leading-order radiation-reaction effects. Such situation was considered in Section 2.4.1, and it leads to the following model of time evolution of the gravitational-wave frequency [see Eq. (2.33) there and Eq. (6.134) in Section 6.2.2]:

$$\omega_{gw}(t; t_{i1}, \mathcal{M}) = \omega_i\left(1 - \frac{64(G\mathcal{M})^{5/3}}{5\sqrt[3]{4}c^5}\omega_i^{8/3}(t - t_{i1})\right)^{-3/8}. \tag{6.200}$$

By optimal filtering of the data in each detector we obtain the ML estimators of the chirp mass $\mathcal{M}$ and the time-of-arrival t_i by maximizing the $\mathcal{F}$-statistic, Eq. (6.153), and then we calculate the ML estimators of amplitude A_0 and phase ξ_0 using Eqs. (6.160) and (6.162). We group the parameters estimated in all N detectors of the network into one vector $\boldsymbol{\eta}$ of $4N$ components:

$$\boldsymbol{\eta} = (A_{01}, \xi_{01}, t_{i1}, \mathcal{M}_1, \ldots, A_{0N}, \xi_{0N}, t_{iN}, \mathcal{M}_N), \tag{6.201}$$

where $\mathcal{M}_I$ is the chirp mass measured in the Ith detector. Note that because of the noise the chirp mass estimated in each detector will be different even though all the detectors are observing the same binary system. Our aim is to estimate the astrophysical parameters of the gravitational waves emitted by the inspiraling binary. They can be grouped into the eight-dimensional vector $\boldsymbol{\mu}$:

$$\boldsymbol{\mu} = (R, t_{\mathrm{a}}, \mathcal{M}, \theta, \phi, \psi, \iota, \phi_{\mathrm{i}}), \tag{6.202}$$

where $t_{\mathrm{a}} \equiv t_{i1}$ is the one independent time of arrival of the signal (which is conventionally equal to the time of arrival of the signal to the detector number 1).

The parameters $\boldsymbol{\mu}$ and $\boldsymbol{\eta}$ are related to each other by means of the following set of $4N$ equations:

$$A_{0I} = A_{0I}(R, \mathcal{M}, \theta, \phi, \psi, \iota), \quad I = 1, \ldots, N, \tag{6.203a}$$

$$\xi_{0I} = \xi_{0I}(\theta, \phi, \psi, \iota, \phi_{\mathrm{i}}), \quad I = 1, \ldots, N, \tag{6.203b}$$

$$t_{i1} = t_{\mathrm{a}}, \quad t_{iJ} = t_{\mathrm{a}} - \tau_{1J}(\theta, \phi), \quad J = 2, \ldots, N, \tag{6.203c}$$

$$\mathcal{M}_I = \mathcal{M}, \quad I = 1, \ldots, N, \tag{6.203d}$$

where the right-hand sides of Eqs. (6.203a) and (6.203b) are given by Eqs. (6.197).

Determination of the position of the source in the sky. By optimal filtering of the data in any three detectors of the network we obtain three amplitudes A_{0I}, three initial phases ξ_{0I}, and also the chirp mass $\mathcal{M}$ of the binary. Then by inverting the six equations (6.197) (for, say, $I = 1, 2, 3$) we can determine the six parameters of the binary: R, θ, ϕ, ψ, ι, and ϕ_{i}. It turns out however that one can obtain much more accurate estimates of the angles θ and ϕ determining the position of the source in the sky from the time delays between the detectors of the network.

Let us first consider a 3-detector network. Then we can use two independent time delays τ_{12} and τ_{13} to determine the position of the gravitational-wave source in the sky. Making use of Eqs. (6.193a) and (6.193b), we compute the cosine and sine functions of the angles θ and ϕ:

$$\sin\theta = \frac{1}{ab_2}\sqrt{\Delta}, \tag{6.204a}$$

$$\cos\theta = \pm\frac{1}{ab_2}\sqrt{(ab_2)^2 - \Delta}, \tag{6.204b}$$

$$\sin\phi = -\frac{b_2\tau_{12}}{\sqrt{\Delta}}, \tag{6.204c}$$

$$\cos\phi = \frac{a\tau_{13} - b_1\tau_{12}}{\sqrt{\Delta}}, \tag{6.204d}$$

where we have introduced

$$\Delta := (b_1\tau_{12} - a\tau_{13})^2 + (b_2\tau_{12})^2. \tag{6.205}$$

The above formulae determine the angle ϕ uniquely, while for the angle θ we have two possible values.

If we have a fourth detector, we can determine the true value of the angle θ by making use of the third independent time delay τ_{14}. It is enough to compute the quantity [see Eq. (6.193c)]

$$\frac{b_2\tau_{12}c_{14} + (b_1\tau_{12} - a\tau_{13})c_{24} - ab_2\tau_{14}}{c_{34}\sqrt{(ab_2)^2 - \Delta}}, \tag{6.206}$$

which is equal to -1 or 1 depending on the sign of the cosine of θ. Thus from the timing of the gravitational-wave signal by four detectors we can determine the position of the binary in the sky uniquely.

Least-squares procedure.   Let us concisely denote the set of Eqs. (6.203) as

$$\boldsymbol{\eta} = \mathbf{f}(\boldsymbol{\mu}), \tag{6.207}$$

where all the three quantities $\boldsymbol{\mu}$, $\boldsymbol{\eta}$, and $\mathbf{f}$ are column vectors with $4N$ components. From filtering the data in individual detectors we only obtain, as a result of the noise in the detectors, some estimators $\hat{\boldsymbol{\eta}}$ of the parameters $\boldsymbol{\eta}$, that always differ from the true values of the parameters by certain random quantities $\boldsymbol{\epsilon}$. We thus have

$$\hat{\eta}_k = \eta_k + \epsilon_k, \quad k = 1, \dots, 4N. \tag{6.208}$$

Those values of the parameters $\boldsymbol{\mu}$ that minimize the function

$$\chi^2 := (\hat{\boldsymbol{\eta}} - \mathbf{f}(\boldsymbol{\mu}))^{\mathsf{T}} \cdot C_{\boldsymbol{\eta}}^{-1} \cdot (\hat{\boldsymbol{\eta}} - \mathbf{f}(\boldsymbol{\mu})), \tag{6.209}$$

are the least-squares (LS) estimators $\hat{\boldsymbol{\mu}}_{\mathrm{LS}}$ of the parameters $\boldsymbol{\mu}$. The matrix $C_{\boldsymbol{\eta}}$ is the covariance matrix for the estimators $\hat{\boldsymbol{\eta}}$. As an approximation to the covariance matrix we can use the inverse of the Fisher matrix for the network of detectors. As noises in the detectors are independent from each other, the Fisher matrix is block-diagonal with each block equal to the Fisher matrix for a given single detector.

Minimization of the function (6.209) leads to a non-linear set of equations that can be solved numerically by an iterative procedure. To perform the iteration we must accept some initial values of the parameters $\boldsymbol{\mu}$. As an initial value for the vector $\boldsymbol{\mu}$ we can use a solution of the set of Eqs. (6.203) with respect to $\boldsymbol{\mu}$. There does not exist a unique way of solving the set of algebraic equations (6.203) with respect to the parameters $\boldsymbol{\mu}$, because the set is overdetermined. One solution can be obtained in the following way. From the three times of arrival t_{iI} we obtain the angle ϕ and the two possible values of the angle θ by means of Eqs. (6.204), then by combining the three equations (6.203a) we compute the angles ψ and ι, choose the true value of the angle θ (out of the two possible values) and after that compute the distance R (see Section IIIC of [196] for details).

The covariance matrix $C_{\boldsymbol{\mu}}$ for the LS estimators of parameters $\boldsymbol{\mu}$ is given by

$$C_{\boldsymbol{\mu}} = (J^{\mathsf{T}} \cdot \Gamma_{\boldsymbol{\eta}} \cdot J)^{-1}, \tag{6.210}$$

where the Jacobi matrix J has components:

$$J_{mn} = \frac{\partial f_m}{\partial \mu_n}, \quad m = 1, \ldots, 4N, \quad n = 1, \ldots, 8. \tag{6.211}$$

All the derivatives $\partial f_m / \partial \mu_n$ can be computed by means of Eqs. (6.203) together with Eqs. (6.193) and (6.197).

We can also obtain the improved estimators $\hat{\boldsymbol{\eta}}_{\mathrm{LS}}$ of the parameters $\boldsymbol{\eta}$ from the LS estimators $\hat{\boldsymbol{\mu}}_{\mathrm{LS}}$ by

$$\hat{\boldsymbol{\eta}}_{\mathrm{LS}} = \mathbf{f}(\hat{\boldsymbol{\mu}}_{\mathrm{LS}}). \tag{6.212}$$

The covariance matrix for the improved estimators is given by

$$\tilde{C}_{\boldsymbol{\eta}} = J \cdot C_{\boldsymbol{\mu}} \cdot J^{\mathsf{T}}. \tag{6.213}$$

Optimal filtering of network data. The procedure presented in the previous section consisted of filtering the data in each detector separately followed by a least-squares procedure to obtain the astrophysical parameters of the binary. Alternatively we can filter the data from all N detectors directly for the astrophysical parameters, taking into account the $N - 1$ constraints relating the angles θ and ϕ (determining the position of the binary in the sky) with time delays between the detectors. To find the optimal procedure we have to solve maximum-likelihood equations directly for the astrophysical parameters and take into account these $N - 1$ constraints. We shall find that the estimates of the distance R and the initial phase ϕ_i can be determined analytically. (We can also find an equation for the angle ι, however this equation is unmanageable). We find that it is optimal to linearly filter the data in each detector. We take the constraints into account in the following way. The filters are parametrized by angles θ and ϕ. For each set of these angles we calculate the time delays. We shift the outputs from the linear filters in the detectors according to the calculated time delays and then add the outputs (with suitable weights). We proceed until we find the maximum of the sum.

The network log likelihood function is given by

$$\ln \Lambda[\mathbf{x}(t); \boldsymbol{\mu}] = \sum_{I=1}^{N} \ln \Lambda_I[x_I(t); \boldsymbol{\mu}]$$

$$= \sum_{I=1}^{N} \left((x_I(t)|s_I(t; \boldsymbol{\mu})) - \frac{1}{2}(s_I(t; \boldsymbol{\mu})|s_I(t; \boldsymbol{\mu})) \right). \quad (6.214)$$

Let us rewrite the response function s_I in the following form:

$$s_I(t; \boldsymbol{\mu}) = \frac{B_{0I}(\boldsymbol{\nu})}{R} \times \Big(h_{cI}(t; \boldsymbol{\nu}) \cos(\chi_{0I}(\boldsymbol{\nu}) - \phi_i)$$

$$+ h_{sI}(t; \boldsymbol{\nu}) \sin(\chi_{0I}(\boldsymbol{\nu}) - \phi_i) \Big), \quad (6.215)$$

where we have singled out the distance R and the initial phase parameter ϕ_i by defining $B_{0I} := A_{0I} R$ and $\chi_{0I} := \xi_{0I} - \phi_i$, and $\boldsymbol{\nu} := (t_a, \mathcal{M}, \theta, \phi, \psi, \iota)$ denotes the remaining six parameters of the binary.

Making use of the stationary phase approximation one can compute the Fourier transform of the functions h_{cI} and h_{sI} and show that

$$C := (h_{cI}|h_{cI}) \cong (h_{sI}|h_{sI}), \quad (h_{cI}|h_{sI}) \cong 0, \quad (6.216)$$

where the quantity C is the same for all detectors (i.e. it does not require the subscript I) and it depends only on the chirp mass $\mathcal{M}$ of the binary. By virtue of the equalities (6.216) the network log likelihood ratio (6.214) for the signals (6.215) can be written in the form

$$\ln \Lambda[\mathbf{x}; R, \phi_i, \boldsymbol{\nu}] = \frac{1}{R} \left(\cos \phi_i \sum_{I=1}^{N} D_I + \sin \phi_i \sum_{I=1}^{N} E_I \right) - \frac{C}{2R^2} \sum_{I=1}^{N} B_{0I}^2, \quad (6.217)$$

where we have introduced some auxiliary notation:

$$D_I := B_{0I}\Big((x_I|h_{cI})\cos\chi_{0I} + (x_I|h_{sI})\sin\chi_{0I}\Big), \tag{6.218a}$$

$$E_I := B_{0I}\Big((x_I|h_{cI})\sin\chi_{0I} - (x_I|h_{sI})\cos\chi_{0I}\Big). \tag{6.218b}$$

Maximization of (6.217) with respect to the distance R gives the following ML estimator of this parameter:

$$\hat{R} = \frac{C\sum_{I=1}^N B_{0I}^2}{\cos\phi_i \sum_{I=1}^N D_I + \sin\phi_i \sum_{I=1}^N E_I}. \tag{6.219}$$

Replacing in (6.217) R by $\hat{R}$ leads to

$$\ln\Lambda[\mathbf{x}; \hat{R}, \phi_i, \boldsymbol{\nu}] = \frac{\Big(\cos\phi_i \sum_{I=1}^N D_I + \sin\phi_i \sum_{I=1}^N E_I\Big)^2}{2C\sum_{I=1}^N B_{0I}^2}. \tag{6.220}$$

The likelihood ratio (6.220) attains its maximal value for the angle ϕ_i determined by the equation

$$\tan\hat{\phi}_i = \frac{\sum_{I=1}^N E_I}{\sum_{I=1}^N D_I}. \tag{6.221}$$

Replacing again in (6.220) the angle ϕ_i by its ML estimator (6.221), we get the likelihood ratio that depends on the remaining six parameters of the binary (collected into the vector $\boldsymbol{\nu}$),

$$\ln\Lambda[\mathbf{x}; \hat{R}, \hat{\phi}_i, \boldsymbol{\nu}] = \frac{\Big(\sum_{I=1}^N D_I\Big)^2 + \Big(\sum_{I=1}^N E_I\Big)^2}{2C\sum_{I=1}^N B_{0I}^2}. \tag{6.222}$$

6.3.2 Periodic signal from a spinning neutron star

Let us assume that a network of N Earth-based detectors is observing the gravitational-wave signal from the same spinning neutron star. The response s_I of the Ith detector ($I = 1, \ldots, N$) has the form given by Eqs. (6.84)–(6.89), i.e. it can be written as

$$s_I(t; \boldsymbol{\theta}) = \sum_{k=1}^4 A_k h_{Ik}(t; \boldsymbol{\xi}), \quad I = 1, \ldots, N, \tag{6.223}$$

where the vector $\boldsymbol{\theta} = (\mathbf{A}, \boldsymbol{\xi})$ collects all parameters of the signal [they split into extrinsic parameters $\mathbf{A} = (A_1, \ldots, A_4)$ defined in Eqs. (6.85) and (6.86) and intrinsic parameters $\boldsymbol{\xi}$]. Since each detector responds to the same gravitational wave, the set of parameters $\boldsymbol{\theta}$ is the same for each detector. But the functions h_{Ik} are different for different detectors, as they depend on the amplitude modulation functions a_I, b_I and on the phase modulation functions ϕ_I. All these functions depend on the location of the detector on Earth. The explicit form of the functions h_{Ik} is as follows [see Eqs. (6.87)]:

$$
\begin{aligned}
h_{I1}(t;\boldsymbol{\xi}) &:= a_I(t;\delta,\alpha)\cos\phi_I(t;\boldsymbol{\xi}), \\
h_{I2}(t;\boldsymbol{\xi}) &:= b_I(t;\delta,\alpha)\cos\phi_I(t;\boldsymbol{\xi}), \\
h_{I3}(t;\boldsymbol{\xi}) &:= a_I(t;\delta,\alpha)\sin\phi_I(t;\boldsymbol{\xi}), \\
h_{I4}(t;\boldsymbol{\xi}) &:= b_I(t;\delta,\alpha)\sin\phi_I(t;\boldsymbol{\xi}),
\end{aligned}
\qquad I = 1,\ldots,N.
\tag{6.224}
$$

We assume that the noises in the detectors are independent, so the cross spectrum matrix for the network is diagonal and the log likelihood ratio for the entire network is the sum of log likelihood ratios for individual detectors, i.e.

$$
\ln\Lambda[\mathbf{x}(t);\boldsymbol{\theta}] \cong 2T_o \sum_{I=1}^{N} \frac{\langle x_I(t)s_I(t;\boldsymbol{\theta})\rangle - \frac{1}{2}\langle s_I(t;\boldsymbol{\theta})^2\rangle}{S_I(f_0)},
\tag{6.225}
$$

where $S_I(f_0)$ is the spectral density of the noise in the Ith detector evaluated for frequency f_0.

It turns out that the ML estimators of the amplitudes, the $\mathcal{F}$-statistic and also the signal-to-noise ratio and the reduced Fisher matrix can be written in the same form as for a single detector if we introduce the following noise-weighted average procedure [191]

$$
\langle \mathbf{q}\rangle_S := \sum_{I=1}^{N} w_I \langle q_I\rangle, \quad \langle \mathbf{q}\,\mathbf{p}\rangle_S := \sum_{I=1}^{N} w_I \langle q_I\,p_I\rangle,
\tag{6.226}
$$

where the weights w_I are defined by

$$
w_I := \frac{\left(S_I(f_0)\right)^{-1}}{S_0^{-1}}, \quad \text{with} \quad \frac{1}{S_0} = \sum_{I=1}^{N} \frac{1}{S_I(f_0)},
\tag{6.227}
$$

and the time averaging operator $\langle\cdot\rangle$ is defined by Eq. (6.91). With these definitions the log likelihood function (6.225) is given by

$$
\ln\Lambda[\mathbf{x}(t);\boldsymbol{\theta}] \cong 2\frac{T_o}{S_0}\left(\langle \mathbf{x}(t)\mathbf{s}(t;\boldsymbol{\theta})\rangle_S - \frac{1}{2}\langle \mathbf{s}(t;\boldsymbol{\theta})^2\rangle_S\right).
\tag{6.228}
$$

Let us collect the amplitude modulation functions a_I, b_I and the signal quadratures h_{Ik} into the column vectors $\mathbf{a}$, $\mathbf{b}$, and $\mathbf{h}_k$ ($k = 1, \ldots, 4$),

$$
\mathbf{a}(t; \delta, \alpha) := \begin{pmatrix} a_1(t; \delta, \alpha) \\ a_2(t; \delta, \alpha) \\ \vdots \\ a_N(t; \delta, \alpha) \end{pmatrix}, \quad \mathbf{b}(t; \delta, \alpha) := \begin{pmatrix} b_1(t; \delta, \alpha) \\ b_2(t; \delta, \alpha) \\ \vdots \\ b_N(t; \delta, \alpha) \end{pmatrix},
$$

$$
\mathbf{h}_k(t; \boldsymbol{\xi}) := \begin{pmatrix} h_{1k}(t; \boldsymbol{\xi}) \\ h_{2k}(t; \boldsymbol{\xi}) \\ \vdots \\ h_{Nk}(t; \boldsymbol{\xi}) \end{pmatrix}, \quad k = 1, \ldots, 4. \tag{6.229}
$$

The ML estimators $\hat{A}_k$ of the amplitudes A_k have exactly the same form as the corresponding values for the single detector with the time average operator $\langle \cdot \rangle$ replaced by a noise-weighted average operator $\langle \cdot \rangle_S$. We thus have [cf. Eqs. (6.100)]

$$
\hat{A}_1 \cong \frac{B \langle \mathbf{x} \mathbf{h}_1 \rangle_S - C \langle \mathbf{x} \mathbf{h}_2 \rangle_S}{D}, \tag{6.230a}
$$

$$
\hat{A}_2 \cong \frac{A \langle \mathbf{x} \mathbf{h}_2 \rangle_S - C \langle \mathbf{x} \mathbf{h}_1 \rangle_S}{D}, \tag{6.230b}
$$

$$
\hat{A}_3 \cong \frac{B \langle \mathbf{x} \mathbf{h}_3 \rangle_S - C \langle \mathbf{x} \mathbf{h}_4 \rangle_S}{D}, \tag{6.230c}
$$

$$
\hat{A}_4 \cong \frac{A \langle \mathbf{x} \mathbf{h}_4 \rangle_S - C \langle \mathbf{x} \mathbf{h}_3 \rangle_S}{D}, \tag{6.230d}
$$

where we have used

$$
A := \langle \mathbf{a}^2 \rangle_S, \quad B := \langle \mathbf{b}^2 \rangle_S, \quad C := \langle \mathbf{ab} \rangle_S, \quad D := AB - C^2. \tag{6.231}
$$

Let us now introduce the data-dependent vectors

$$
\mathbf{F}_a := \begin{pmatrix} \displaystyle\int_0^{T_o} x_1(t)\, a_1(t) \exp[-i\phi_1(t)]\, dt \\ \vdots \\ \displaystyle\int_0^{T_o} x_N(t)\, a_N(t) \exp[-i\phi_N(t)]\, dt \end{pmatrix}, \tag{6.232a}
$$

$$
\mathbf{F}_b := \begin{pmatrix} \displaystyle\int_0^{T_o} x_1(t)\, b_1(t) \exp[-i\phi_1(t)]\, dt \\ \vdots \\ \displaystyle\int_0^{T_o} x_N(t)\, b_N(t) \exp[-i\phi_N(t)]\, dt \end{pmatrix}. \tag{6.232b}
$$

With their help the $\mathcal{F}$-statistic can be written in the form

$$\mathcal{F} \cong \frac{2}{S_0 T_{\mathrm{o}}} \frac{B|\mathbf{F}_a|^2 + A|\mathbf{F}_b|^2 - 2C\Re(\mathbf{F}_a\mathbf{F}_b^*)}{D}. \qquad (6.233)$$

The signal-to-noise ratio ρ and the reduced Fisher information matrix $\tilde{\Gamma}$ are given by formulae analogous to those given in Eqs. (6.106) and (6.120b) for single detector case:

$$\rho^2 \cong h_0^2 \frac{T_{\mathrm{o}}}{S_0} (\alpha_1 A + \alpha_2 B + 2\alpha_3 C), \qquad (6.234\mathrm{a})$$

$$\tilde{\Gamma}_{AB} \cong \frac{\alpha_1 \, \mathbf{m}_{AB}^1 + \alpha_2 \, \mathbf{m}_{AB}^2 + 2\alpha_3 \, \mathbf{m}_{AB}^3}{\alpha_1 A + \alpha_2 B + 2\alpha_3 C}, \qquad (6.234\mathrm{b})$$

where the quantities $\mathbf{m}_{AB}^r$ (in bold) are defined in exactly the same way as the quantities m_{AB}^r [in plain text; they are defined in Eq. (6.118) with the aid of Eqs. (6.116b) and (6.119)] but with the averages $\langle\cdot\rangle$ replaced by averages $\langle\cdot\rangle_S$ and with the spectral density S_0 being the harmonic average defined in Eq. (6.227) [191].

6.3.3 Detection of periodic signals by LISA

In this section we consider the ML detection of almost monochromatic gravitational-wave signals in the data collected by the space-borne LISA detector. Such signals are created by gravitational waves emitted by compact binary systems.

From the LISA detector we obtain three independent data streams. The independence of these data streams means that their cross-spectrum matrix $\mathbf{S}$ has a maximum rank equal to 3. We can therefore combine the data streams in such a way that their cross-spectrum matrix is diagonal [193]. For the so called *Michelson observables* X, Y, and Z (in Appendix D we give the explicit analytic formulae describing the responses of the observables X, Y, and Z to almost periodic gravitational waves from a binary system) the following combinations have the diagonal cross-spectrum matrix:

$$A(t) := \frac{Z(t) - X(t)}{\sqrt{2}}, \qquad (6.235\mathrm{a})$$

$$E(t) := \frac{X(t) - 2Y(t) + Z(t)}{\sqrt{6}}, \qquad (6.235\mathrm{b})$$

$$T(t) := \frac{X(t) + Y(t) + Z(t)}{\sqrt{3}}. \qquad (6.235\mathrm{c})$$

Let us denote the data streams corresponding to the gravitational-wave responses A, E, and T by x_A, x_E, and x_T, respectively, and let their noise spectral densities be S_A, S_E, and S_T, respectively. The data streams and their noises we arrange into the column vectors $\mathbf{x}$ and $\mathbf{n}$,

$$\mathbf{x}(t) := \begin{pmatrix} x_A(t) \\ x_E(t) \\ x_T(t) \end{pmatrix}, \quad \mathbf{n}(t) := \begin{pmatrix} n_A(t) \\ n_E(t) \\ n_T(t) \end{pmatrix}. \tag{6.236}$$

We can thus write

$$\mathbf{x}(t) = \mathbf{n}(t) + \mathbf{s}(t), \quad \text{where} \quad \mathbf{s}(t) := \begin{pmatrix} A(t) \\ E(t) \\ T(t) \end{pmatrix}. \tag{6.237}$$

The gravitational-wave signal, $\mathbf{s}$, we are looking for can be written in the following form:

$$\mathbf{s}(t) = 2\omega L \sin(\omega L) \sum_{k=1}^{4} A_k \, \mathbf{h}_k(t), \quad \mathbf{h}_k(t) := \begin{pmatrix} h_{Ak}(t) \\ h_{Ek}(t) \\ h_{Tk}(t) \end{pmatrix}, \quad k = 1, \ldots, 4,$$

$$\tag{6.238}$$

where ω is the angular frequency of gravitational wave and L is the (approximately constant) armlength of the LISA detector. The component functions h_{Ak}, h_{Ek}, and h_{Tk} can be expressed in terms of the quadratures h_{Xk}, h_{Yk}, and h_{Zk} [given in Eqs. (D.6) of Appendix D] of the Micheslon variables X, Y, and Z:

$$h_{Ak}(t) := \frac{h_{Zk}(t) - h_{Xk}(t)}{\sqrt{2}}, \tag{6.239a}$$

$$h_{Ek}(t) := \frac{h_{Xk}(t) - 2h_{Yk}(t) + h_{Zk}(t)}{\sqrt{6}}, \tag{6.239b}$$

$$h_{Tk}(t) := \frac{h_{Xk}(t) + h_{Yk}(t) + h_{Zk}(t)}{\sqrt{3}}. \tag{6.239c}$$

The logarithm of the likelihood ratio for the vectorial data stream $\mathbf{x}$ of Eq. (6.237) reads

$$\ln \Lambda[\mathbf{x}] = (\mathbf{x}|\mathbf{s}) - \frac{1}{2}(\mathbf{s}|\mathbf{s}) = (x_A|A)_A - \frac{1}{2}(A|A)_A$$

$$+ (x_E|E)_E - \frac{1}{2}(E|E)_E + (x_T|T)_T - \frac{1}{2}(T|T)_T, \tag{6.240}$$

where the scalar products $(\cdot | \cdot)_I$ $(I = A, E, T)$ are defined as

$$(x|y)_I := 4\Re \int_0^\infty \frac{\tilde{x}(f)\,\tilde{y}(f)^*}{S_I(f)}\,df, \quad I = A, E, T. \tag{6.241}$$

We assume that the spectral densities S_A, S_E, and S_T are approximately constant over the frequency bandwidths of the gravitational-wave responses A, E, and T, so we can employ the following approximate equalities:

$$(x|y)_I \cong 2\frac{T_o}{S_I(f_0)}\langle xy \rangle, \quad I = A, E, T, \tag{6.242}$$

where $[0, T_o]$ is the observational interval and the time-averaging operator $\langle \cdot \rangle$ is defined in Eq. (6.91).

It is convenient to introduce a noise-weighted average operator, which allows to write the $\mathcal{F}$-statistic and the ML estimators of the signal's amplitudes in a compact form. For any two vectorial quantities $\mathbf{p}$ and $\mathbf{q}$,

$$\mathbf{p}(t) := \begin{pmatrix} p_A(t) \\ p_E(t) \\ p_T(t) \end{pmatrix}, \quad \mathbf{q}(t) := \begin{pmatrix} q_A(t) \\ q_E(t) \\ q_T(t) \end{pmatrix}, \tag{6.243}$$

where the components (p_A, q_A), (p_E, q_E), and (p_T, q_T) belong to the variable respectively A, E, and T, the noise-weighted average operator $\langle \cdot \rangle_S$ is defined as follows,

$$\langle \mathbf{p}\,\mathbf{q} \rangle_S := w_A \langle p_A\,q_A \rangle + w_E \langle p_E\,q_E \rangle + w_T \langle p_T\,q_T \rangle, \tag{6.244}$$

where the weights w_I $(I = A, E, T)$ are defined by

$$w_I := \frac{S_I^{-1}}{S^{-1}}, \quad I = A, E, T, \quad \text{with} \quad S^{-1} = S_A^{-1} + S_E^{-1} + S_T^{-1}. \tag{6.245}$$

With the definitions introduced above, the log likelihood ratio (6.240) reads

$$\ln \Lambda[\mathbf{x}] \cong 2\frac{T_o}{S}\left(\langle \mathbf{x}\,\mathbf{s} \rangle_S - \frac{1}{2}\langle \mathbf{s}^2 \rangle_S\right). \tag{6.246}$$

The ML estimators $\hat{A}_k$ of the amplitude parameters A_k can be found analytically. The ML equations for $\hat{A}_k$,

$$\frac{\partial \ln \Lambda}{\partial A_k} = 0, \quad k = 1, \ldots, 4, \tag{6.247}$$

can be written, by virtue of Eqs. (6.246) and (6.238), in the form

$$\sum_{l=1}^{4} \mathcal{M}_{kl} A_l = \frac{\mathcal{N}_k}{2\omega L \sin(\omega L)}, \quad k = 1, \ldots, 4, \tag{6.248}$$

where the quantities $\mathcal{M}_{kl}$ and $\mathcal{N}_k$ are given by

$$\mathcal{M}_{kl} := \langle \mathbf{h}_k\, \mathbf{h}_l \rangle_S, \quad \mathcal{N}_k := \langle \mathbf{x}\, \mathbf{h}_k \rangle_S. \tag{6.249}$$

The 4×4 matrix $\mathcal{M}$ made of the numbers $\mathcal{M}_{kl}$ can be approximately written in the following form (see Section IVA of Ref. [32]):

$$\mathcal{M} \cong \begin{pmatrix} U & Q & 0 & P \\ Q & V & -P & 0 \\ 0 & -P & U & Q \\ P & 0 & Q & V \end{pmatrix}, \tag{6.250}$$

where the quantities U, V, Q, and P are defined through equations

$$\langle \mathbf{h}_1\, \mathbf{h}_1 \rangle_S \cong \langle \mathbf{h}_3\, \mathbf{h}_3 \rangle_S = \frac{1}{2}U, \tag{6.251a}$$

$$\langle \mathbf{h}_2\, \mathbf{h}_2 \rangle_S \cong \langle \mathbf{h}_4\, \mathbf{h}_4 \rangle_S = \frac{1}{2}V, \tag{6.251b}$$

$$\langle \mathbf{h}_1\, \mathbf{h}_2 \rangle_S \cong \langle \mathbf{h}_3\, \mathbf{h}_4 \rangle_S = \frac{1}{2}Q, \tag{6.251c}$$

$$\langle \mathbf{h}_1\, \mathbf{h}_4 \rangle_S \cong -\langle \mathbf{h}_2\, \mathbf{h}_3 \rangle_S = \frac{1}{2}P, \tag{6.251d}$$

and where we have taken into account the approximate equalities:

$$\langle \mathbf{h}_1\, \mathbf{h}_3 \rangle \cong \langle \mathbf{h}_2\, \mathbf{h}_4 \rangle \cong 0. \tag{6.252}$$

The analytic solution to Eqs. (6.248) [with the aid of Eq. (6.250)] gives the following expressions for the ML estimators $\hat{A}_k$ of the amplitudes A_k:

$$\hat{A}_1 \cong \frac{V\mathcal{N}_1 - Q\mathcal{N}_2 - P\mathcal{N}_4}{2D\omega L \sin(\omega L)}, \tag{6.253a}$$

$$\hat{A}_2 \cong \frac{-Q\mathcal{N}_1 + U\mathcal{N}_2 + P\mathcal{N}_3}{2D\omega L \sin(\omega L)}, \tag{6.253b}$$

$$\hat{A}_3 \cong \frac{P\mathcal{N}_2 + V\mathcal{N}_3 - Q\mathcal{N}_4}{2D\omega L \sin(\omega L)}, \tag{6.253c}$$

$$\hat{A}_4 \cong \frac{-P\mathcal{N}_1 - Q\mathcal{N}_3 + U\mathcal{N}_4}{2D\omega L \sin(\omega L)}, \tag{6.253d}$$

where $D := UV - P^2 - Q^2$. Replacing in the log likelihood function (6.246) the amplitudes A_k by their ML estimators $\hat{A}_k$ yields the

F-statistic. Using Eqs. (6.249) and (6.253), we find

$$
\mathcal{F} \cong \frac{T_o}{S} \sum_{k=1}^{4} \sum_{l=1}^{4} \left(\mathcal{M}^{-1}\right)_{kl} \mathcal{N}_k \mathcal{N}_l
$$

$$
\cong \frac{T_o}{SD} \Big(V \left(\mathcal{N}_1^2 + \mathcal{N}_3^2\right) + U \left(\mathcal{N}_2^2 + \mathcal{N}_4^2\right)
$$

$$
- 2Q \left(\mathcal{N}_1 \mathcal{N}_2 + \mathcal{N}_3 \mathcal{N}_4\right) - 2P \left(\mathcal{N}_1 \mathcal{N}_4 - \mathcal{N}_2 \mathcal{N}_3\right) \Big). \tag{6.254}
$$

By introducing a number of complex quantities we can write the ML estimators of the amplitudes and the $\mathcal{F}$-statistic in more compact forms. We thus introduce the complex amplitudes

$$
A_u := A_1 + \mathrm{i} A_3, \quad A_v := A_2 + \mathrm{i} A_4, \tag{6.255}
$$

a complex quantity W,

$$
W := Q + \mathrm{i} P, \tag{6.256}
$$

and two more complex and data-dependent functions:

$$
\mathcal{N}_u := \mathcal{N}_1 + \mathrm{i} \mathcal{N}_3, \quad \mathcal{N}_v := \mathcal{N}_2 + \mathrm{i} \mathcal{N}_4. \tag{6.257}
$$

We can then write the ML estimators of the complex amplitudes A_u and A_v in the following form:

$$
\hat{A}_u \cong 2 \frac{V \mathcal{N}_u - W^* \mathcal{N}_v}{2 D \omega L \sin(\omega L)}, \quad \hat{A}_v \cong 2 \frac{U \mathcal{N}_v - W \mathcal{N}_u}{2 D \omega L \sin(\omega L)}, \tag{6.258}
$$

and the $\mathcal{F}$-statistic (6.254) can be written as

$$
\mathcal{F} \cong \frac{T_o}{SD} \Big(V |\mathcal{N}_u|^2 + U |\mathcal{N}_v|^2 - 2\Re \left(W \mathcal{N}_u \mathcal{N}_v^*\right) \Big). \tag{6.259}
$$

6.4 Detection of stochastic signals

In this section we discuss the problem of detecting the stochastic background of gravitational waves by means of Earth-based laser interferometric detectors. Our exposition closely follows Refs. [114, 115] (see also [201, 202, 203]). We consider here the situation (which we expect to hold in reality) when the stochastic gravitational waves we are looking for have amplitudes much smaller than the amplitude of the intrinsic instrumental noise in a gravitational-wave detector. To detect such weak stochastic background one has to correlate the outputs of *two* or *more* detectors. We also assume that the stochastic background is isotropic, unpolarized, stationary, and Gaussian (see Section IIB of [115] for a discussion of these assumptions).

Let us consider the case of two gravitational-wave detectors. Their output data streams x_I ($I = 1, 2$) contain noises n_I and possibly gravitational-wave signals s_I,

$$x_I(t) = n_I(t) + s_I(t), \quad I = 1, 2. \tag{6.260}$$

The noises are assumed to be stationary, zero-mean, and Gaussian stochastic processes with *one-sided* spectral densities S_I ($I = 1, 2$) defined through the relations

$$\mathrm{E}\left\{\tilde{n}_I^*(f)\, \tilde{n}_I(f')\right\} = \frac{1}{2}\delta(f - f')\, S_I(|f|), \quad I = 1, 2. \tag{6.261}$$

Let us recall that one-sided spectral densities are defined only for frequencies $f \geq 0$ and this is the reason why in the right-hand side of the above equation we put the modulus of f in the argument of S_I.

The detectors' responses, s_I, to a stochastic gravitational-wave background are superpositions of responses to waves coming from different directions with different polarizations,

$$s_I(t) = \sum_A \oint_{\mathbb{S}^2} d\Omega\, F_{AI}(\theta, \phi)\, h_A(t, \mathbf{x}_{0I}; \theta, \phi), \quad I = 1, 2, \tag{6.262}$$

where the sum is over two wave polarizations $A = +, \times$, the integration is over Euclidean two-sphere $\mathbb{S}^2$, $h_A(t, \mathbf{x}_{0I}; \theta, \phi)$ represents the waves propagating in the direction defined by the spherical angles (θ, ϕ), F_{AI} are the Ith detector's beam-pattern functions, $\mathbf{x}_{0I}$ is the position vector of a characteristic point of the Ith detector (say the central station of the interferometric detector). The functions F_{+I}, $F_{\times I}$ and the vectors $\mathbf{x}_{0I}$ are actually functions of time due to the Earth's motion with respect to the cosmological rest frame. They can be treated as constants, however, since the velocity of the Earth with respect to the cosmological rest frame is small compared to the speed of light and the distance that the detectors move during their *correlation time* (equal to the light travel time between the detectors) is small (see Ref. [204] for more details).

In correlating data from two widely separated detectors we must take into account a reduction in sensitivity due to the separation time delay between the detectors and the non-parallel alignment of the detector arms. These effects are most apparent in the frequency domain and the *overlap reduction function* γ, first calculated in closed-form by Flanagan ([202], see also Ref. [115]) quantifies these two effects. This is a dimensionless function of frequency f, which is determined by the relative positions and orientations of a pair of detectors. Explicitly,

$$\gamma(f) := \frac{5}{8\pi}\sum_A \oint_{\mathbb{S}^2} d\Omega\, F_{A1}(\theta, \phi)\, F_{A2}(\theta, \phi)\, \exp\left(2\pi i f \frac{\mathbf{n} \cdot \Delta \mathbf{x}}{c}\right), \tag{6.263}$$

where $\mathbf{n}$ is a unit vector describing a direction on the Euclidean two-sphere $\mathbb{S}^2$ [$\mathbf{n}$ is parametrized by the standard angles θ and ϕ and has components given in Eq. (2.81)], $\mathbf{\Delta x} := \mathbf{x}_{01} - \mathbf{x}_{02}$ is the separation vector between the two detector sites, and the sum is over the two polarization states $A = +, \times$ of the wave. The overlap reduction function γ equals unity for coincident and coaligned detectors and it decreases below unity when the detectors are shifted apart (so there is a phase shift between the signals in the two detectors), or rotated out of coalignment (so the detectors are most sensitive to different polarizations). Later in the current section we will see that γ arises naturally when evaluating the expectation value of the product of the gravitational strains at two different detectors, when they are driven by an isotropic and unpolarized stochastic background of gravitational radiation.

To get a better feeling for the meaning of γ, let us look at each term in Eq. (6.263) separately [115]. (i) The overall normalization factor $5/(8\pi)$ is chosen so that for a pair of coincident and coaligned detectors $\gamma(f) = 1$ for all frequencies f. (ii) The sum over polarizations A is appropriate for an unpolarized stochastic background. (iii) The integral over the two-sphere is an isotropic average over all directions $\mathbf{n}$ of the incoming radiation. (iv) The exponential phase factor is the phase shift arising from the time delay between the two detectors for radiation arriving along the direction $\mathbf{n}$. In the limit $f \to 0$, this phase shift also goes to zero, and the two detectors become effectively coincident. (v) The quantity $\sum_A F_{A1} F_{A2}$ is the sum of products of the pattern functions of the two detectors. For coaligned detectors $F_{A1} = F_{A2}$ and

$$\sum_A \oint_{\mathbb{S}^2} d\Omega \, (F_A(\theta, \phi))^2 = \frac{8\pi}{5}. \tag{6.264}$$

In this case the overlap reduction function is equal to 1.

We shall now calculate the signal-to-noise ratio ρ that we can achieve using the filter given by Eq. (3.121) and we shall obtain the *optimal* filter by maximizing the signal-to-noise ratio. Optimal filtering for stochastic background consists of calculating the following cross-correlation between the outputs x_1 and x_2 of the two detectors:

$$\mathcal{C}[x_1, x_2] := \int_0^{T_o} dt \int_0^{T_o} dt' \, x_1(t) \, x_2(t') \, Q(t, t'), \tag{6.265}$$

where $[0, T_o]$ is the observational interval and $Q(t, t')$ is a filter function. Because we are assuming in this section that the statistical properties of the stochastic gravitational-wave background and intrinsic detectors' noises are both stationary, the best choice of the filter function $Q(t, t')$

can depend only upon the time difference $\Delta t := t - t'$. The goal is thus to find the optimal choice of the filter function $Q(t - t') = Q(t, t')$.

The filter function Q falls off rapidly to zero for time delays $\Delta t = t - t'$ whose magnitude is large compared to the light travel time d/c (d is the distance between the detectors) between the two detector sites. Since a typical observation time T_o will be $\gg d/c$, we are justified in changing the limits in one of the integrations in (6.265) to obtain

$$C[x_1, x_2] \cong \int_0^{T_o} \mathrm{d}t \int_{-\infty}^{+\infty} \mathrm{d}t' \, x_1(t) \, x_2(t') \, Q(t - t'). \tag{6.266}$$

This change of limits simplifies the mathematical analysis that follows. We first rewrite Eq. (6.266) in the frequency domain. To do this we express the functions x_1, x_2, and Q by their Fourier transforms $\tilde{x}_1$, $\tilde{x}_2$, and $\tilde{Q}$, respectively. We obtain

$$C[x_1, x_2] \cong \int_{-\infty}^{\infty} \mathrm{d}f \int_{-\infty}^{\infty} \mathrm{d}f' \, \delta_{T_o}(f - f') \, \tilde{x}_1^*(f) \, \tilde{x}_2(f') \, \tilde{Q}(f'), \tag{6.267}$$

where δ_{T_o} is the finite-time approximation to the Dirac delta function δ defined by (let us recall that $\operatorname{sinc} x := \sin x / x$)

$$\delta_{T_o}(x) := \int_0^{T_o} \exp(-2\pi i x t) \, \mathrm{d}t = T_o \, \exp(-\pi i x T_o) \operatorname{sinc}(\pi x T_o). \tag{6.268}$$

Note also that for a real Q, $\tilde{Q}(-f) = \tilde{Q}^*(f)$.

We *define* the signal-to-noise ratio ρ for the filter Q as the ratio of the gravitational-wave contribution $\mu_{\mathrm{gw}}(C)$ to the expectation value $\mu(C)$ of the correlation C and its standard deviation $\sigma(C)$:

$$\rho := \frac{\mu_{\mathrm{gw}}(C)}{\sigma(C)}. \tag{6.269}$$

The quantity $\mu_{\mathrm{gw}}(C)$ is defined as follows:

$$\mu_{\mathrm{gw}}(C) := \mathrm{E}\{C[s_1, s_2]\}. \tag{6.270}$$

We further assume that the intrinsic noises in the two detectors are statistically independent of each other and of the gravitational-wave strains, therefore

$$\mu_{\mathrm{gw}}(C) = \mathrm{E}\{C[x_1, x_2]\} = \mu(C). \tag{6.271}$$

From Eq. (6.267) follows

$$\mu_{\mathrm{gw}}(C) \cong \int_{-\infty}^{\infty} \mathrm{d}f \int_{-\infty}^{\infty} \mathrm{d}f' \, \delta_{T_o}(f - f') \, \mathrm{E}\left\{\tilde{s}_1^*(f) \, \tilde{s}_2(f')\right\} \tilde{Q}(f'). \tag{6.272}$$

Making use of Eq. (2.80) it is easy to compute the Fourier transforms $\tilde{s}_I$ of the gravitational-wave responses s_I from Eq. (6.262). They read

$$\tilde{s}_I(f) = \sum_A \oint_{\mathbb{S}^2} d\Omega \, F_{A1}(\theta, \phi) \, \tilde{h}_A(f; \theta, \phi) \, \exp\left(-2\pi i f \frac{\mathbf{n} \cdot \mathbf{x}_{0I}}{c}\right), \quad I = 1, 2.$$
$$(6.273)$$

Making use of (6.273) we compute

$$\mathrm{E}\left\{\tilde{s}_1^*(f) \, \tilde{s}_2(f')\right\} = \sum_A \sum_{A'} \oint_{\mathbb{S}^2} d\Omega \oint_{\mathbb{S}^2} d\Omega' \, \mathrm{E}\{\tilde{h}_A^*(f; \theta, \phi) \, \tilde{h}_{A'}(f'; \theta', \phi')\}$$

$$\times F_{A1}(\theta, \phi) \, F_{A'2}(\theta', \phi') \, \exp\left(2\pi i \left(f \frac{\mathbf{n} \cdot \mathbf{x}_{01}}{c} - f' \frac{\mathbf{n}' \cdot \mathbf{x}_{02}}{c}\right)\right)$$

$$= \frac{1}{5} \delta(f - f') \, S_{\mathrm{gw}}(|f|) \, \gamma(|f|)$$

$$= \frac{3H_0^2}{20\pi^2} \delta(f - f') \, |f|^{-3} \, \Omega_{\mathrm{gw}}(|f|) \, \gamma(|f|), \qquad (6.274)$$

where we have used Eqs. (2.85) and (6.263) to obtain the second equality and Eq. (2.90) to obtain the third equality; S_{gw} [introduced in Eq. (2.85)] denotes here the *one-sided* spectral density of the stochastic gravitational-wave background and Ω_{gw} [defined in Eq. (2.88)] is the ratio of the gravitational-wave energy density per logarithmic frequency interval to the closure density of the universe. Substituting (6.274) into (6.272) yields

$$\mu_{\mathrm{gw}}(\mathcal{C}) \cong \frac{3H_0^2}{20\pi^2} T_{\mathrm{o}} \int_{-\infty}^{\infty} df \, |f|^{-3} \, \Omega_{\mathrm{gw}}(|f|) \, \gamma(|f|) \, \tilde{Q}(f). \qquad (6.275)$$

The factor of T_{o} on the right-hand side arises from evaluating $\delta_{T_{\mathrm{o}}}(0) = T_{\mathrm{o}}$. To calculate the standard deviation $\sigma(\mathcal{C})$,

$$\sigma(\mathcal{C})^2 = \mathrm{E}\left\{\mathcal{C}[x_1, x_2]^2\right\} - \left(\mathrm{E}\left\{\mathcal{C}[x_1, x_2]\right\}\right)^2, \qquad (6.276)$$

we will use the assumption that the intrinsic noises in the two detectors have much larger magnitudes than the stochastic gravitational-wave background. We can thus replace in (6.276) the outputs x_I by pure noises

n_I and compute

$$\sigma(\mathcal{C})^2 \cong \mathrm{E}\{\mathcal{C}[n_1, n_2]^2\}$$

$$\cong \int_{-\infty}^{\infty} \mathrm{d}f \int_{-\infty}^{\infty} \mathrm{d}f' \int_{-\infty}^{\infty} \mathrm{d}k \int_{-\infty}^{\infty} \mathrm{d}k' \, \delta_{T_o}(f - f') \, \delta_{T_o}(k - k')$$

$$\times \mathrm{E}\{\tilde{n}_1^*(f) \, \tilde{n}_2(f') \, \tilde{n}_1^*(k) \, \tilde{n}_2(k')\} \, \tilde{Q}(f') \, \tilde{Q}(k')$$

$$\cong \int_{-\infty}^{\infty} \mathrm{d}f \int_{-\infty}^{\infty} \mathrm{d}f' \int_{-\infty}^{\infty} \mathrm{d}k \int_{-\infty}^{\infty} \mathrm{d}k' \, \delta_{T_o}(f - f') \, \delta_{T_o}(k - k')$$

$$\times \mathrm{E}\{\tilde{n}_1^*(f) \, \tilde{n}_1(-k)\} \, \mathrm{E}\{\tilde{n}_2^*(-f') \, \tilde{n}_2(k')\} \, \tilde{Q}(f') \, \tilde{Q}(k'), \quad (6.277)$$

where we have used the statistical independence and reality of the noises n_I to obtain the last line. Using the definition (6.261) of the spectral densities S_I of the noises in the two detectors, Eq. (6.277) yields

$$\sigma(\mathcal{C})^2 \cong \frac{1}{4} \int_{-\infty}^{\infty} \mathrm{d}f \int_{-\infty}^{\infty} \mathrm{d}f' \, \delta_{T_o}(f - f') \, \delta_{T_o}(f' - f)$$

$$\times S_1(|f|) \, S_2(|f'|) \, \tilde{Q}(f') \, \tilde{Q}^*(f')$$

$$\cong \frac{T_o}{4} \int_{-\infty}^{\infty} \mathrm{d}f \, S_1(|f|) \, S_2(|f|) \, |\tilde{Q}(f)|^2, \quad (6.278)$$

where we have replaced one of the finite-time delta functions δ_{T_o} by an ordinary Dirac delta function δ, and evaluated the other at $f = f'$ to obtain the last line.

The problem now is to find the filter function $\tilde{Q}$ that *maximizes* the signal-to-noise ratio (6.269) with $\mu_{\mathrm{gw}}(\mathcal{C})$ and $\sigma(\mathcal{C})$ obtained above in Eqs. (6.275) and (6.278), respectively. The solution is simple if we first introduce a *positive-definite inner product* (x, y) for any pair of complex-valued functions x and y [115]. The product is defined by

$$(x, y) := \int_{-\infty}^{\infty} \mathrm{d}f \, x^*(f) \, y(f) \, S_1(|f|) \, S_2(|f|). \quad (6.279)$$

For any complex number λ, $(x, y_1 + \lambda y_2) = (x, y_1) + \lambda(x, y_2)$. In addition, $(x, y) = (y, x)^*$ and, since $S_I(|f|) > 0$, $(x, x) \geq 0$ and $(x, x) = 0$ if and only if $x(f) = 0$. Thus (x, y) satisfies all of the properties of an ordinary dot product of vectors in a three-dimensional Euclidean space. In terms of this inner product the mean value $\mu_{\mathrm{gw}}(\mathcal{C})$ and the variance $\sigma(\mathcal{C})^2$ can be written as

$$\mu_{\mathrm{gw}}(\mathcal{C}) = \frac{3H_0^2}{20\pi^2} T_o \left(\tilde{Q}, \frac{\gamma(|f|) \, \Omega_{\mathrm{gw}}(|f|)}{|f|^3 \, S_1(|f|) \, S_2(|f|)} \right), \quad (6.280\mathrm{a})$$

$$\sigma(\mathcal{C})^2 \cong \frac{T_o}{4} (\tilde{Q}, \tilde{Q}). \quad (6.280\mathrm{b})$$

The problem is to choose $\tilde{Q}$ so that it maximizes the signal-to-noise ratio or, equivalently, the squared signal-to-noise ratio,

$$\rho^2 = \frac{\mu_{\mathrm{gw}}(\mathcal{C})^2}{\sigma(\mathcal{C})^2} \simeq \left(\frac{3H_0^2}{10\pi^2}\right)^2 T_{\mathrm{o}} \frac{\left(\tilde{Q}, \dfrac{\gamma(|f|)\,\Omega_{\mathrm{gw}}(|f|)}{|f|^3\,S_1(|f|)\,S_2(|f|)}\right)^2}{(\tilde{Q}, \tilde{Q})}. \qquad (6.281)$$

Suppose we are given a fixed three-dimensional vector $\mathbf{y}$, and are asked to find the three-dimensional vector $\mathbf{x}$ that maximizes the ratio $(\mathbf{x} \cdot \mathbf{y})^2/(\mathbf{x} \cdot \mathbf{x})$. Since this ratio is proportional to the squared cosine of the angle between the two vectors, it is maximized by choosing $\mathbf{x}$ to point in the same direction as $\mathbf{y}$. The problem of maximizing (6.281) is identical. The solution is thus

$$\tilde{Q}(f) = \lambda \frac{\gamma(|f|)\,\Omega_{\mathrm{gw}}(|f|)}{|f|^3\,S_1(|f|)\,S_2(|f|)}, \qquad (6.282)$$

where λ is a real constant normalization factor.

We see that the optimal filter $\tilde{Q}$ depends, through the overlap reduction function γ, on the locations and orientations of the two detectors, it also depends on the spectral densities of intrinsic noises in the detectors and the spectrum Ω_{gw} of the stochastic gravitational-wave background. The function Ω_{gw} we do not know a priori. This means that in data analysis we need to use a *set* of optimal filters which somehow parametrize all reasonable functional forms of the spectrum Ω_{gw}. For example, within the bandwidth of interest for the ground-based interferometers, it is reasonable to assume that the spectrum is given by a power-law [115]

$$\Omega_{\mathrm{gw}}(f) = \Omega_\alpha f^\alpha, \quad \text{where} \quad \Omega_\alpha = \text{const.} \qquad (6.283)$$

We could then construct a set of optimal filters $\tilde{Q}_\alpha$ (say, for $\alpha = -4$, $-7/2, \cdots, 7/2, 4$). In addition we can choose the normalization constants λ_α such that the following condition is fulfilled [115]:

$$\mu_{\mathrm{gw}}(\mathcal{C}) = \Omega_\alpha T_{\mathrm{o}}. \qquad (6.284)$$

With this choice of normalization the optimal filter functions $\tilde{Q}_\alpha$ do not depend on the constants Ω_α. They are thus completely specified by the value of the exponent α, the overlap reduction function γ, and the noise power spectra S_1, S_2 of the two detectors. When the normalization condition (6.284) is fulfilled, the explicit form of the filter functions $\tilde{Q}_\alpha$ read

$$\tilde{Q}_\alpha(f) = \frac{20\pi^2}{3H_0^2} \left(\frac{\gamma(|f|)\,|f|^{\alpha-3}}{S_1(|f|)\,S_2(|f|)}, \frac{\gamma(|f|)\,|f|^{\alpha-3}}{S_1(|f|)\,S_2(|f|)}\right)^{-1} \frac{\gamma(|f|)\,|f|^{\alpha-3}}{S_1(|f|)\,S_2(|f|)}. \qquad (6.285)$$

In data analysis we would analyze the outputs of the two detectors for each of these filters separately.

To calculate the maximized value of the signal-to-noise ratio for a given pair of detectors, we substitute (6.282) into (6.281). Taking the square root gives

$$\rho \cong \frac{3H_0^2}{10\pi^2} \sqrt{T_0 \int_{-\infty}^{\infty} \frac{\gamma^2(|f|)\,\Omega_{\text{gw}}^2(|f|)}{f^6\,S_1(|f|)\,S_2(|f|)}\,\mathrm{d}f}. \qquad (6.286)$$

7

Data analysis tools

In this chapter we describe a number of data analysis tools needed to perform a search for one important class of gravitational-wave signals – periodic signals. Typical sources for such signals are rotating neutron stars. These tools were primarily developed for the analysis of real data from the resonant bar detectors EXPLORER and NAUTILUS [190, 205], however they can be applied to a search for periodic signals in a narrow frequency band (typically of a few hertz bandwidth) in data from any detector. Moreover the methods presented in this chapter can be used in a search for gravitational-wave signals from white-dwarf binaries in the planned LISA detector [32]. The methodology outlined here can also be applied to search for other types of deterministic gravitational-wave signals.

In Section 7.1 we introduce an approximate linear model of a periodic gravitational-wave signal. For this model the Fisher matrix is constant. In Section 7.2 we present the construction of a grid of templates using the linear signal model approximation. In Section 7.3 we introduce several numerical algorithms that increase the computational speed of a search and its accuracy. In Section 7.4 we present several tools for the analysis of the candidate signals obtained in a search.

7.1 Linear signal model

Let us consider the following gravitational-wave signal

$$s(t, \boldsymbol{\theta}) = A_0 \cos[\phi(t, \phi_0, \boldsymbol{\xi})], \qquad (7.1)$$

$$\phi(t, \phi_0, \boldsymbol{\xi}) = \sum_{k=0}^{M} \xi_k m_k(t) + \phi_0, \qquad (7.2)$$

where A_0 and ϕ_0 are the constant amplitude and the initial phase, respectively. The vector $\boldsymbol{\theta}$ collects all the signal's parameters, $\boldsymbol{\theta} := (A_0, \phi_0, \boldsymbol{\xi})$, with the vector $\boldsymbol{\xi}$ comprising the $M + 1$ parameters of the signal's phase. The above signal model is called *linear* because it has the property that its phase $\phi(t)$ is a linear function of the parameters. Assuming that the signal (7.1) is buried in a stationary Gaussian noise and that the spectral density is constant and equal to S_0 over the bandwidth of the signal, the $\mathcal{F}$-statistic can approximately be written as

$$\mathcal{F} \cong \frac{2}{S_0 T_o} \left| \left\langle x(t) \exp \left(-\mathrm{i} \sum_{k=0}^{M} \xi_k m_k(t) \right) \right\rangle \right|^2, \tag{7.3}$$

where $x(t)$ are the data, T_o is the observation time and $\langle \cdot \rangle$ is the averaging operator defined by Eq. (6.91).

Using Eqs. (6.18) and (6.20) we can compute the optimal signal-to-noise ratio ρ and the components of the Fisher information matrix Γ for the signal (7.1). We assume that the phase $\phi(t)$ is a rapidly changing function of time, therefore the following equations are approximately fulfilled:

$$\langle \cos^2(\phi(t, \phi_0, \boldsymbol{\xi})) m_k(t) m_l(t) \rangle \cong \frac{1}{2} \langle m_k(t) m_l(t) \rangle, \tag{7.4a}$$

$$\langle \sin^2(\phi(t, \phi_0, \boldsymbol{\xi})) m_k(t) m_l(t) \rangle \cong \frac{1}{2} \langle m_k(t) m_l(t) \rangle, \tag{7.4b}$$

$$\langle \cos(\phi(t, \phi_0, \boldsymbol{\xi})) \sin(\phi(t, \phi_0, \boldsymbol{\xi})) m_k(t) m_l(t) \rangle \cong 0. \tag{7.4c}$$

Making use of Eqs. (7.4) we find that the signal-to-noise ratio ρ equals

$$\rho \cong A_0 \sqrt{\frac{T_o}{S_0}}, \tag{7.5}$$

and the components of the Fisher information matrix are

$$\Gamma_{A_0 A_0} \cong \frac{\rho^2}{A_0^2}, \quad \Gamma_{\phi_0 \phi_0} \cong \rho^2, \tag{7.6a}$$

$$\Gamma_{A_0 \phi_0} \cong \Gamma_{A_0 \xi_k} \cong 0, \quad k = 0, \ldots, M, \tag{7.6b}$$

$$\Gamma_{\phi_0 \xi_k} \cong \rho^2 \langle m_k \rangle, \quad k = 0, \ldots, M, \tag{7.6c}$$

$$\Gamma_{\xi_k \xi_\ell} \cong \rho^2 \langle m_k m_\ell \rangle, \quad k, \ell = 0, \ldots, M. \tag{7.6d}$$

We see that the Fisher matrix above does not depend on the values of the phase parameters. This is a consequence of the linearity of the phase in the parameters. There is only a simple dependence on the amplitude parameter contained in the signal-to-noise ratio ρ.

The components of the reduced Fisher matrix $\tilde{\Gamma}$ [see Eq. (6.31)] for the case of the linear model are given by

$$\tilde{\Gamma}_{k\ell} = \langle m_k m_\ell \rangle - \langle m_k \rangle \langle m_\ell \rangle; \qquad (7.7)$$

they are constants, independent of the values of the signal parameters. The autocovariance function [defined by Eq. (6.37)] approximately takes the form

$$\mathcal{C}(\tau) \cong \left\langle \cos \left(\sum_k \tau_k\, m_k(t) \right) \right\rangle^2 + \left\langle \sin \left(\sum_k \tau_k\, m_k(t) \right) \right\rangle^2, \qquad (7.8)$$

where $\tau := \xi - \xi'$. Thus for a linear model its autocovariance function only depends on the difference of the parameters at two points and not on the parameters themselves. The autocovariance $\mathcal{C}$ attains its maximum value equal 1 when $\tau = 0$. Let us consider Taylor expansion of $\mathcal{C}$ around the maximum up to terms quadratic in τ [see Eq. (6.43) for the general case],

$$\mathcal{C}(\tau) \cong 1 + \sum_k \frac{\partial \mathcal{C}(\tau)}{\partial \tau_k}\bigg|_{\tau=0} \tau_k + \frac{1}{2} \sum_{k,\ell} \frac{\partial^2 \mathcal{C}(\tau)}{\partial \tau_k \partial \tau_l}\bigg|_{\tau=0} \tau_k \tau_\ell. \qquad (7.9)$$

As $\mathcal{C}$ attains its maximum for $\tau = 0$, we have

$$\frac{\partial \mathcal{C}(\tau)}{\partial \tau_k}\bigg|_{\tau=0} = 0. \qquad (7.10)$$

Introducing the symmetric matrix G [see Eq. (6.45)] with elements

$$G_{k\ell} := -\frac{1}{2} \frac{\partial^2 \mathcal{C}(\tau)}{\partial \tau_k \partial \tau_\ell}\bigg|_{\tau=0} \qquad (7.11)$$

one easily finds that $G = \tilde{\Gamma}$, where $\tilde{\Gamma}$ is the reduced Fisher matrix given by Eq. (7.7). Thus the Taylor expansion (7.9) can be written as

$$\mathcal{C}(\tau) \cong 1 - \sum_{k,\ell} \tilde{\Gamma}_{k\ell}\, \tau_k \tau_\ell. \qquad (7.12)$$

7.1.1 Examples

Polynomial phase signal. In this case the vector ξ has the components $\xi := (\omega_0, \omega_1, \ldots, \omega_s)$, and the functions $m_k(t)$ are polynomials of degree $k + 1$ in time, $m_k(t) := t^{k+1}/(k + 1)!$ for $k = 0, 1, \ldots, s$. The parameter ω_k is called the kth spin down. A special case of this signal is the monochromatic signal where $s = 0$.

The polynomial signal is a useful model for gravitational waves from a rotating neutron star of a known sky position or for an arbitrary neutron star when the observation time T_o is short compared with 1 day. It can also be a useful model for a neutron star in a binary system when the observation time is short compared with the binary period. An analysis of the Fisher matrix for the polynomial phase model can be found in Appendix C of [206].

Linear model I. This is an approximate model of the response of a ground-based detector to a gravitational wave from a rotating neutron star (see Section 6.2.1). Let us associate the following coordinate system with the SSB reference frame. The x axis of the system is parallel to the x axis of the celestial coordinate system, the z axis is perpendicular to the ecliptic and has direction of the orbital angular momentum vector of the Earth. In this SSB coordinate system the vector $\mathbf{n}_0$ pointing towards the source has the components

$$\mathbf{n}_0 = \begin{pmatrix} 1 & 0 & 0 \\ 0 & \cos\varepsilon & \sin\varepsilon \\ 0 & -\sin\varepsilon & \cos\varepsilon \end{pmatrix} \begin{pmatrix} \cos\alpha\cos\delta \\ \sin\alpha\cos\delta \\ \sin\delta \end{pmatrix}, \qquad (7.13)$$

where ε is the obliquity of the ecliptic, δ and α are respectively the declination and the right ascension of the source. The position vector $\mathbf{r}_d$ of the detector with respect to the SSB has the components

$$\mathbf{r}_d = \begin{pmatrix} R_{\mathrm{ES}}^x \\ R_{\mathrm{ES}}^y \\ 0 \end{pmatrix} + \begin{pmatrix} 1 & 0 & 0 \\ 0 & \cos\varepsilon & \sin\varepsilon \\ 0 & -\sin\varepsilon & \cos\varepsilon \end{pmatrix} \begin{pmatrix} R_{\mathrm{E}}^x \\ R_{\mathrm{E}}^y \\ R_{\mathrm{E}}^z \end{pmatrix}, \qquad (7.14)$$

where $(R_{\mathrm{ES}}^x, R_{\mathrm{ES}}^y, 0)$ are the components of the vector joining the SSB with the center of the Earth in the SSB coordinate system, and $(R_{\mathrm{E}}^x, R_{\mathrm{E}}^y, R_{\mathrm{E}}^z)$ are the components of the vector joining the center of the Earth and the detector's location in the celestial coordinate system.

Let us neglect the component r_{d}^z of the vector $\mathbf{r}_d$ perpendicular to the ecliptic. In this case the scalar product $\mathbf{n}_0 \cdot \mathbf{r}_d(t)$ is given by

$$\mathbf{n}_0 \cdot \mathbf{r}_d(t) = n_0^x r_{\mathrm{d}}^x + n_0^y r_{\mathrm{d}}^y, \qquad (7.15)$$

where explicitly we have

$$n_0^x := \omega_0 \cos\alpha \cos\delta, \qquad (7.16a)$$

$$n_0^y := \omega_0 (\sin\alpha \cos\delta \cos\varepsilon + \sin\delta \sin\varepsilon), \qquad (7.16b)$$

$$r_{\mathrm{d}}^x(t) := R_{\mathrm{ES}}^x(t) + R_{\mathrm{E}}^x(t), \qquad (7.16c)$$

$$r_{\mathrm{d}}^y(t) := R_{\mathrm{ES}}^y(t) + R_{\mathrm{E}}^y(t) \cos\varepsilon. \qquad (7.16d)$$

Thus after neglecting all spin downs in the phase modulation due to motion of the detector with respect to the SSB, the phase of the gravitational-wave signal from a rotating star [see Eq. (6.89)] can be approximated by a linear model with $M = s + 3$ parameters, which we collect into the vector $\boldsymbol{\xi} := (\omega_0, \omega_1, \ldots, \omega_s, \alpha_1, \alpha_2)$ such that $\xi_k := \omega_k$ for $k = 0, 1, \ldots, s$, $\xi_{s+1} := \alpha_1$, and $\xi_{s+2} := \alpha_2$. The known functions of time m_k are: $m_k(t) := t^{k+1}/(k+1)!$ for $k = 0, 1, \ldots, s$, $m_{s+1}(t) := \mu_1(t)$, and $m_{s+2}(t) := \mu_2(t)$. The parameters α_1 and α_2 are defined by

$$\alpha_1 := n_0^y, \tag{7.17a}$$

$$\alpha_2 := n_0^x, \tag{7.17b}$$

and the known functions of time $\mu_1(t)$ and $\mu_2(t)$ are given by

$$\mu_1(t) := r_d^y(t), \tag{7.18a}$$

$$\mu_2(t) := r_d^x(t). \tag{7.18b}$$

We obtain a linear model by neglecting the amplitude modulation of the signal.

Linear model II. This is another approximate model of the response of a ground-based detector to a gravitational wave from a rotating neutron star. The phase modulation of the response due to the motion of the detector with respect to the SSB consists of two contributions, one which comes from the motion of the Earth barycenter with respect to the SSB, and the other which is due to the diurnal motion of the detector with respect to the Earth barycenter. The first contribution has a period of one year and thus for observation times of a few days can be well approximated by a few terms of the Taylor expansion. The second term has a period of 1 sidereal day and to a very good accuracy can be approximated by a circular motion. This leads to a linear phase model with $M = s + 3$ parameters $\boldsymbol{\xi} := (p_0, p_1, \ldots, p_s, A, B)$ such that $\xi_k := p_k$ for $k = 0, 1, \ldots, s$, $\xi_{s+1} := A$, and $\xi_{s+2} := B$. The functions m_k in this model are: $m_k(t) := t^{k+1}/(k+1)!$ for $k = 0, 1, \ldots, s$, $m_{s+1}(t) := \cos(\Omega_r t)$, and $m_{s+2}(t) := \sin(\Omega_r t)$, where Ω_r is the rotational angular velocity of the Earth. The constant parameters A and B are defined by

$$A := \omega_0 \, r_E \cos\phi \cos\delta \cos(\alpha - \phi_r), \tag{7.19}$$

$$B := \omega_0 \, r_E \cos\phi \cos\delta \sin(\alpha - \phi_r), \tag{7.20}$$

where ω_0 is the angular frequency of the gravitational-wave signal, r_E is the Earth radius, ϕ is the detector's geographical latitude, and ϕ_r is the deterministic phase, which defines the position of the Earth in its diurnal motion at $t = 0$. As in linear model I we neglect all the spin downs in the phase modulation due to the motion of the detector with respect to the SSB. Note that the parameters $p_0, p_1, \ldots, p_s$ contain not only the contribution from frequency ω_0 and the spin down parameters but also contributions from coefficients of the Taylor expansion of the orbital motion.

7.2 Grid of templates in the parameter space

To construct a grid of templates in the parameter space we can use a linear model that approximates the signal well. For a linear model the reduced Fisher matrix is independent of the values of the parameters. This allows the construction of a uniform grid with cells that are independent of the values of the parameters. To test whether a linear model is a good approximation one can perform a Monte Carlo simulation comparing the reduced Fisher matrix for a true signal to that of the linear model (see [189]).

7.2.1 Covering problem

The problem of constructing a grid in the parameter space is equivalent to the so called *covering problem* [207]. In the case of a linear model using the Taylor expansion (7.12) the general equation (6.66) takes the form

$$\sum_{k,\ell} \tilde{\Gamma}_{k\ell}\, \tau_k\, \tau_\ell = 1 - \mathcal{C}_0, \tag{7.21}$$

where $\tilde{\Gamma}$ is the reduced Fisher matrix and $\mathcal{C}_0$ is some chosen value of the correlation function. Thus the covering problem in our case is the problem to cover $(M+1)$-dimensional parameter space with equal hyperellipsoids given by Eq. (7.21) in such a way that any point of the space belongs to at least one ellipsoid. Moreover, we look for an optimal covering, i.e. the one having the smallest possible number of grid points per unit volume. The *covering thickness* Θ is commonly defined as the average number of ellipsoids that contain a point in the space. The optimal covering would have minimal possible thickness.

Let us introduce in the parameter space a new set of coordinates $\mathbf{x} = (x_0, \ldots, x_M)$, defined by the equality

$$\tau = \mathsf{M}\mathbf{x}, \tag{7.22}$$

where the transformation matrix M is given by

$$\mathsf{M} = \mathsf{U}_0 \cdot \mathsf{D}_0^{-1}. \tag{7.23}$$

Here D_0 is the diagonal matrix whose diagonal components are square roots of eigenvalues of the reduced Fisher matrix $\tilde{\Gamma}$, and U_0 is a matrix whose columns are eigenvectors of $\tilde{\Gamma}$, normalized to unity. One can show that U_0 is an orthogonal matrix, $U_0^{-1} = U_0^{\mathsf{T}}$ (superscript "T" denotes matrix transposition) and

$$\tilde{\Gamma} = U_0 \cdot D_0^2 \cdot U_0^{\mathsf{T}}. \tag{7.24}$$

Hyperellipse (7.21) in coordinates $\mathbf{x}$ reduces to the $(M+1)$-dimensional sphere of radius $R := \sqrt{1 - C_0}$. Therefore, the optimal grid can be expressed by means of a sphere covering.

In general, the thinnest possible sphere coverings are known only in dimensions 1 and 2. In dimensions up to 5 the thinnest *lattice* coverings are known, while in many higher dimensions the thinnest known coverings are lattices [207]. From this point on we consider only lattice coverings. The best lattice coverings for dimension k up to 5 are the A_k^* coverings. A_2^* is the well-known hexagonal covering and A_3^* is the body-centered-cubic (bcc) covering. The thickness of A_k^* covering is given by

$$\Theta_{A_k^*} = V_k \sqrt{k+1} \left(\frac{k(k+2)}{12(k+1)} \right)^{k/2}, \tag{7.25}$$

where V_k is the volume of the k-dimensional sphere of unit radius:

$$V^k = \frac{\pi^{k/2}}{(k/2)!}. \tag{7.26}$$

It is useful to compare $\Theta_{A_k^*}$ with the thickness $\Theta_{\mathbf{Z}^k}$ of the hypercubic lattice $\mathbf{Z}^k$

$$\Theta_{\mathbf{Z}^k} = V_k \frac{k^{k/2}}{2^k}. \tag{7.27}$$

We have plotted the two thicknesses in Fig. 7.1. We see that the optimal grid is around 1.3 times thinner than the hypercubic one in the two-dimensional case, 2.8 thinner in the four-dimensional case and as much as 18 times thinner in eight dimensions.

7.2.2 The covering problem with constraints

The general well-known solutions of the covering problem cannot always be adapted to calculations of the $\mathcal{F}$-statistic. In particular we would like to use the FFT algorithm in order to reduce the computational time. This requires the nodes of the grid to coincide with the Fourier frequencies. Thus we need to solve the covering problem subject to a constraint: one of its basis vectors needs to lie on the frequency axis and have a given

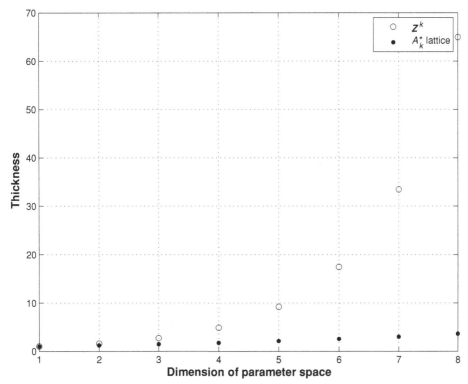

Fig. 7.1.   Comparison of the thickness of the hypercubic lattice with that of the A_k^* lattice as the function of the dimension k of the parameter space.

length. In other words, we look for the optimal covering with one of the lattice vectors fixed. We denote it by

$$\mathbf{a}_0 = (\Delta p_0, 0, 0, \ldots), \tag{7.28}$$

where Δp_0 is the fixed frequency resolution of our procedure. As far as we know, the general solution to the covering problem with a constraint is not known. Starting from the hypercubic covering (i.e. having all lattice vectors orthogonal), we shall construct the covering suitable for further calculations, satisfying the constraint (7.28).

We define a lattice as a set of all linear combinations of its basis vectors $\mathbf{a}_i$ with integer coefficients:

$$\Lambda = \left\{ \sum_i k_i \mathbf{a}_i : k_i \in \mathbb{Z} \right\}. \tag{7.29}$$

Given lattice Λ, its *fundamental parallelotope* is the set of points of the form $\sum_i \theta_i \mathbf{a}_i$, with $0 \leq \theta_i < 1$. Fundamental parallelotope is one example of *elementary cell*. The thickness Θ of a lattice covering is equal to the ratio of the volume of one hyperellipsoid to the volume of fundamental parallelotope.

For any lattice point $\mathbf{P}_i \in \Lambda$, the *Voronoi cell* around $\mathbf{P}_i$ is defined as

$$V(\mathbf{P}_i) = \{\boldsymbol{\tau} : C(\boldsymbol{\tau} - \mathbf{P}_i) \geq C(\boldsymbol{\tau} - \mathbf{P}_j) \text{ for all } j \neq i\}, \qquad (7.30)$$

where $C(\boldsymbol{\tau})$ is the Taylor expansion (7.12) of the autocovariance function. All the Voronoi cells of any lattice Λ are congruent, disjoint, and their union is the whole space. Voronoi cell is another example of elementary cell and is sometimes called Wigner–Seitz cell or Brillouin zone. The Voronoi cell of Λ is inscribed into the correlation ellipsoid (7.21).

Let Λ be *any* lattice with basis vectors $(\mathbf{a}_0, \mathbf{a}_1, \dots)$. The square of minimal match of Λ is

$$\mathrm{MM}^2(\Lambda) = \inf_{\boldsymbol{\tau} \in V(\mathbf{P}_i)} C(\boldsymbol{\tau} - \mathbf{P}_i), \qquad (7.31)$$

where $\mathbf{P}_i$ can be any lattice point.

Let $\boldsymbol{\zeta} \in V(\mathbf{P}_i)$ be the value for which the minimum in (7.31) is found. The function $C(\boldsymbol{\tau} - \mathbf{P}_i)$ has at the point $\boldsymbol{\zeta}$ its absolute minimum inside the Voronoi cell $V(\mathbf{P}_i)$, and $\boldsymbol{\zeta}$ is a *deep hole* of Λ. Note that the deep hole must be one of the vertices of the Voronoi cell. It makes Voronoi cells especially useful for calculating minimal match of a given lattice.

We can now outline the construction of a hypercubic grid in the parameter space. Given the value of $\mathcal{C}_0$, we look for the thinnest possible lattice covering Λ (subject to constraint on $\mathbf{a}_0$), satisfying

$$\mathrm{MM}^2(\Lambda) = \mathcal{C}_0. \qquad (7.32)$$

Due to the constraint (7.28) the first basis vector of Λ needs to be $\mathbf{a}_0$. As a first approximation of the remaining basis vectors $\mathbf{a}_1, \dots, \mathbf{a}_M$, we take unit vectors parallel to the main axes of the correlation ellipsoid (7.21), excluding the one that forms the smallest angle with the frequency axis and is replaced by $\mathbf{a}_0$.

By examining the vertices of the Voronoi cell we find the deep hole $\boldsymbol{\zeta}$ of this lattice. We then express $\boldsymbol{\zeta}$ as the combination

$$\boldsymbol{\zeta} = \alpha_0 \mathbf{a}_0 + \sum_{i=1}^{M} \alpha_i \mathbf{a}_i. \qquad (7.33)$$

Since the choice of the basis was arbitrary, $C(\boldsymbol{\zeta}) \neq \mathcal{C}_0$. We then shrink the lattice by the factor μ along all basis vectors except $\mathbf{a}_0$:

$$\mathbf{a}_0' = \mathbf{a}_0, \qquad (7.34\text{a})$$

$$\mathbf{a}_i' = \mu \mathbf{a}_i, \text{ for } i = 1, \dots, M, \qquad (7.34\text{b})$$

where μ satisfies the condition

$$C(\zeta') = C_0, \tag{7.35}$$

with

$$\zeta' = \alpha_0 \mathbf{a}_0 + \mu \sum_{i=1}^{M} \alpha_i \mathbf{a}_i. \tag{7.36}$$

The set of vectors $\{\mathbf{a}'_0, \ldots, \mathbf{a}'_M\}$ is the basis of the newly defined lattice Λ'.

Note that in general ζ' is *not* a deep hole of Λ'. Replacing Λ with Λ', we then iterate the procedure outlined in (7.33)–(7.36). As we proceed with iterations, values of μ are found to converge to unity. We end up with the lattice Λ, satisfying the condition (7.32).

7.3 Numerical algorithms to calculate the $\mathcal{F}$-statistic

7.3.1 The FFT algorithm

The *discrete Fourier transform* (DFT) $X(k)$ of a time series x_j is given by

$$X(k) = \sum_{j=1}^{N} x_j e^{-2\pi i (j-1)(k-1)/N}, \quad k = 1, \ldots, N. \tag{7.37}$$

The frequencies $(k-1)/N$ are called *Fourier frequencies* and DFT components calculated at Fourier frequencies are called *Fourier bins*. The inverse DFT is defined as

$$x_j = \frac{1}{N} \sum_{k=1}^{N} X(k) e^{2\pi i (j-1)(k-1)/N}. \tag{7.38}$$

The DFT is a periodic function with period equal to N [i.e. $X(k + lN) = X(k)$ for any integer l].

The DFT can be thought of as an $N \times N$ matrix that transforms N-point time series into N Fourier bins. Thus in general to calculate the N Fourier bins we need of the order of $\mathcal{O}(N^2)$ complex multiplications and additions. When number of data points N is a power of 2 there exists an algorithm called the *fast Fourier transform* (FFT) [208, 209] that calculates N Fourier bins with $\mathcal{O}(N \log_2 N)$ operations. The algorithm relies on a factorization of the $N \times N$ matrix in the DFT such that factors have many zeros and repeated factors. The FFT allows great saving in computational time to calculate DFTs. For N of the order of 2^{20}, which is typical in gravitational-wave data analysis, the saving is of the order of 5×10^4. Yet another advantage of calculating DFT using FFT is its numerical precision. The rms round-off error of the FFT is of the order of $\mathcal{O}(\epsilon\sqrt{\log_2 N})$ where ϵ is the machine floating-point relative precision comparing to $\mathcal{O}(\epsilon\sqrt{N})$ for the direct calculation of the DFT.

The FFT is useful in calculating the convolutions. Discrete convolution y_r of two time series x_m and h_n is given by

$$y_r = \sum_{m=1}^{N} x_m h_{r-m} \qquad (7.39)$$

DFT of a discrete convolution is given by a product of DFTs of the two time series x_m and h_n. Thus in order to calculate the convolution (7.39) we first calculate FFTs of the two time series, we multiply them and we take the inverse FFT.

7.3.2 Resampling

For continuous sources, like gravitational waves from rotating neutron stars observed by ground-based detectors (see Section 6.2.1), the detection statistic $\mathcal{F}$ [see Eq. (6.103)] involves integrals given by Eqs. (6.104). Let us consider the integral (6.104a) [the same arguments will apply to the integral (6.104b)]. The phase $\phi(t)$ [see Eq. (6.89)] can be written as

$$\phi(t) = \omega_0[t + \phi_{\mathrm{m}}(t)] + \phi_{\mathrm{s}}(t), \qquad (7.40)$$

where

$$\phi_{\mathrm{m}}(t; \mathbf{n}_0) := \frac{\mathbf{n}_0 \cdot \mathbf{r}_{\mathrm{d}}(t)}{c}, \qquad (7.41a)$$

$$\phi_{\mathrm{s}}(t; \tilde{\xi}) := \sum_{k=1}^{s} \omega_k \frac{t^{k+1}}{(k+1)!} + \frac{\mathbf{n}_0 \cdot \mathbf{r}_{\mathrm{d}}(t)}{c} \sum_{k=1}^{s} \omega_k \frac{t^k}{k!}, \qquad (7.41b)$$

and where $\tilde{\xi}$ comprises all parameters of the phase $\phi(t)$ but frequency ω_0. The functions $\phi_{\mathrm{m}}(t)$ and $\phi_{\mathrm{s}}(t)$ do not depend on the frequency ω_0. We can write the integral (6.104a) as

$$F_a = \int_0^{T_0} x(t)\, a(t)\, e^{-i\phi_{\mathrm{s}}(t)} \exp\left\{ -i\omega_0[t + \phi_{\mathrm{m}}(t)] \right\} \mathrm{d}t. \qquad (7.42)$$

We see that the integral (7.42) can be interpreted as a Fourier transform (and computed efficiently with an FFT), if $\phi_{\mathrm{m}} = 0$. In order to convert Eq. (7.42) to a Fourier transform we introduce a new time variable t_{b}, so called *barycentric time* [26, 31, 134],

$$t_{\mathrm{b}} := t + \phi_{\mathrm{m}}(t; \mathbf{n}_0). \qquad (7.43)$$

In the new time coordinate the integral (7.42) is approximately given by (see Ref. [31], Section IIID)

$$F_a \cong \int_0^{T_0} x[t(t_\mathrm{b})]a[t(t_\mathrm{b})]\mathrm{e}^{-\mathrm{i}\phi_\mathrm{s}[t(t_\mathrm{b})]}\mathrm{e}^{-\mathrm{i}\omega_0 t_\mathrm{b}}\,\mathrm{d}t_\mathrm{b}. \qquad (7.44)$$

Thus in order to compute the integral (7.42), we first multiply the data $x(t)$ by the function $a(t)\exp[-\mathrm{i}\phi_\mathrm{s}(t;\tilde{\xi})]$ for each set of the intrinsic parameters $\tilde{\xi}$ and then resample the resulting function according to Eq. (7.43). At the end we perform the FFT.

We consider two numerical interpolation methods in order to obtain the resampled function. We assume that the original data is a time series x_k, $k = 1,\ldots,N$ sampled at uniform intervals. The fastest method is the *nearest neighbor interpolation* also called *stroboscopic resampling* [26]. In this method we obtain the value of the time series x_k at barycentric time t_b by taking the value y_{k_o} such that k_o is the nearest integer to t_b. The second method has two steps. The first step consists of upsampling the time series and the second step consists of interpolating the upsampled time series to barycentric time using splines [210]. To perform the first step we use an interpolation method based on the Fourier transform. We Fourier transform the original time series, pad the Fourier transform series with an appropriate number of zeros and then transform it back to a time domain by an inverse Fourier transform. Thus we obtain an interpolated time series with points inserted between the original points. If we have a time series with N points and pad its DFT with N zeros, by inverse transform we obtain a $2N$-point time series. To perform the Fourier transforms we use the FFT algorithm. The second step consists in applying splines to interpolate the upsampled time series to barycentric time for number of points equal to the number of original data points. Thus if the original time series contains N points the final interpolated time series contains also N points.

We have compared the performance of the two interpolation methods and we have also compared these methods with an exact matched filter. To carry out the comparison we have used noise-free waveforms given in Eqs. (6.84)–(6.89) with one spin down. We have calculated the $\mathcal{F}$-statistic using the two interpolation methods and exact matched-filtering method. In the matched-filtering method we have assumed that we know the frequency of the signal and thus the Doppler modulation due to the motion of the detector. However, we have used FFT to calculate the $\mathcal{F}$-statistic for the whole frequency band. The results are shown in Fig. 7.2.

We find that the two-step method is much more accurate than the stroboscopic resampling. We have performed a Monte Carlo simulation consisting of 1000 trials and we have found that the rms error divided by maximum of the $\mathcal{F}$-statistic in the second method was 0.1% whereas

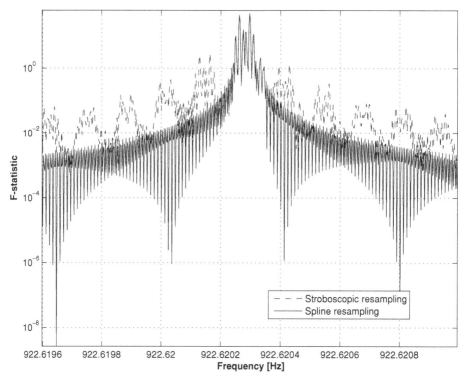

Fig. 7.2.   Comparison of the two interpolation methods. We see that the two-step interpolation method that uses Fourier and spline interpolation very accurately reproduces a perfectly matched filter. The difference between the spline resampling and the matched filter would not be visible on the plot.

in the first, fastest method it was 5%. The nearest neighbor interpolation leads to a greater signal-to-noise ratio loss than the spline interpolation and also, very importantly, to elevated sidelobes of the $\mathcal{F}$-statistic. In the presence of noise this can lead to a loss of the parameter estimation accuracy if the noise elevates the sidelobes above the main maximum.

7.3.3 Fourier interpolation

Let us consider the DFT $X(k)$ of a time series x_j given by formula (7.37). The dimensionless quantities $(k-1)/N$ are called the *Fourier frequencies* and DFT components calculated at Fourier frequencies are called *Fourier bins*. The Fourier frequencies give dimensionless normalized frequency. The true frequency is obtained by multiplying the Fourier frequency by the sampling frequency of the time series. Substituting x_j in Eq. (7.37) with the inverse transform [see Eq. (7.38)] and rearranging the summation gives

$$X(k) = \frac{1}{N} \sum_{l=1}^{N} X(l) \sum_{j=1}^{N} e^{-2\pi i\,(j-1)(k-l)/N}. \tag{7.45}$$

The summation over the index j can be explicitly computed yielding

$$X(k) = \frac{1}{N} \sum_{l=1}^{N} X(l)\, e^{-\pi i\,(k-l)(1-1/N)} \frac{\sin\left(\pi(k-l)\right)}{\sin\left(\pi(k-l)/N\right)}. \tag{7.46}$$

When the number of data points N is large the above expression can be approximated by

$$X(k) = \sum_{l=1}^{N} X(l)\, e^{-\pi i\,(k-l)} \frac{\sin\left(\pi(k-l)\right)}{\pi(k-l)}. \tag{7.47}$$

When the signal power is concentrated in the Fourier component $X(l)$ and l does not coincide with the Fourier frequency k, the application of a discrete N-point Fourier transform leads to signal-to-noise ratio lowered by factor $|D(k-l)|$, where

$$D(k-l) = e^{-\pi i\,(k-l)} \frac{\sin\left(\pi(k-l)\right)}{\pi(k-l)}.$$

The worst case is when $k - l = 1/2$. Then $|D(1/2)| = 2/\pi$, which amounts to over 36% loss of signal-to-noise ratio. To improve this we append the time series with N zeros (this is called the *zero-padding* procedure). We then have

$$X(k) = \sum_{l=1}^{N} X(l)\, e^{-\pi i\,(k/2-l)} \frac{\sin\left(\pi(k/2-l)\right)}{\pi(k/2-l)}, \tag{7.48}$$

and signal-to-noise ratio loss in the worst case is $1 - |D(1/4)| = 2\sqrt{2}/\pi$, which is only around 10%. However this zero-padding procedure involves computing two times longer FFT, which increases the computing time more than twice. We can obtain an approximate interpolation of the FFT at points $k = l + 1/2$ with the help of Eq. (7.47). We just use the two neighboring bins $l = k$ and $l = k + 1$ in the series (7.47) and we correct for the SNR loss factor of $|D(1/4)| = 2\sqrt{2}/\pi$. As a result we obtain an approximate formula

$$X(k + 1/2) \cong \left(X(k+1) - X(k)\right)/\sqrt{2}. \tag{7.49}$$

The interpolation defined by Eq. (7.49) is called *interbinning*. We have tested the above interpolation formula and we have compared it to padding the time series with zeros for the case of a monochromatic signal.

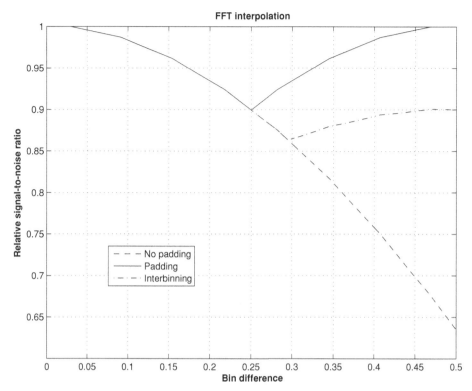

Fig. 7.3. Comparison of the interbinning interpolation procedure with zero padding. The dashed line shows the decrease of the signal-to-noise ratio as a function of the difference between the frequency of the signal and the nearest Fourier frequency. We see that in the worst case when the true signal frequency is in the middle of the Fourier bin, the signal-to-noise ratio is only $2/\pi \simeq 0.64$ of the true signal-to-noise ratio. By using twice as long FFT with zero padding of data the worst case improves to $2\sqrt{2}/\pi \simeq 0.90$ of the true signal-to-noise ratio (continuous line). The use of a simple interpolation formula [given in Eq. (7.49)] that requires only one complex addition per Fourier bin brings down the worst case fraction only by 5% to $\simeq 0.86$ (dotted-dashed line).

The results are shown in Fig. 7.3. The loss of signal-to-noise ratio is only 5% less than using zero padding and a twice as long FFT.

7.3.4 Optimization algorithms

In order to detect signals we search for threshold crossings of the $\mathcal{F}$-statistic over the intrinsic parameter space. Once we have a threshold crossing we need to find the precise location of the maximum of $\mathcal{F}$ in order to estimate accurately parameters of the signal. A satisfactory procedure is a two-step procedure. The first step is a *coarse search* where we evaluate

$\mathcal{F}$ on a coarse grid in the parameter space and locate threshold crossings. The second step, called *fine search*, is a refinement around the region of the parameter space where the maximum identified by the coarse search is located.

There are two methods to perform the fine search. One is to refine the grid around the threshold crossing found by the coarse search [188, 211, 212, 213], and the other is to use an optimization routine to find the maximum of $\mathcal{F}$ [32, 189]. As initial value to the optimization routine we input the values of the parameters found by the coarse search. There are many maximization algorithms available. As our maximum finding routines we can use ones that do not require calculations of derivatives of $\mathcal{F}$. In one-dimensional case this is Brent's algorithm [214], and in multidimensional case this is Nelder–Mead algorithm [215, 216]. Efficient algorithms to find the maxima of the optimal statistic in multidimensional parameter space are also provided by Markov chain Monte Carlo methods [217, 218, 219].

Comparison with the Cramèr–Rao bound. In order to test the performance of the maximization method of the $\mathcal{F}$-statistic it is useful to perform Monte Carlo simulations of the parameter estimation and compare the variances of the estimators with the variances calculated from the Fisher matrix. Such simulations were performed for various gravitational-wave signals [189, 220, 221]. In these simulations we observe that above a certain signal-to-noise ratio, that we shall call the threshold signal-to-noise ratio, the results of the Monte Carlo simulations agree very well with the calculations of the rms errors from the inverse of the Fisher matrix. However below the threshold signal-to-noise ratio they differ by a large factor. This threshold effect is well-known in signal processing [128].

There exist more refined theoretical bounds on the rms errors that explain this effect and they were also studied in the context of the gravitational-wave signal from a coalescing binary [150]. Here we present a simple model that explains the deviations from the covariance matrix and reproduces well the results of the Monte Carlo simulations. The model makes use of the concept of the elementary cell of the parameter space that we introduced in Section 6.1.3. The calculation given below is a generalization of the calculation of the rms error for the case of a monochromatic signal given by Rife and Boorstyn [222].

When the values of parameters of the template that correspond to the maximum of the functional $\mathcal{F}$ fall within the cell in the parameter space where the signal is present the rms error is satisfactorily approximated by the inverse of the Fisher matrix. However sometimes as a result of noise the global maximum is in the cell where there is no signal. We then say that an *outlier* has occurred. In the simplest case we can assume that the probability density of the values of the outliers is uniform over the search interval of a parameter and then the rms error is given by

$$\sigma^2_{\text{out}} = \frac{\Delta^2}{12},\tag{7.50}$$

where Δ is the length of the search interval for a given parameter. The probability that an outlier occurs will be the higher the lower the signal-to-noise ratio. Let q be the probability that an outlier occurs. Then the total variance σ^2 of the estimator of a parameter is the weighted sum of the two errors

$$\sigma^2 = \sigma^2_{\text{out}}\, q + \sigma^2_{\text{CR}}(1 - q),\tag{7.51}$$

where σ_{CR} is the rms error calculated from the covariance matrix for a given parameter. One can show [189] that the probability q can be approximated by the following formula:

$$q = 1 - \int_0^\infty p_1(\rho, \mathcal{F}) \left(\int_0^{\mathcal{F}} p_0(y)\,\mathrm{d}y \right)^{N_c - 1} \mathrm{d}\mathcal{F},\tag{7.52}$$

where p_0 and p_1 are probability density functions of respectively false alarm and detection given by Eqs. (6.32a) and (6.32b) and where N_c is the number of cells in the parameter space. Equation (7.52) is in good but not perfect agreement with the rms errors obtained from the Monte Carlo simulations (see [189]). There are clearly also other reasons for deviations from the Cramèr–Rao bound. One important effect (see [150]) is that the functional $\mathcal{F}$ has many local subsidiary maxima close to the global one. Thus for a low signal-to-noise ratio the noise may promote the subsidiary maximum to a global one thereby biasing the estimation of the parameters.

7.4 Analysis of the candidates

7.4.1 Coincidences

Let us assume that we analyze the same parameter space L times and that each analysis is statistically independent. For example this may be a search of the same time interval for impulses using stretches of data in that interval from L different detectors or it may be a search of the same frequency band for monochromatic signals using L non-overlapping time stretches of data from the same detector. As the parameter spaces are the same each can be divided into the same number N_{cell} of independent cells, as explained in Section 6.1.3. Suppose that the lth analysis produces N_l triggers. We would like to test the null hypothesis that the coincidences are the result of the noise only. The probability for a candidate event to fall into any given coincidence cell is equal to $1/N_{\text{cell}}$. Thus probability

ϵ_l that a given coincidence cell is populated with one or more candidate events is given by

$$\epsilon_l = 1 - \left(1 - \frac{1}{N_{\text{cell}}}\right)^{N_l}. \tag{7.53}$$

We can also consider only the independent triggers i.e. such that there is no more than one trigger within one cell. If we obtain more than one trigger within a given cell we choose the one which has the highest signal-to-noise ratio. In this case

$$\epsilon_l = \frac{N_l}{N_{\text{cell}}}. \tag{7.54}$$

The probability p_F that any given coincidence cell contains candidate events from $\mathcal{C}_{\max}$ or more distinct data segments is given by a generalized binomial distribution

$$p_F = \sum_{n=\mathcal{C}_{\max}}^{L} \frac{1}{n!(L-n)!} \sum_{\sigma \in \Pi(L)} \epsilon_{\sigma(1)} \cdots \epsilon_{\sigma(n)} (1 - \epsilon_{\sigma(n+1)}) \cdots (1 - \epsilon_{\sigma(L)}), \tag{7.55}$$

where $\sum_{\sigma \in \Pi(L)}$ is the sum over all the permutations of the L data sequences. Finally the probability P_F that there are $\mathcal{C}_{\max}$ or more coincidences in one or more of the N_{cell} cells is

$$P_F = 1 - (1 - p_F)^{N_{\text{cell}}}. \tag{7.56}$$

The expected number of cells with $\mathcal{C}_{\max}$ or more coincidences is given by

$$N_F = N_{\text{cell}} \, p_F. \tag{7.57}$$

Thus by choosing a certain false alarm probability P_F we can calculate the threshold number $\mathcal{C}_{\max}$ of coincidences. If we obtain more than $\mathcal{C}_{\max}$ coincidences in our search we reject the null hypothesis that coincidences are due to noise only at the significance level of P_F.

7.4.2 Signal-to-noise gain

Let us suppose that we have a persistent gravitational-wave signal that is continuously present in the data. An example of such a signal is a periodic gravitational-wave signal originating from a rotating neutron star (and considered in detail in Section 2.5). In such a case a candidate signal present for a certain observation time T_o should also be present during any other observational time T_o'. Thus we can verify a candidate obtained in search of data x by testing whether it is present in data y. It is useful to choose the duration T_o' of data y to be longer than the duration T_o of data x. If the candidate is a real signal, theoretically its signal-to-noise ratio should increase as a square root of the observation time.

Whatever the length of the observation time the $\mathcal{F}$-statistic has a χ^2 distribution with four degrees of freedom. Thus to test for a significant increase of the signal-to-noise ratio we can consider the ratio F of the $\mathcal{F}$-statistic $\mathcal{F}_{T'_0}$ for observation time T'_0 to the $\mathcal{F}$-statistic $\mathcal{F}_{T_0}$ for observation time T_0. Moreover we assume that the two observation times do not overlap so that the two data sets x and y are statistically independent and consequently so are the two corresponding $\mathcal{F}$-statistics. We can test the null hypothesis that data is noise only. Under the null hypothesis the ratio F has Fisher–Snedecor $F(4, 4)$ distribution as it is the ratio of two χ^2 distributions with four degrees of freedom. In general Fisher–Snedecor $F_{(m,n)}$ distribution is defined as

$$F_{(m,n)} = \frac{\chi_m^2/m}{\chi_n^2/n}, \tag{7.58}$$

where χ_m^2 and χ_n^2 are two independent χ^2 distributions with m and n degrees of freedom respectively. The probability density $p_{(m,n)}(F)$ of $F_{(m,n)}$ distribution is given by

$$p_{(m,n)}(F) = \frac{\Gamma((m+n)/2)}{\Gamma(m/2)\Gamma(n/2)} \left(\frac{m}{n}\right)^{m/2} \frac{F^{(m-2)/2}}{\left(1 + \frac{m}{n}F\right)^{(m+n)/2}}, \tag{7.59}$$

where Γ is the gamma function. Probability density function $p_{(4,4)}$ and cumulative density function $P_{(4,4)}$ for the Fisher–Snedecor $F(4, 4)$ distribution are explicitly given by

$$p_{(4,4)}(F) = \frac{6F}{(1+F)^4}, \quad P_{(4,4)}(F) = \frac{F^2(3+F)}{(1+F)^3}. \tag{7.60}$$

In Fig. 7.4 we have plotted the false alarm probability equal to $1 - P_{(4,4)}$ for the null hypothesis that data is noise only.

7.4.3 Upper limits

The detection of a signal is signified by a large value of the $\mathcal{F}$-statistic that is unlikely to arise from the noise-only distribution. If instead the value of $\mathcal{F}$ is consistent with pure noise with high probability we can place an upper limit on the strength of the signal. One way of doing this is to take the loudest event obtained in the search and solve the equation

$$\mathcal{P} = P_D(\rho_{\text{ul}}, \mathcal{F}_0) \tag{7.61}$$

for signal-to-noise ratio ρ_{ul}, where P_D is the detection probability given by Eq. (6.35), $\mathcal{F}_0$ is the value of the $\mathcal{F}$-statistic corresponding to the

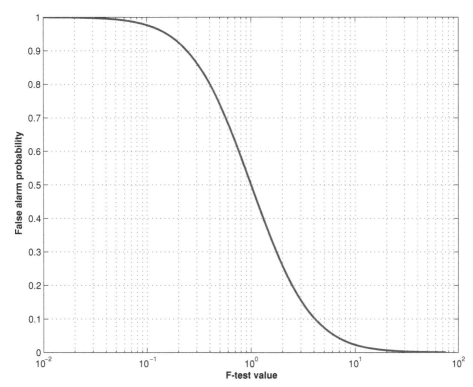

Fig. 7.4. False alarm probability: the probability density that the ratio F exceeds a certain value when the data is noise only.

loudest event and $\mathcal{P}$ is a chosen confidence [205, 223]. Then ρ_{ul} is the desired upper limit with confidence $\mathcal{P}$. For several independent searches the relation between the confidence $\mathcal{P}$ and upper limit ρ_{ul} is given by

$$\mathcal{P} = 1 - \prod_{s=1}^{L} \left(1 - P_D(\rho_{\mathrm{ul}}, \mathcal{F}_{0s}) \right), \tag{7.62}$$

where $\mathcal{F}_{0s}$ is the threshold corresponding to the loudest event in the sth search and L is the number of searches. Here $\mathcal{P}$ is the probability that a signal of signal-to-noise ratio ρ_{ul} crosses the threshold $\mathcal{F}_{0s}$ for at least one of the L independent searches.

When gravitational-wave data do not conform to a Gaussian probability density assumed in Eq. (6.35), a more accurate upper limit can be imposed by injecting the signals into the detector's data and thereby estimating the probability of detection P_D [224]. An improved upper limit can be obtained using the Bayesian methodology presented in Section 3.6.7 (see [225] for a detailed exposition).

Appendix A
The chirp waveform

In this appendix we give the explicit forms of the different functions needed to describe the time evolution of the phase of gravitational waves emitted by a coalescing binary system. We restrict ourselves here to the situation when binary members move along *quasi-circular* orbits and we take into account all post-Newtonian effects up to the 3.5 post-Newtonian order [i.e. up to the terms of the order $(v/c)^7$ beyond the predictions of the Newtonian theory of gravitation].

We start from the function $\mathcal{F}$ defined in Eq. (2.49). After expanding the integrand in Eq. (2.49) in power series with respect to x and performing the integration, we get the following expression for the function $\mathcal{F}$:

$$
\begin{aligned}
\mathcal{F}(x) = \frac{5}{256\,\eta}\frac{1}{x^4}\Bigg\{ &1 + \left(\frac{743}{252} + \frac{11}{3}\eta\right)x - \frac{32}{5}\pi\,x^{3/2} \\
&+ \left(\frac{3\,058\,673}{508\,032} + \frac{5\,429}{504}\eta + \frac{617}{72}\eta^2\right)x^2 \\
&+ \left(-\frac{7\,729}{252} + \frac{13}{3}\eta\right)\pi\,x^{5/2} \\
&+ \left(-\frac{10\,052\,469\,856\,691}{23\,471\,078\,400} + \frac{128}{3}\pi^2 + \frac{6\,848}{105}\gamma_{\mathrm{E}} + \frac{3\,424}{105}\ln(16\,x)\right) \\
&+ \left(\frac{3\,147\,553\,127}{3\,048\,192} - \frac{451}{12}\pi^2\right)\eta - \frac{15\,211}{1\,728}\eta^2 + \frac{25\,565}{1\,296}\eta^3\bigg)x^3 \\
&+ \left(-\frac{15\,419\,335}{127\,008} - \frac{75\,703}{756}\eta + \frac{14\,809}{378}\eta^2\right)\pi\,x^{7/2} + \mathcal{O}(x^8)\Bigg\},
\end{aligned}
$$

(A.1)

where γ_E denotes Euler constant which can be defined as the limit

$$\gamma_E := \lim_{n \to \infty} \left(\sum_{k=1}^{n} \frac{1}{k} - \ln n \right) \cong 0.577\,216. \tag{A.2}$$

We put the result of Eq. (A.1) into Eq. (2.48) and such obtained formula we invert perturbatively to get x as a function of $\hat{t}$. The explicit form of the final expression reads [here we use instead of $\hat{t}$ the time variable τ defined in Eq. (2.50)]

$$x(\hat{t}) = \mathcal{T}(\tau(\hat{t})), \tag{A.3}$$

where the function $\mathcal{T}$ is defined as follows:

$$\mathcal{T}(\tau) = \frac{1}{4}\tau^{-1/4} \left\{ 1 + \left(\frac{743}{4\,032} + \frac{11}{48}\eta \right) \tau^{-1/4} - \frac{1}{5}\pi\,\tau^{-3/8} \right.$$

$$+ \left(\frac{19\,583}{254\,016} + \frac{24\,401}{193\,536}\eta + \frac{31}{288}\eta^2 \right) \tau^{-1/2}$$

$$+ \left(-\frac{11\,891}{53\,760} + \frac{109}{1\,920}\eta \right) \pi\,\tau^{-5/8}$$

$$+ \left(\frac{1}{6}\pi^2 - \frac{10\,052\,469\,856\,691}{6\,008\,596\,070\,400} + \frac{107}{420}\gamma_E - \frac{107}{3\,360}\ln\left(\frac{\tau}{256}\right) \right)$$

$$+ \left(\frac{3\,147\,553\,127}{780\,337\,152} - \frac{451}{3\,072}\pi^2 \right)\eta - \frac{15\,211}{442\,368}\eta^2 + \frac{25\,565}{331\,776}\eta^3 \right)\tau^{-3/4}$$

$$+ \left(-\frac{113\,868\,647}{433\,520\,640} - \frac{31\,821}{143\,360}\eta + \frac{294\,941}{3\,870\,720}\eta^2 \right)\pi\,\tau^{-7/8} + \mathcal{O}(\tau^{-1}) \right\}. \tag{A.4}$$

Next we study the function $\mathcal{G}$ introduced in Eq. (2.54). After expanding the integrand in Eq. (2.54) in a power series with respect to x and performing the integration, we get the following expression for the function $\mathcal{G}$:

$$\mathcal{G}(x) = -\frac{x^{-5/2}}{32\eta} \left\{ 1 + \left(\frac{3\,715}{1\,008} + \frac{55}{12}\eta \right)x - 10\pi\,x^{3/2} \right.$$

$$+ \left(\frac{15\,293\,365}{1\,016\,064} + \frac{27\,145}{1\,008}\eta + \frac{3\,085}{144}\eta^2 \right)x^2$$

$$+ \left(\frac{38\,645}{1\,344} - \frac{65}{16}\eta \right)\pi\,x^{5/2}\ln x$$

$$+\left(\frac{12\,348\,611\,926\,451}{18\,776\,862\,720}-\frac{160}{3}\pi^2-\frac{1\,712}{21}\gamma_E-\frac{856}{21}\ln(16\,x)\right)$$

$$+\left(\frac{2\,255}{48}\pi^2-\frac{15\,737\,765\,635}{12\,192\,768}\right)\eta+\frac{76\,055}{6\,912}\eta^2-\frac{127\,825}{5\,184}\eta^3\right)x^3$$

$$+\left(\frac{77\,096\,675}{2\,032\,128}+\frac{378\,515}{12\,096}\eta-\frac{74\,045}{6\,048}\eta^2\right)\pi\,x^{7/2}+\mathcal{O}(x^8)\bigg\}.\qquad\text{(A.5)}$$

We put Eqs. (A.3) and (A.4) into Eq. (2.53), then we expand the resultant formula into the PN series. We get

$$\phi(\tau)=\phi_a-\mathcal{G}(x_a)-\frac{\tau^{5/8}}{\eta}\left\{1+\left(\frac{3\,715}{8\,064}+\frac{55}{96}\eta\right)\tau^{-1/4}-\frac{3}{4}\pi\,\tau^{-3/8}\right.$$

$$+\left(\frac{9\,275\,495}{14\,450\,688}+\frac{284\,875}{258\,048}\eta+\frac{1\,855}{2\,048}\eta^2\right)\tau^{-1/2}$$

$$+\left(\frac{47\,561}{64\,512}+\frac{67}{768}\eta+\left(-\frac{38\,645}{172\,032}+\frac{65}{2\,048}\eta\right)\ln(256\tau)\right)\pi\,\tau^{-5/8}$$

$$+\left(\frac{831\,032\,450\,749\,357}{57\,682\,522\,275\,840}-\frac{53}{40}\pi^2-\frac{107}{56}\gamma_E+\frac{107}{448}\ln\left(\frac{\tau}{256}\right)\right.$$

$$+\left(\frac{2\,255}{2\,048}\pi^2-\frac{126\,510\,089\,885}{4\,161\,798\,144}\right)\eta+\frac{154\,565}{1\,835\,008}\eta^2-\frac{1\,179\,625}{1\,769\,472}\eta^3\right)\tau^{-3/4}$$

$$+\left(\frac{188\,516\,689}{173\,408\,256}+\frac{488\,825}{516\,096}\eta-\frac{141\,769}{516\,096}\eta^2\right)\pi\,\tau^{-7/8}+\mathcal{O}(\tau^{-1})\bigg\}.\qquad\text{(A.6)}$$

Stationary phase approximation. Let us consider the time-dependent function of the form

$$s(t)=A(t)\cos\Phi(t),\qquad\text{(A.7)}$$

where A is the amplitude which varies on the time scale much longer than the time scale at which the phase Φ is varying:

$$\frac{|\dot{A}(t)|}{|A(t)|}\ll|\dot{\Phi}(t)|,\qquad\text{(A.8)}$$

where the overdot denotes differentiation with respect to the time t. It is often useful to have an analytic formula for the Fourier transform $\tilde{s}$ of the function s. If, apart from the relation (A.8), the following condition is also fulfilled

$$|\ddot{\Phi}(t)| \ll \left(\dot{\Phi}(t)\right)^2, \tag{A.9}$$

then one can employ the *stationary phase approximation* to evaluate the Fourier transform of s. The result is

$$\tilde{s}(f) \cong \sqrt{\frac{\pi}{2\ddot{\Phi}(t_f)}}\, A(t_f)\, \exp\left(i\left(\Psi_f(t_f) - \frac{\pi}{4}\right)\right), \quad f \geq 0, \tag{A.10}$$

where the function Ψ_f is defined as

$$\Psi_f(t) := 2\pi f t - \Phi(t), \tag{A.11}$$

and where t_f is a function of the Fourier transform variable f implicitly defined by the relation

$$\frac{1}{2\pi}\dot{\Phi}(t_f) = f, \tag{A.12}$$

i.e. t_f is the time at which the instantaneous frequency of emitted radiation is equal to the Fourier transform variable f.

Fourier transform of the chirp waveform. As an example of the waveform of the kind given in Eq. (A.7), whose Fourier transform can be computed within the stationary phase approximation, we consider here the response function of the Earth-based interferometer to gravitational waves from a coalescing compact binary. This response function is described by Eqs. (6.154)–(6.155) and (6.141), it reads

$$s(t) = A_0\, \Theta(t - t_\mathrm{i})\, \omega_{\mathrm{gw}}(t)^{2/3} \cos\left(\phi_{\mathrm{gw}}(t) - \phi_\mathrm{i} - \xi_0\right), \tag{A.13}$$

where the step function Θ is defined in Eq. (6.124), the constant factor A_0 in the waveform's amplitude is given in Eq. (6.156), ϕ_i is the initial phase of the polarization waveforms h_+ and $h_\times$, and the constant angle ξ_0 [defined through Eqs. (6.157)] is the initial phase of the response function s. The instantaneous phase ϕ_{gw} of the waveform (A.13) is related to the instantaneous phase ϕ of the orbital motion through equations [see Eq. (6.138) and the discussion preceding it]

$$\phi_{\mathrm{gw}}(t) = 2\phi(t), \quad t_a = t_\mathrm{i}, \quad \phi_a := \phi(t_a) = \frac{1}{2}\phi_\mathrm{i}, \quad \dot{\phi}_a := \dot{\phi}(t_a) = \frac{1}{2}\omega_\mathrm{i}. \tag{A.14}$$

Equation (A.12) applied to the waveform (A.13) leads to the relation

$$x_f := x(t_f) = \frac{(\pi G M f)^{2/3}}{c^2}, \tag{A.15}$$

where we have employed the connection (2.41) between the dimensionless variable x and the orbital angular frequency ϕ. Let us also introduce

$$x_i := \frac{1}{c^2}\left(GM\frac{\omega_i}{2}\right)^{2/3}. \tag{A.16}$$

We expect that the amplitude of the waveform (A.13) is very small, therefore quite often the post-Newtonian corrections to the time evolution of the gravitational-wave angular frequency ω_{gw} in the amplitude $A_0\,\omega_{gw}(t)^{2/3}$ can be neglected. It means that the time evolution of ω_{gw} can be treated here in the leading-order approximation, which is given by [see Eq. (2.38)]

$$\omega_{gw}(t) = \frac{5^{3/8}c^{15/8}}{4\,(GM)^{5/8}}\,(t_c - t)^{-3/8}, \tag{A.17}$$

where t_c defines the (formal) moment of coalescence of the binary system. Equation (A.17) implies that

$$\dot{\omega}_{gw} = \frac{12\sqrt[3]{2}\,(GM)^{5/3}}{5\,c^5}\,\omega_{gw}^{11/3}. \tag{A.18}$$

Applying the formula (A.10) to the waveform (A.13) and making use of Eqs. (A.17), (A.18) and (6.156) one gets

$$\tilde{s}(f) \cong A\,f^{-7/6}\exp\left(i\left(\Psi_f(t_f) - \frac{\pi}{4}\right)\right), \tag{A.19}$$

where the constant A is equal

$$A = \frac{1}{2\pi^{2/3}}\sqrt{\frac{5}{6}}\frac{(GM)^{5/6}}{c^{3/2}R}\sqrt{\left(F_+\frac{1+\cos^2\iota}{2}\right)^2 + (F_\times\cos\iota)^2}. \tag{A.20}$$

Making use of Eqs. (2.48), (2.53) and replacing the time variable t by its dimensionless version $\hat{t}$ [defined in Eq. (2.47)] one can compute the phase Ψ_f of the Fourier-domain waveform (A.19) as a function of the dimensionless parameter x_f up to the 3.5PN order of approximation. It is given by

$$\Psi_f(x_f) = \xi_0 + 2\,\hat{t}_i\,x_f^{3/2} + 2\big(\mathcal{F}(x_i) - \mathcal{F}(x_f)\big)x_f^{3/2} + 2\big(\mathcal{G}(x_i) - \mathcal{G}(x_f)\big)$$

$$= \xi_0 + 2\,\hat{t}_i\,x_f^{3/2} + \frac{3}{128\,\eta}\frac{1}{x_f^{5/2}}\,\Xi(x_f), \tag{A.21}$$

where $\Xi(x)$ is the polynomial of the two variables $x^{1/2}$ and $\ln x$ with coefficients which depend on the ratio $r := x/x_i$:

$$\Xi(x) = a_0(r) + k_2\,a_2(r)\,x + k_3\,a_3(r)\,x^{3/2} + k_4\,a_4(r)\,x^2 + k_5\,a_5(r)\,x^{5/2}$$

$$+ \big(a_{61}(r)(k_{61} + k_{62}\ln x) + a_{62}(r)\big)x^3 + k_7\,a_7(r)\,x^{7/2}. \tag{A.22}$$

The coefficients $a_n(r)$ are given by

$$a_0(r) = 1 - \frac{8}{3}r^{5/2} + \frac{5}{3}r^4, \tag{A.23a}$$

$$a_2(r) = 1 - 2r^{3/2} + r^3, \tag{A.23b}$$

$$a_3(r) = 1 - \frac{5}{3}r + \frac{2}{3}r^{5/2}, \tag{A.23c}$$

$$a_4(r) = 1 - \frac{4}{3}\sqrt{r} + \frac{1}{3}r^2, \tag{A.23d}$$

$$a_5(r) = 1 - r^{3/2} + \frac{3}{2}\ln r, \tag{A.23e}$$

$$a_{61}(r) = -\frac{2}{3\sqrt{r}} + 1 - \frac{r}{3}, \tag{A.23f}$$

$$a_{62}(r) = \frac{6\,848}{63}\left(r - \frac{1}{\sqrt{r}} - \left(\frac{r}{2} + \frac{1}{\sqrt{r}}\right)\ln r\right), \tag{A.23g}$$

$$a_7(r) = -\frac{2\sqrt{r}}{3} + 1 - \frac{1}{3r}. \tag{A.23h}$$

The coefficients k_n depend only on the symmetric mass ratio η:

$$k_2 = \frac{3\,715}{756} + \frac{55}{9}\eta, \tag{A.24a}$$

$$k_3 = -16\pi, \tag{A.24b}$$

$$k_4 = \frac{15\,293\,365}{508\,032} + \frac{27\,145}{504}\eta + \frac{3\,085}{72}\eta^2, \tag{A.24c}$$

$$k_5 = \left(\frac{38\,645}{756} - \frac{65}{9}\eta\right)\pi, \tag{A.24d}$$

$$k_{61} = \frac{11\,583\,231\,236\,531}{4\,694\,215\,680} - \frac{6\,848}{21}\gamma_E - \frac{640}{3}\pi^2 - \frac{13\,696}{21}\ln 2$$
$$+ \left(\frac{2\,255}{12}\pi^2 - \frac{15\,737\,765\,635}{3\,048\,192}\right)\eta + \frac{76\,055}{1\,728}\eta^2 - \frac{127\,825}{1\,296}\eta^3, \tag{A.24e}$$

$$k_{62} = -\frac{3\,424}{21}, \tag{A.24f}$$

$$k_7 = \left(\frac{77\,096\,675}{254\,016} + \frac{378\,515}{1\,512}\eta - \frac{74\,045}{756}\eta^2\right)\pi. \tag{A.24g}$$

Appendix B
Proof of the Neyman–Pearson lemma

In this appendix we give the proof of the Neyman–Pearson lemma [124].

1. *Existence.* Let us define the function

$$T(x) := \begin{cases} \dfrac{p_1(x)}{p_0(x)}, & \text{for } p_0(x) > 0, \\ +\infty, & \text{for } p_0(x) = 0, \end{cases} \tag{B.1}$$

and let us consider the following function related to the random variable $T(X)$:

$$\alpha(\lambda) := P_0\big(T(X) > \lambda\big). \tag{B.2}$$

From this definition of the random variable T, it follows that the function $1 - \alpha(\lambda)$ is a cumulative distribution function, therefore the function $\alpha(\lambda)$ itself is a non-increasing and right-continuous function. This implies that for any $\alpha \in (0, 1)$ there exists λ_0 such that

$$\alpha(\lambda_0) \leq \alpha \leq \alpha(\lambda_0 - 0), \tag{B.3}$$

where $\alpha(\lambda_0 - 0)$ denotes the left-hand limit,

$$\alpha(\lambda_0 - 0) := \lim_{\lambda \to \lambda_0^-} \alpha(\lambda).$$

Since $\alpha(\lambda_0 - 0) = \alpha(\lambda_0)$ if and only if $P_0(T(X) = \lambda_0) = 0$, i.e. if and only if $P_0(p_1(X) = \lambda_0\, p_0(X)) = 0$, we can define a function

$$
\phi(x) := \begin{cases} 1, & \text{when } p_1(x) > \lambda_0\, p_0(x), \\[2mm] \dfrac{\alpha - \alpha(\lambda_0 - 0)}{\alpha(\lambda_0) - \alpha(\lambda_0 - 0)}, & \text{when } p_1(x) = \lambda_0\, p_0(x), \\[2mm] 0, & \text{when } p_1(x) < \lambda_0\, p_0(x). \end{cases} \qquad \text{(B.4)}
$$

We have

$$
E_0\{\phi(X)\} = P_0\Big(p_1(X) > \lambda_0\, p_0(X)\Big)
$$

$$
+ \frac{\alpha - \alpha(\lambda_0 - 0)}{\alpha(\lambda_0) - \alpha(\lambda_0 - 0)} P_0\Big(p_1(X) = \lambda_0\, p_0(X)\Big)
$$

$$
= \alpha(\lambda_0 - 0) + \frac{\alpha - \alpha(\lambda_0 - 0)}{\alpha(\lambda_0) - \alpha(\lambda_0 - 0)}\big(\alpha(\lambda_0) - \alpha(\lambda_0 - 0)\big)
$$

$$
= \alpha. \qquad \text{(B.5)}
$$

Consequently the test ϕ defined in Eq. (B.4) fulfills the conditions (3.75) and (3.76).

The value λ_0 is defined uniquely except for the case when $\alpha(\lambda) = \alpha$ for a certain interval $[\lambda', \lambda'']$. However, we can exclude this case because both probability distributions P_0 and P_1 assign it a zero probability.

2. *Sufficiency.* Let ϕ be a test satisfying the conditions (3.76) and (3.77) and let ϕ^* be a different test such that

$$
E_0\{\phi^*(X)\} \le \alpha. \qquad \text{(B.6)}
$$

We have

$$
E_1\{\phi(X)\} - E_1\{\phi^*(X)\} = \int \big(\phi(x) - \phi^*(x)\big)\, p_1(x)\, dx
$$

$$
= \int_{\{x:\, \phi(x) > \phi^*(x)\}} \big(\phi(x) - \phi^*(x)\big)\, p_1(x)\, dx
$$

$$
+ \int_{\{x:\, \phi(x) < \phi^*(x)\}} \big(\phi(x) - \phi^*(x)\big)\, p_1(x)\, dx. \qquad \text{(B.7)}
$$

If $\phi(x) > \phi^*(x)$ then $\phi(x) > 0$ and consequently by the definition of test ϕ we must have $p_1(x) > \lambda\, p_0(x)$. Thus the first integral on the right of Eq. (B.7) is not less than

$$\lambda \int_{\{x:\, \phi(x)>\phi^*(x)\}} \big(\phi(x) - \phi^*(x)\big)\, p_0(x)\, \mathrm{d}x. \tag{B.8}$$

Similarly on the set $\{x : \phi(x) > \phi^*(x)\}$ we have $\phi(x) < 1$, i.e. $p_1(x) \leq \lambda\, p_0(x)$. It follows that the second integral is not less than

$$\lambda \int_{\{x:\, \phi(x)<\phi^*(x)\}} \big(\phi(x) - \phi^*(x)\big)\, p_0(x)\, \mathrm{d}x. \tag{B.9}$$

Thus

$$E_1\{\phi(X)\} - E_1\{\phi^*(X)\} \geq \lambda \int \big(\phi(x) - \phi^*(x)\big)\, p_0(x)\, \mathrm{d}x$$

$$= \lambda\Big(E_0\{\phi(X)\} - E_0\{\phi^*(X)\}\Big)$$

$$= \lambda\Big(\alpha - E_0\{\phi^*(X)\}\Big) \geq 0. \tag{B.10}$$

Thus the test ϕ is at least as powerful as an arbitrarily chosen test ϕ^*.

3. *Necessity.* Let ϕ be the most powerful test at the significance level α and let ϕ^* be a test of the same power and fulfilling the condition (3.76). Let us consider the set

$$C := \{x : \phi(x) \neq \phi^*(x), p_1(x) \neq \lambda\, p_0(x)\}. \tag{B.11}$$

To prove the thesis we have to show that the measure $\int_C \mathrm{d}x$ of the set C is 0. By the argument used in the proof of the previous item we find that on the set C the product

$$\big(\phi^*(x) - \phi(x)\big)\big(p_1(x) - \lambda\, p_0(x)\big) \tag{B.12}$$

is positive (when $p_1 > \lambda\, p_0$ then by assumption $\phi^* = 1$, thus we must have $\phi = 0$, when $p_1 < \lambda\, p_0$ then $\phi^* = 0$ and $\phi = 1$), i.e.

$$\int \big(\phi^*(x) - \phi(x)\big)\big(p_1(x) - \lambda\, p_0(x)\big)\, \mathrm{d}x$$

$$= \int_C \big(\phi^*(x) - \phi(x)\big)\big(p_1(x) - \lambda\, p_0(x)\big)\, \mathrm{d}x > 0, \tag{B.13}$$

when $\int_C \mathrm{d}x > 0$. But then

$$\int \big(\phi^*(x) - \phi(x)\big) p_1(x)\, \mathrm{d}x > \lambda \int \big(\phi^*(x) - \phi(x)\big) p_0(x)\, \mathrm{d}x = 0 \tag{B.14}$$

and ϕ^* is more powerful than ϕ, which contradicts the assumption. Consequently we must have $\int_C \mathrm{d}x = 0$.

Appendix C
Detector's beam-pattern functions

In this appendix we present explicit analytic formulae for the detector's *beam-pattern functions* for the three types of gravitational-wave detectors: space-borne LISA detector (see e.g. Section II of [32]), a ground-based interferometric detector (see e.g. Section II of [31]), and a ground-based resonant-bar detector (see e.g. Appendix A of [190]).

The basic building blocks for the responses of different detectors to a *plane* gravitational wave are the scalar quantities Φ_a given by (see Section 5.4 for more details)

$$\Phi_a(t) := \frac{1}{2}\, \mathsf{n}_a(t)^{\mathsf{T}} \cdot \mathsf{H}\big(t - t_{\mathrm{r}}(t)\big) \cdot \mathsf{n}_a(t), \quad a = 1, 2, 3, \qquad (\mathrm{C.1})$$

where n_a are the column 3×1 matrices made of the components [computed in the *TT* or *"wave"* *reference frame* with coordinates $(t, x_{\mathrm{w}}, y_{\mathrm{w}}, z_{\mathrm{w}})$] of the unit 3-vectors $\mathbf{n}_a$ directed along the arms of the detector (so the subscript a labels different detector's arms, see Fig. 5.1), H is the 3×3 matrix of gravitational-wave-induced perturbation of space metric computed in the TT reference frame, and $t_{\mathrm{r}}(t)$ is some retardation. We assume here that the gravitational wave is propagating in the $+z_{\mathrm{w}}$ direction, so the components of the matrix H are given in Eq. (5.30).

Let us introduce the *detector's proper reference frame* with coordinates $(t, \hat{x}, \hat{y}, \hat{z})$. As this reference frame is comoving with the detector, the unit 3-vectors $\mathbf{n}_a$ directed along the detecor's arms have (approximately) constant components with respect to this frame. We arrange these components into the column 3×1 matrices $\hat{\mathsf{n}}_a$. The matrices $\hat{\mathsf{n}}_a$ and n_a (made from the components of the *same* 3-vector $\mathbf{n}_a$, but the components are computed with respect to *different* reference frames) are related to each other through the equation

$$\hat{n}_a = O(t) \cdot n_a(t), \qquad\qquad (C.2)$$

where O is the 3×3 *orthogonal* matrix of transformation from the wave to the detector's proper-reference-frame coordinates. An inversion of the relation (C.2) reads

$$n_a(t) = O(t)^T \cdot \hat{n}_a. \qquad\qquad (C.3)$$

Making use of Eq. (C.3), one can rewrite the quantities Φ_a in the form

$$\Phi_a(t) = \frac{1}{2} \hat{n}_a^T \cdot \hat{H}\big(t - t_r(t)\big) \cdot \hat{n}_a, \qquad\qquad (C.4)$$

where $\hat{H}$ is the matrix of gravitational-wave-induced perturbation of space metric computed in the detector's proper reference frame; it is related to the matrix H by the formula

$$\hat{H}\big(t - t_r(t)\big) = O(t) \cdot H\big(t - t_r(t)\big) \cdot O(t)^T. \qquad\qquad (C.5)$$

C.1 LISA detector

The quantities Φ_a introduced in Eq. (C.1) depend linearly on the components of the matrix H, therefore they can be written as linear combinations of the wave polarizations h_+ and $h_\times$:

$$\Phi_a(t) = F_{a+}(t)\, h_+\big(t - t_r(t)\big) + F_{a\times}(t)\, h_\times\big(t - t_r(t)\big). \qquad (C.6)$$

In the case of the LISA detector, the functions F_{a+} and $F_{a\times}$ are called the *beam-pattern functions* for that detector.

We choose the origin of the LISA proper reference frame (which is comoving with the spacecraft configuration) to coincide with the guiding center of the spacecraft configuration. The orthogonal matrix of transformation O from Eq. (C.2) can be written as

$$O(t) = O_2^{-1}(t) \cdot O_1, \qquad\qquad (C.7)$$

where O_1 is the orthogonal matrix of transformation from wave to *ecliptic* coordinates and O_2 is the orthogonal matrix of transformation from the detector's proper-reference-frame to ecliptic coordinates. The matrix O_1 depends on the ecliptic latitude β and the ecliptic longitude λ of the gravitational-wave source, and on the polarization angle ψ of the gravitational wave. It reads

$$\mathsf{O}_1 = \begin{pmatrix} \sin\lambda\cos\psi - \cos\lambda\sin\beta\sin\psi \\ -\cos\lambda\cos\psi - \sin\lambda\sin\beta\sin\psi \\ \cos\beta\sin\psi \end{pmatrix}$$

$$\begin{matrix} -\sin\lambda\sin\psi - \cos\lambda\sin\beta\cos\psi & -\cos\lambda\cos\beta \\ \cos\lambda\sin\psi - \sin\lambda\sin\beta\cos\psi & -\sin\lambda\cos\beta \\ \cos\beta\cos\psi & -\sin\beta \end{matrix} \Bigg). \quad (\text{C.8})$$

In comoving coordinates the 3-vector joining the LISA guiding center with the ath spacecraft has components

$$\hat{x}_a = \frac{L}{\sqrt{3}}\left(-\cos 2\sigma_a,\ \sin 2\sigma_a,\ 0\right)^{\mathsf{T}}, \quad a = 1, 2, 3, \quad (\text{C.9})$$

where the angles σ_a equal

$$\sigma_a := -\frac{3}{2}\pi + \frac{2}{3}(a-1)\pi, \quad a = 1, 2, 3. \quad (\text{C.10})$$

Let us also introduce the functions

$$\eta(t) := \Omega t + \eta_0, \quad (\text{C.11a})$$

$$\xi(t) := -\Omega t + \xi_0, \quad (\text{C.11b})$$

where $\Omega := 2\pi/(1\,\text{yr})$. The function η gives the *true anomaly* of the motion of the LISA guiding center around the Sun, and the function ξ returns the phase of the motion of each spacecraft around the guiding center. The matrix O_2 of transformation from LISA proper-reference-frame to ecliptic coordinates reads

$$\mathsf{O}_2(t) = \begin{pmatrix} \sin\eta(t)\cos\xi(t) - \sin\zeta\cos\eta(t)\sin\xi(t) \\ -\cos\eta(t)\cos\xi(t) - \sin\zeta\sin\eta(t)\sin\xi(t) \\ \cos\zeta\sin\xi(t) \end{pmatrix}$$

$$\begin{matrix} -\sin\eta(t)\sin\xi(t) - \sin\zeta\cos\eta(t)\cos\xi(t) & -\cos\zeta\cos\eta(t) \\ \cos\eta(t)\sin\xi(t) - \sin\zeta\sin\eta(t)\cos\xi(t) & -\cos\zeta\sin\eta(t) \\ \cos\zeta\cos\xi(t) & -\sin\zeta \end{matrix} \Bigg),$$

$$(\text{C.12})$$

where ζ is the inclination angle of the spacecraft configuration plane with respect to the ecliptic. For the LISA detector we have $\zeta = -\pi/6$.

Making use of Eqs. (C.4)–(C.12) one can show that the beam-pattern functions can be written as

$$F_{a+}(t) = u_a(t)\cos 2\psi + v_a(t)\sin 2\psi, \quad (\text{C.13a})$$

$$F_{a\times}(t) = v_a(t)\cos 2\psi - u_a(t)\sin 2\psi, \quad (\text{C.13b})$$

where the *amplitude modulation functions* u_a and v_a are given by

$$u_a(t) = U_0 \cos(-2\gamma_a) + U_1 \cos(\delta(t) - 2\gamma_a) + U_2 \cos(2\delta(t) - 2\gamma_a)$$
$$+ U_3 \cos(3\delta(t) - 2\gamma_a) + U_4 \cos(4\delta(t) - 2\gamma_a)$$
$$+ \left(\frac{1}{4} - \frac{3}{8}\cos^2\zeta\right)\cos^2\beta - \frac{1}{2}\sin\beta\cos\beta\cos\zeta\sin\zeta\cos\delta(t)$$
$$+ \frac{1}{4}\cos^2\zeta\left(1 - \frac{1}{2}\cos^2\beta\right)\cos 2\delta(t), \qquad \text{(C.14a)}$$

$$v_a(t) = V_0 \sin(-2\gamma_a) + V_1 \sin(\delta(t) - 2\gamma_a)$$
$$+ V_3 \sin(3\delta(t) - 2\gamma_a) + V_4 \sin(4\delta(t) - 2\gamma_a)$$
$$- \frac{1}{2}\cos\zeta\sin\zeta\cos\beta\sin\delta(t) + \frac{1}{4}\cos^2\zeta\sin\beta\sin 2\delta(t), \qquad \text{(C.14b)}$$

where we have defined

$$\delta(t) := \lambda - \eta_0 - \Omega t, \qquad \text{(C.15a)}$$
$$\gamma_a := \lambda - \eta_0 - \xi_0 - \sigma_a, \qquad \text{(C.15b)}$$

and where the constant coefficients U_k and V_k (for $k = 0, \ldots, 4$) are given by

$$U_0 = \frac{1}{16}(1 + \sin^2\beta)(1 - \sin\zeta)^2, \qquad \text{(C.16a)}$$

$$U_1 = -\frac{1}{8}\sin 2\beta\cos\zeta(1 - \sin\zeta), \qquad \text{(C.16b)}$$

$$U_2 = \frac{3}{8}\cos^2\beta\cos^2\zeta, \qquad \text{(C.16c)}$$

$$U_3 = \frac{1}{8}\sin 2\beta\cos\zeta(1 + \sin\zeta), \qquad \text{(C.16d)}$$

$$U_4 = \frac{1}{16}(1 + \sin^2\beta)(1 + \sin\zeta)^2, \qquad \text{(C.16e)}$$

$$V_0 = -\frac{1}{8}\sin\beta(1 - \sin\zeta)^2, \qquad \text{(C.16f)}$$

$$V_1 = \frac{1}{4}\cos\beta\cos\zeta(1 - \sin\zeta), \qquad \text{(C.16g)}$$

$$V_3 = \frac{1}{4}\cos\beta\cos\zeta(1 + \sin\zeta), \qquad \text{(C.16h)}$$

$$V_4 = \frac{1}{8}\sin\beta(1 + \sin\zeta)^2. \qquad \text{(C.16i)}$$

C.2 Earth-based detectors

In the case of Earth-based detectors we represent the matrix O [which describes the transformation from the wave to the detector's proper-reference-frame coordinates, see Eq. (C.2)] as

$$\mathsf{O}(t) = \mathsf{O}_3 \cdot \mathsf{O}_2(t) \cdot \mathsf{O}_1^\mathsf{T}, \qquad (\text{C.17})$$

where O_1 is the matrix of transformation from wave to *celestial* coordinates, O_2 is the matrix of transformation from celestial to *cardinal* coordinates, and O_3 is the matrix of transformation from cardinal to the detector's proper-reference-frame coordinates.

In celestial coordinates the z axis coincides with the Earth's rotation axis and points toward the North pole, the x and y axes lie in the Earth's equatorial plane, and the x axis points toward the vernal point. Under such conventions the matrix O_1 of transformation from wave to celestial coordinates is as follows

$$\mathsf{O}_1 = \begin{pmatrix} \sin\alpha\cos\psi - \cos\alpha\sin\delta\sin\psi & -\cos\alpha\cos\psi - \sin\alpha\sin\delta\sin\psi & \cos\delta\sin\psi \\ -\sin\alpha\sin\psi - \cos\alpha\sin\delta\cos\psi & \cos\alpha\sin\psi - \sin\alpha\sin\delta\cos\psi & \cos\delta\cos\psi \\ -\cos\alpha\cos\delta & -\sin\alpha\cos\delta & -\sin\delta \end{pmatrix}, \qquad (\text{C.18})$$

where the angles α and δ are respectively the right ascension and the declination of the gravitational-wave source, and ψ is the polarization angle of the wave.

In cardinal coordinates the (x, y) plane is tangential to the surface of the Earth at the detector's location with x axis in the North–South direction and y axis in the West–East direction, the z axis is along the Earth's radius pointing toward zenith. The matrix O_2 of transformation from cardinal to detector coordinates has components

$$\mathsf{O}_2(t) = \begin{pmatrix} \sin\phi\cos(\phi_r + \Omega_r t) & \sin\phi\sin(\phi_r + \Omega_r t) & -\cos\phi \\ -\sin(\phi_r + \Omega_r t) & \cos(\phi_r + \Omega_r t) & 0 \\ \cos\phi\cos(\phi_r + \Omega_r t) & \cos\phi\sin(\phi_r + \Omega_r t) & \sin\phi \end{pmatrix}, \qquad (\text{C.19})$$

where the angle ϕ is the geodetic latitude of the detector's site, Ω_r is the rotational angular velocity of the Earth, and the phase ϕ_r defines the position of the Earth in its diurnal motion at $t = 0$ (so the sum $\phi_r + \Omega_r t$ essentially coincides with the local sidereal time of the detector's site).

C.2.1 Interferometric detector

The noise-free response function s of the interferometric detector in the long-wavelength approximation is equal to the difference between the wave-induced relative length changes of the two interferometer arms. It can be computed from the formula (see Section 5.4.2 for a derivation)

$$s(t) = \frac{1}{2} \left(\hat{n}_1^\mathsf{T} \cdot \hat{H}(t) \cdot \hat{n}_1 - \hat{n}_2^\mathsf{T} \cdot \hat{H}(t) \cdot \hat{n}_2 \right), \tag{C.20}$$

where $\hat{n}_1$ and $\hat{n}_2$ denote 3×1 column matrices made of the proper-reference-frame components of the unit vectors $\mathbf{n}_1$ and $\mathbf{n}_2$ parallel to the arm number 1 and 2, respectively (the order of arms is defined such that the vector $\mathbf{n}_1 \times \mathbf{n}_2$ points *outwards* from the surface of the Earth). The z_d axis of the detector's coordinates $(x_\mathrm{d}, y_\mathrm{d}, z_\mathrm{d})$ coincides with the z axis of the cardinal coordinates, the x_d axis is along the first interferometer arm, then the y_d axis is along the second arm if the arms are at a right angle. The matrix O_3 of transformation from cardinal coordinates to detector coordinates has the form

$$O_3 = \begin{pmatrix} -\sin(\gamma - \zeta/2) & \cos(\gamma - \zeta/2) & 0 \\ -\cos(\gamma - \zeta/2) & -\sin(\gamma - \zeta/2) & 0 \\ 0 & 0 & 1 \end{pmatrix}, \tag{C.21}$$

where ζ is the angle between the interferometer arms (usually $90°$) and the angle γ determines the orientation of the detector's arms with respect to local geographical directions: γ is measured counter-clockwise from East to the bisector of the interferometer arms. The components of the unit vectors $\mathbf{n}_1$ and $\mathbf{n}_2$ in the detector's frame are as follows

$$\hat{n}_1 = (1, 0, 0)^\mathsf{T}, \quad \hat{n}_2 = (\cos\zeta, \sin\zeta, 0)^\mathsf{T}. \tag{C.22}$$

To find the explicit formulae for the beam-pattern functions F_+ and $F_\times$ one has to combine Eqs. (C.17)–(C.22). After extensive algebraic manipulations we arrive at the expressions

$$F_+(t) = \sin\zeta \left(a(t) \cos 2\psi + b(t) \sin 2\psi \right), \tag{C.23a}$$

$$F_\times(t) = \sin\zeta \left(b(t) \cos 2\psi - a(t) \sin 2\psi \right), \tag{C.23b}$$

where $a(t)$ and $b(t)$ are the *amplitude modulation functions*:

$$a(t) = \frac{1}{4} \sin 2\gamma \, (1 + \sin^2\phi) \, (1 + \sin^2\delta) \cos 2\eta(t)$$

$$- \frac{1}{2} \cos 2\gamma \, \sin\phi \, (1 + \sin^2\delta) \sin 2\eta(t)$$

$$+ \frac{1}{4} \sin 2\gamma \, \sin 2\phi \, \sin 2\delta \, \cos\eta(t)$$

$$-\frac{1}{2}\cos 2\gamma \cos\phi \sin 2\delta \sin\eta(t)$$

$$+\frac{3}{4}\sin 2\gamma \cos^2\phi \cos^2\delta, \tag{C.24a}$$

$$b(t) = \cos 2\gamma \sin\phi \sin\delta \cos 2\eta(t)$$

$$+\frac{1}{2}\sin 2\gamma \left(1+\sin^2\phi\right)\sin\delta \sin 2\eta(t)$$

$$+\cos 2\gamma \cos\phi \cos\delta \cos\eta(t)$$

$$+\frac{1}{2}\sin 2\gamma \sin 2\phi \cos\delta \sin\eta(t), \tag{C.24b}$$

and where we have defined

$$\eta(t) := \alpha - \phi_r - \Omega_r t. \tag{C.25}$$

C.2.2 Resonant-bar detector

The noise-free response of the resonant-bar detector can be computed from the formula (see Section 5.4.3)

$$s(t) = \hat{n}^{\mathsf{T}} \cdot \hat{H}(t) \cdot \hat{n}, \tag{C.26}$$

where the column matrix $\hat{n}$ is made of the proper-reference-frame components of the unit vector along the symmetry axis of the bar.

In the detector's proper reference frame we introduce Cartesian coordinates (x_d, y_d, z_d) with the z_d axis along the Earth's radius pointing toward zenith, and the x_d axis along the bar's axis of symmetry. In these coordinates the unit vector along the symmetry axis of the bar has components

$$\hat{n} = (1, 0, 0)^{\mathsf{T}}. \tag{C.27}$$

The matrix O_3 of transformation from cardinal coordinates to detector coordinates has the form

$$O_3 = \begin{pmatrix} -\sin\gamma & \cos\gamma & 0 \\ -\cos\gamma & -\sin\gamma & 0 \\ 0 & 0 & 1 \end{pmatrix}, \tag{C.28}$$

where the angle γ determines the orientation of the bar with respect to local geographical directions: it is measured counter-clockwise from the East to the bar's axis of symmetry.

To find the explicit formulae for the beam-pattern functions F_+ and $F_\times$ we have to combine Eqs. (C.26)–(C.28) together with Eqs. (C.17)–(C.19).

After some algebraic manipulations we arrive at the expressions

$$F_+(t) = a(t)\cos 2\psi + b(t)\sin 2\psi, \tag{C.29a}$$

$$F_\times(t) = b(t)\cos 2\psi - a(t)\sin 2\psi, \tag{C.29b}$$

where $a(t)$ and $b(t)$ are the *amplitude modulation functions*:

$$a(t) = \frac{1}{2}\left(\cos^2\gamma - \sin^2\gamma\,\sin^2\phi\right)\left(1 + \sin^2\delta\right)\cos 2\eta(t)$$

$$+ \frac{1}{2}\sin 2\gamma\,\sin\phi\,(1 + \sin^2\delta)\,\sin 2\eta(t)$$

$$- \frac{1}{2}\sin^2\gamma\,\sin 2\phi\,\sin 2\delta\,\cos\eta(t)$$

$$+ \frac{1}{2}\sin 2\gamma\,\cos\phi\,\sin 2\delta\,\sin\eta(t)$$

$$+ \frac{1}{2}\left(1 - 3\sin^2\gamma\,\cos^2\phi\right)\cos^2\delta, \tag{C.30a}$$

$$b(t) = -\sin 2\gamma\,\sin\phi\,\sin\delta\,\cos 2\eta(t)$$

$$+ \left(\cos^2\gamma - \sin^2\gamma\,\sin^2\phi\right)\sin\delta\,\sin 2\eta(t)$$

$$- \sin 2\gamma\,\cos\phi\,\cos\delta\,\cos\eta(t)$$

$$- \sin^2\gamma\,\sin 2\phi\,\cos\delta\,\sin\eta(t), \tag{C.30b}$$

here $\eta(t)$ is given by Eq. (C.25).

Appendix D

Response of the LISA detector to an almost monochromatic wave

In this appendix we summarize the analytic formulae for the LISA detector response of the *first-generation TDI Michelson observables* X, Y, and Z to gravitational waves emitted by moderately chirping binary systems. In terms of the delayed one-way Doppler frequency shifts y_{ab} $(a, b = 1, 2, 3)$ the observable X is given by [226]

$$X = (y_{31} + y_{13,2}) + (y_{21} + y_{12,3}),_{22} - (y_{21} + y_{12,3}) - (y_{31} + y_{13,2}),_{33}.$$

$$(D.1)$$

The observables Y and Z can be obtained from Eq. (D.1) by cyclical permutation of the spacecraft indices.

Wave polarization functions for gravitational waves emitted by moderately chirping binary systems have the following form (see discussion in Section IID of [32]):

$$h_+(t) = h_0 \frac{1 + \cos^2 \iota}{2} \cos\left(\omega t + \frac{1}{2}\dot{\omega} t^2 + \phi_0\right), \qquad (D.2a)$$

$$h_\times(t) = h_0 \cos \iota \, \sin\left(\omega t + \frac{1}{2}\dot{\omega} t^2 + \phi_0\right), \qquad (D.2b)$$

where ω is the angular frequency of the gravitational wave evaluated at $t = 0$, $\dot{\omega}$ is approximately constant angular frequency drift, ϕ_0 is a constant initial phase, ι is the angle between the normal to the orbital plane and the direction of propagation of the gravitational wave, and h_0 is the approximately constant amplitude of the waveforms [see Eq. (2.40)],

$$h_0 = \frac{4(G\mathcal{M})^{5/3}}{c^4 D} \left(\frac{\omega}{2}\right)^{2/3}. \qquad (D.3)$$

229

Here $\mathcal{M} := (m_1 m_2)^{3/5}/(m_1 + m_2)^{1/5}$ is the chirp mass of the binary (m_1 and m_2 are the individual masses of the binary's components) and D is the luminosity distance to the source. The frequency drift may be either due to the gravitational radiation reaction or as a result of the tidal interaction between the components of the binary system. In the case of the gravitational-radiation-reaction driven evolution the frequency drift $\dot{\omega}$ is approximately given by [see Eq. (A.18)]

$$\dot{\omega} = \frac{48}{5}\left(\frac{G\mathcal{M}}{2\,c^3}\right)^{5/3} \omega^{11/3}. \tag{D.4}$$

Making use of Eqs. (D.2) and several times of Eq. (5.44) for the Doppler shifts y_{ab} (computed for different spacecraft labels a, b and different retardations), the Michelson observable X from Eq. (D.1) can be written as a linear combination of the four time-dependent functions h_{Xk} ($k = 1, \ldots, 4$; see Appendix C of [32]):

$$X(t) = 2\,\omega L\,\sin(\omega L) \sum_{k=1}^{4} A_k\, h_{Xk}(t), \tag{D.5}$$

where L is the (approximately constant) distance between the spacecraft in the LISA detector and A_k ($k = 1, \ldots, 4$) are constant amplitudes [they are given by Eqs. (E.1) of Appendix E]. The functions h_{Xk} read (here $x := \omega L$ and $\operatorname{sinc} y := \sin y / y$)

$$\begin{pmatrix} h_{X1}(t) \\ h_{X2}(t) \end{pmatrix} = \begin{pmatrix} u_2(t) \\ v_2(t) \end{pmatrix} \left\{ \operatorname{sinc}\left(\frac{x}{2}(1 + c_2(t))\right) \cos\left(\phi(t) - x\left(d_2(t) + \frac{3}{2}\right)\right) \right.$$

$$\left. + \operatorname{sinc}\left(\frac{x}{2}(1 - c_2(t))\right) \cos\left(\phi(t) - x\left(d_2(t) + \frac{5}{2}\right)\right) \right\}$$

$$- \begin{pmatrix} u_3(t) \\ v_3(t) \end{pmatrix} \left\{ \operatorname{sinc}\left(\frac{x}{2}(1 + c_3(t))\right) \cos\left(\phi(t) - x\left(d_3(t) + \frac{5}{2}\right)\right) \right.$$

$$\left. + \operatorname{sinc}\left(\frac{x}{2}(1 - c_3(t))\right) \cos\left(\phi(t) - x\left(d_3(t) + \frac{3}{2}\right)\right) \right\}, \tag{D.6a}$$

$$\begin{pmatrix} h_{X3}(t) \\ h_{X4}(t) \end{pmatrix} = \begin{pmatrix} u_2(t) \\ v_2(t) \end{pmatrix} \left\{ \operatorname{sinc}\left(\frac{x}{2}(1 + c_2(t))\right) \sin\left(\phi(t) - x\left(d_2(t) + \frac{3}{2}\right)\right) \right.$$

$$\left. + \operatorname{sinc}\left(\frac{x}{2}(1 - c_2(t))\right) \sin\left(\phi(t) - x\left(d_2(t) + \frac{5}{2}\right)\right) \right\}$$

$$- \begin{pmatrix} u_3(t) \\ v_3(t) \end{pmatrix} \left\{ \operatorname{sinc}\left(\frac{x}{2}(1+c_3(t))\right) \sin\left(\phi(t) - x\left(d_3(t) + \frac{5}{2}\right)\right) \right.$$

$$+ \operatorname{sinc}\left(\frac{x}{2}(1-c_3(t))\right) \sin\left(\phi(t) - x\left(d_3(t) + \frac{3}{2}\right)\right) \right\}.$$

$$\text{(D.6b)}$$

The functions h_{Xk} depend on several other functions of the time t: ϕ, u_a, v_a, c_a, and d_a ($a = 1, 2, 3$). The phase modulation function ϕ is given by (see Section IIE of [32])

$$\phi(t) = \omega t + \frac{1}{2}\dot{\omega} t^2 + (\omega + \dot{\omega} t) R \cos\beta \cos\left(\Omega t + \eta_0 - \lambda\right), \quad \text{(D.7)}$$

where the angles β and λ are respectively the latitude and the longitude of the source in ecliptic coordinates, $\Omega := 2\pi/(1\,\text{yr})$, η_0 is the position of the constellation on the orbit around the Sun at time $t = 0$, and R is 1 astronomical unit. The amplitude modulation functions u_a and v_a are explicitly given in Eqs. (C.14) of Appendix C.1. The time-dependent modulation functions c_a and d_a are given by

$$c_a(t) = \frac{3}{4}\cos\beta\sin\gamma_a - \frac{\sqrt{3}}{2}\sin\beta\sin(\delta(t) - \gamma_a)$$

$$+ \frac{1}{4}\cos\beta\sin(2\delta(t) - \gamma_a), \quad \text{(D.8a)}$$

$$d_a(t) = \frac{\sqrt{3}}{8}\cos\beta\sin\gamma_{2a} - \frac{1}{4}\sin\beta\sin(\delta(t) - \gamma_{2a})$$

$$+ \frac{\sqrt{3}}{24}\cos\beta\sin(2\delta(t) - \gamma_{2a}), \quad \text{(D.8b)}$$

where the function δ and the constants γ_a, γ_{2a}, σ_a ($a = 1, 2, 3$) read

$$\delta(t) := \lambda - \eta_0 - \Omega t, \quad \text{(D.9a)}$$

$$\gamma_a := \lambda - \eta_0 - \xi_0 - \sigma_a, \quad \text{(D.9b)}$$

$$\gamma_{2a} := \lambda - \eta_0 - \xi_0 + 2\sigma_a, \quad \text{(D.9c)}$$

$$\sigma_a := -\frac{3}{2}\pi + \frac{2}{3}(a-1)\pi. \quad \text{(D.9d)}$$

The constant ξ_0 determines the initial orientation of the LISA constellation at time $t = 0$. For synthetic LISA [227] conventions we have

$$\eta_0 = 0, \quad \xi_0 = 3\pi/2. \quad \text{(D.10)}$$

In the long-wavelength approximation the gravitational-wave responses are obtained by taking the leading-order terms of the generic expressions in the limit of $\omega L \to 0$. Let us denote by M the array of Michelson responses (X, Y, Z), i.e. let $M_1 \equiv X$, $M_2 \equiv Y$, and $M_3 \equiv Z$. In the long-wavelength approximation we get

$$M_i(t) \cong 4\sqrt{3}(\omega L)^2 \Big\{ -a_i(t)\big[A_1 \cos \phi(t) + A_3 \sin \phi(t)\big]$$

$$+ b_i(t)\big[A_2 \cos \phi(t) + A_4 \sin \phi(t)\big] \Big\}, \quad i = 1, 2, 3,$$

$$\text{(D.11)}$$

where the modulation functions a_i and b_i $(i = 1, 2, 3)$ equal

$$a_i(t) = U_0 \sin\left(2\gamma_{Li}\right) + U_1 \sin\left(2\gamma_{Li} - \delta(t)\right) + U_2 \sin\left(2\gamma_{Li} - 2\delta(t)\right)$$

$$+ U_3 \sin\left(2\gamma_{Li} - 3\delta(t)\right) + U_4 \sin\left(2\gamma_{Li} - 4\delta(t)\right), \quad \text{(D.12a)}$$

$$b_i(t) = V_0 \cos\left(2\gamma_{Li}\right) + V_1 \cos\left(2\gamma_{Li} - \delta(t)\right)$$

$$+ V_3 \cos\left(2\gamma_{Li} - 3\delta(t)\right) + V_4 \cos\left(2\gamma_{Li} - 4\delta(t)\right). \quad \text{(D.12b)}$$

Here U_k and V_k $(k = 0, \ldots, 4)$ are given by Eqs. (C.16) of Appendix C.1 and we have defined

$$\gamma_{Li} := \lambda - \eta_0 - \xi_0 + \frac{1}{2}\sigma_i + \frac{5}{4}\pi, \quad i = 1, 2, 3. \quad \text{(D.13)}$$

Appendix E
Amplitude parameters of periodic waves

Both for the case of the response of the ground-based detector to gravitational waves from a rotating neutron star and for the case of the response of the LISA detector to gravitational waves emitted by moderately chirping binary system (e.g. made of white dwarfs, see Ref. [32]), the four constant amplitude (or extrinsic) parameters A_k ($k = 1, \ldots, 4$) are given by the same set of algebraic equations,

$$A_1 = h_{0+} \cos 2\psi \cos \phi_0 - h_{0\times} \sin 2\psi \sin \phi_0, \tag{E.1a}$$

$$A_2 = h_{0+} \sin 2\psi \cos \phi_0 + h_{0\times} \cos 2\psi \sin \phi_0, \tag{E.1b}$$

$$A_3 = -h_{0+} \cos 2\psi \sin \phi_0 - h_{0\times} \sin 2\psi \cos \phi_0, \tag{E.1c}$$

$$A_4 = -h_{0+} \sin 2\psi \sin \phi_0 + h_{0\times} \cos 2\psi \cos \phi_0, \tag{E.1d}$$

where ψ is the polarization angle of the gravitational wave and ϕ_0 is the constant initial phase parameter. The amplitudes h_{0+}, $h_{0\times}$ of the individual wave polarizations are

$$h_{0+} = h_0 \frac{1 + \cos^2 \iota}{2}, \tag{E.2a}$$

$$h_{0\times} = h_0 \cos \iota, \tag{E.2b}$$

where the angle ι is the angle between the line of sight and the rotation axis of the star or the orbital angular momentum vector of the binary system.

One can invert Eqs. (E.1) to obtain formulae for the parameters h_{0+}, $h_{0\times}$, ϕ_0, and ψ as functions of the amplitudes A_k ($k = 1, \ldots, 4$). Let us introduce quantities

$$A := A_1^2 + A_2^2 + A_3^2 + A_4^2, \tag{E.3a}$$

$$D := A_1 A_4 - A_2 A_3. \tag{E.3b}$$

Then the amplitudes h_{0+} and $h_{0\times}$ can be uniquely determined from the relations (we assume here, without loss of generality, that $h_{0+} \geq 0$)

$$h_{0+} = \sqrt{\frac{1}{2}\left(A + \sqrt{A^2 - 4D^2}\right)}, \tag{E.4a}$$

$$h_{0\times} = \text{sign}(D)\sqrt{\frac{1}{2}\left(A - \sqrt{A^2 - 4D^2}\right)}. \tag{E.4b}$$

The initial phase ϕ_0 and the polarization angle ψ can be obtained from the following equations:

$$\tan 2\phi_0 = \frac{2(A_1 A_3 + A_2 A_4)}{A_3^2 + A_4^2 - A_1^2 - A_2^2}, \tag{E.5a}$$

$$\tan 4\psi = \frac{2(A_1 A_2 + A_3 A_4)}{A_1^2 + A_3^2 - A_2^2 - A_4^2}. \tag{E.5b}$$

Also Eqs. (6.86) can be solved for the amplitude h_0 and the angle ι. The result is

$$h_0 = h_{0+} + \sqrt{h_{0+}^2 - h_{0\times}^2}, \tag{E.6a}$$

$$\iota = \arccos\left(\frac{h_{0\times}}{h_0}\right). \tag{E.6b}$$

References

[1] J. M. Weisberg and J. H. Taylor. The relativistic binary pulsar B1913+16: Thirty years of observations and analysis. In F. A. Rasio and I. H. Stairs, eds., *Binary Radio Pulsars, Astronomical Society of the Pacific Conference Series*, vol. 328, p. 25, 2005.

[2] B. Allen, J. D. E. Creighton, É. É. Flanagan, and J. D. Romano. Robust statistics for deterministic and stochastic gravitational waves in non-Gaussian noise: Frequentist analyses. *Phys. Rev. D*, **65**:122002, 1–18, 2002.

[3] B. Allen, J. D. E. Creighton, É. É. Flanagan, and J. D. Romano. Robust statistics for deterministic and stochastic gravitational waves in non-Gaussian noise. II. Bayesian analyses. *Phys. Rev. D*, **67**:122002, 1–13, 2003.

[4] É. Chassande-Mottin and A. Pai. Best chirplet chain: Near-optimal detection of gravitational wave chirps. *Phys. Rev. D*, **73**:042003, 1–23, 2006.

[5] A. Pai, É. Chassande-Mottin, and O. Rabaste. Best network chirplet chain: Near-optimal coherent detection of unmodeled gravitational wave chirps with a network of detectors. *Phys. Rev. D*, **77**:062005, 1–22, 2008.

[6] R. Flaminio, L. Massonnet, B. Mours, S. Tissot, D. Verkindt, and M. Yvert. Fast trigger algorithms for binary coalescences. *Astropart. Phys.*, **2**, 235–248, 1994.

[7] A. Królak and P. Trzaskoma. Application of wavelet analysis to estimation of parameters of the gravitational-wave signal from a coalescing binary. *Class. Quantum Grav.*, **13**, 813–830, 1996.

[8] J.-M. Innocent and B. Torrèsani. Wavelet transform and binary coalescence detection. In A. Królak, ed., *Mathematics of Gravitation. Part II. Gravitational Wave Detection*. Warszawa: Banach Center Publications, pp. 179–208, 1997.

[9] F. Acernese, M. Alshourbagy, P. Amico, *et al.* Search for gravitational waves associated with GRB 050915a using the Virgo detector. *Class. Quantum Grav.*, **25**:225001, 1–20, 2008.

[10] B. Krishnan, A. M. Sintes, M. A. Papa, B. F. Schutz, S. Frasca, and C. Palomba. Hough transform search for continuous gravitational waves. *Phys. Rev. D*, **70**:082001, 1–23, 2004.

[11] B. Abbott, R. Abbott, R. Adhikari, *et al.* First all-sky upper limits from LIGO on the strength of periodic gravitational waves using the Hough transform. *Phys. Rev. D*, **72**:122004, 1–22, 2005.

[12] P. R. Brady and T. Creighton. Searching for periodic sources with LIGO. II. Hierarchical searches. *Phys. Rev. D*, **61**:082001, 1–20, 2000.

[13] B. Abbott, R. Abbott, R. Adhikari, *et al.* All-sky search for periodic gravitational waves in LIGO S4 data. *Phys. Rev. D*, **77**:022001, 1–38, 2008.

[14] J. Weber. Evidence for discovery of gravitational radiation. *Phys. Rev. Lett.*, **22**, 1320–1324, 1969.

[15] P. Kafka. Optimal detection of signals through linear devices with thermal noise sources and application to the Munich-Frascati Weber-type gravitational wave detectors. In V. De Sabbata and J. Weber, eds., *Topics in Theoretical and Experimental Gravitation Physics, NATO ASI Series*, vol. B27. Dordrecht: Reidel Publishing Company, p. 161, 1977.

[16] K. S. Thorne. Gravitational radiation. In S. W. Hawking and W. Israel, eds., *Three Hundred Years of Gravitation*. Cambridge: Cambridge University Press, pp. 330–458, 1987.

[17] B. F. Schutz. Determining the nature of the Hubble constant. *Nature*, **323**:310, 1986.

[18] A. Królak and B. F. Schutz. Coalescing binaries – probe to the universe. *Gen. Rel. Grav.*, **19**, 1163–1171, 1987.

[19] B. F. Schutz, ed. *Gravitational Wave Data Analysis. Proceedings of the NATO Advanced Research Workshop held at St. Nicholas, Cardiff, Wales, July 6–9, 1987, NATO ASI Series*, vol. C253. Dordrecht: Kluwer, 1989.

[20] L. S. Finn. Observing binary inspiral in gravitational radiation: one interferometer. *Phys. Rev. D*, **47**, 2198–2219, 1993.

[21] A. Królak, J. A. Lobo, and B. J. Meers. Estimation of the parameters of the gravitational-wave signal of a coalescing binary system. *Phys. Rev. D*, **48**, 3451–3462, 1993.

[22] C. Cutler and É. É. Flanagan. Gravitational waves from merging compact binaries: how accurately can one extract the binary's parameters from the inspiral waveform? *Phys. Rev. D*, **49**, 2658–2697, 1994.

[23] A. Królak, D. Kokkotas, and G. Schäfer. On estimation of the post-Newtonian parameters in the gravitational-wave emission of a coalescing binary. *Phys. Rev. D*, **52**, 2089–2111, 1995.

[24] R. Balasubramanian and S. V. Dhurandhar. Estimation of parameters of gravitational waves from coalescing binaries. *Phys. Rev. D*, **57**, 3408–3422, 1998.

[25] K. G. Arun, B. R. Iyer, B. S. Sathyaprakash, and P. A. Sundararajan. Parameter estimation of inspiralling compact binaries using 3.5 post-Newtonian gravitational wave phasing: the nonspinning case. *Phys. Rev. D*, **71**:084008, 1–16, 2005.

[26] B. F. Schutz. Data processing, analysis and storage for interferometric antennas. In D. G. Blair, ed., *The Detection of Gravitational Waves*. Cambridge: Cambridge University Press, pp. 406–452, 1991.

[27] C. Cutler, T. A. Apostolatos, L. Bildsten, *et al.* The last three minutes: Issues in gravitational wave measurements of coalescing compact binaries. *Phys. Rev. Lett.*, **70**, 2984–2987, 1993.

[28] T. A. Apostolatos. Search templates for gravitational waves from precessing, inspiraling binaries. *Phys. Rev. D*, **52**, 605–620, 1995.

[29] B. J. Owen. Search templates for gravitational waves from inspiraling binaries: Choice of template spacing. *Phys. Rev. D*, **53**, 6749–6761, 1996.

[30] P. R. Brady, T. Creighton, C. Cutler, and B. F. Schutz. Searching for periodic sources with LIGO. *Phys. Rev. D*, **57**, 2101–2116, 1998.

[31] P. Jaranowski, A. Królak, and B. F. Schutz. Data analysis of gravitational-wave signals from spinning neutron stars: The signal and its detection. *Phys. Rev. D*, **58**:063001, 1–24, 1998.

[32] A. Królak, M. Tinto, and M. Vallisneri. Optimal filtering of the LISA data. *Phys. Rev. D*, **70**:022003, 1–24, 2004.

[33] A. Królak, M. Tinto, and M. Vallisneri. Erratum: Optimal filtering of the LISA data (*Phys. Rev. D*, **70**:022003, 2004). *Phys. Rev. D*, **76**:069901(E) 1, 2007.

[34] C. W. Misner, K. S. Thorne, and J. A. Wheeler. *Gravitation*. San Francisco, CA: Freeman, 1973.

[35] B. F. Schutz. *A First Course in General Relativity*. Cambridge: Cambridge University Press, 1985.

[36] S. Carroll. *Spacetime and Geometry. An Introduction to General Relativity*. San Francisco, CA: Addison Wesley, 2004.

[37] M. Maggiore. *Gravitational Waves*, vol. 1. *Theory and Experiments*. Oxford: Oxford University Press, 2007.

[38] K. S. Thorne. The theory of gravitational radiation: An introductory review. In N. Deruelle and T. Piran, eds., *Gravitational Radiation, NATO ASI*. Amsterdam: North-Holland, pp. 1–57, 1983.

[39] R. A. Isaacson. Gravitational radiation in the limit of high frequency. I. The linear approximation and geometrical optics. *Phys. Rev.*, **166**, 1263–1271, 1968.

[40] R. A. Isaacson. Gravitational radiation in the limit of high frequency. II. Nonlinear terms and the effective stress tensor. *Phys. Rev.*, **166**, 1272–1280, 1968.

[41] A. Einstein. Näherungsweise Integration der Feldgleichungen der Gravitation. *Preussische Akademie der Wissenschaften, Sitzungsberichte der physikalisch-mathematischen Klasse*, pp. 688–696, 1916.

[42] A. Einstein. Gravitationswellen. *Preussische Akademie der Wissenschaften, Sitzungsberichte der physikalisch-mathematischen Klasse*, pp. 154–167, 1918.

[43] P. C. Peters. Gravitational radiation and the motion of two point masses. *Phys. Rev.*, **136**, B1224–B1232, 1964.

[44] T. Damour. The problem of motion in Newtonian and Einsteinian gravity. In S. W. Hawking and W. Israel, eds., *Three Hundred Years of Gravitation*, Cambridge: Cambridge University Press, pp. 128–198, 1987.

[45] J. Hough, B. J. Meers, G. P. Newton, *et al.* Gravitational wave astronomy – potential and possible realisation. *Vistas Astron.*, **30**, 109–134, 1986.

[46] J. A. Marck and J. P. Lasota, eds. *Relativistic Gravitation and Gravitational Radiation: Proceedings of the Les Houches School on Astrophysical Sources of Gravitational Radiation.* Cambridge: Cambridge University Press, 1997.

[47] C. Cutler and K. S. Thorne. An overview of gravitational-wave sources. In N. T. Bishop and S. D. Maharaj, eds., *General Relativity and Gravitation. Proceedings of the 16th International Conference*, Durban, South Africa, 15–21 July 2001. Singapore: World Scientific, pp. 72–111, 2002.

[48] S. L. Shapiro and S. A. Teukolsky. *Black Holes, White Dwarfs, and Neutron Stars. The Physics of Compact Objects.* New York: Wiley, 1983.

[49] H.-Th. Janka, K. Langanke, A. Marek, G. Martínez-Pinedo, and B. Müller. Theory of core-collapse supernovae. *Phys. Rep.*, **442**, 38–74, 2007.

[50] C. D. Ott, A. Burrows, L. Dessart, and E. Livne. A new mechanism for gravitational-wave emission in core-collapse supernovae. *Phys. Rev. Lett.*, **96**:201102, 1–4, 2006.

[51] H. Dimmelmeier, C. D. Ott, H.-Th. Janka, A. Marek, and E. Müller. Generic gravitational-wave signals from the collapse of rotating stellar cores. *Phys. Rev. Lett.*, **98**:251101, 1–4, 2007.

[52] C. L. Fryer and K. C. B. New. Gravitational waves from gravitational collapse. *Liv. Rev. Relat.*, **6**, 2003.

[53] B. Abbott, R. Abbott, R. Adhikari, *et al.* Beating the spin-down limit on gravitational wave emission from the Crab pulsar. *Ap. J.*, **683**, L45–L49, 2008.

[54] M. Maggiore. Gravitational wave experiments and early universe cosmology. *Phys. Rep.*, **331**, 283–367, 2000.

[55] A. Barrau, A. Gorecki, and J. Grain. An original constraint on the Hubble constant: $h > 0.74$. *Mon. Not. R. Astron. Soc.*, **389**, 919–924, 2008.

[56] T. Damour. Gravitational radiation and the motion of compact bodies. In N. Deruelle and T. Piran, eds., *Gravitational Radiation, NATO ASI.* Amsterdam: North-Holland, pp. 59–144, 1983.

[57] F. Pretorius. Gravitational waves and the death spiral of compact objects: Neutron star and black hole mergers in numerical simulations. In M. Colpi *et al.*, eds. *Physics of Relativistic Objects in Compact Binaries: from Birth to Coalescence, Astrophysics and Space Science Library*, vol. 359. Berlin: Springer-Verlag, pp. 305–369, 2009.

[58] G. Schäfer. The gravitational quadrupole radiation-reaction force and the canonical formalism of ADM. *Ann. Phys. (N. Y.)*, **161**, 81–100, 1985.

[59] L. Blanchet. Gravitational radiation from post-Newtonian sources and inspiralling compact binaries. *Liv. Rev. Relat.*, **9**, 2006.

[60] M. Sasaki and H. Tagoshi. Analytic black hole perturbation approach to gravitational radiation. *Liv. Rev. Relat.*, **6**, 2003.

[61] T. Futamase and Y. Itoh. The post-Newtonian approximation for relativistic compact binaries. *Liv. Rev. Relat.*, **10**, 2007.

[62] M. E. Pati and C. M. Will. Post-Newtonian gravitational radiation and equations of motion via direct integration of the relaxed Einstein equations: Foundations. *Phys. Rev. D*, **62**:124015, 1–28, 2000.

[63] M. E. Pati and C. M. Will. Post-Newtonian gravitational radiation and equations of motion via direct integration of the relaxed Einstein equations. II. Two-body equations of motion to second post-Newtonian order, and radiation-reaction to 3.5 post-Newtonian order. *Phys. Rev. D*, **65**:104008, 1–21, 2002.

[64] C. M. Will. Post-Newtonian gravitational radiation and equations of motion via direct integration of the relaxed Einstein equations. III. Radiation reaction for binary systems with spinning bodies. *Phys. Rev. D*, **71**:084027, 1–15, 2005.

[65] Han Wang and C. M. Will. Post-Newtonian gravitational radiation and equations of motion via direct integration of the relaxed Einstein equations. IV. Radiation reaction for binary systems with spin–spin coupling. *Phys. Rev. D*, **75**:064017, 1–8, 2007.

[66] Th. Mitchell and C. M. Will. Post-Newtonian gravitational radiation and equations of motion via direct integration of the relaxed Einstein equations. V. Evidence for the strong equivalence principle to second post-Newtonian order. *Phys. Rev. D*, **75**:124025, 1–12, 2007.

[67] W. D. Goldberger and I. Z. Rothstein. Effective field theory of gravity for extended objects. *Phys. Rev. D*, **73**:104029, 1–22, 2006.

[68] P. Jaranowski and G. Schäfer. Third post-Newtonian higher order ADM Hamilton dynamics for two-body point-mass systems. *Phys. Rev. D*, **57**, 7274–7291, 1998.

[69] P. Jaranowski and G. Schäfer. Erratum: third post-Newtonian higher order ADM Hamilton dynamics for two-body point-mass systems (*Phys. Rev. D*, **57**, 7274, 1998). *Phys. Rev. D*, **63**:029902(E), 1, 2000.

[70] P. Jaranowski and G. Schäfer. Binary black-hole problem at the third post-Newtonian approximation in the orbital motion: Static part. *Phys. Rev. D*, **60**:124003, 1–7, 1999.

[71] T. Damour, P. Jaranowski, and G. Schäfer. Dynamical invariants for general relativistic two-body systems at the third post-Newtonian approximation. *Phys. Rev. D*, **62**:044024, 1–17, 2000.

[72] T. Damour, P. Jaranowski, and G. Schäfer. Poincaré invariance in the ADM Hamiltonian approach to the general relativistic two-body problem. *Phys. Rev. D*, **62**:021501(R), 1–5, 2000.

[73] T. Damour, P. Jaranowski, and G. Schäfer. Erratum: Poincaré invariance in the ADM Hamiltonian approach to the general relativistic two-body problem (*Phys. Rev. D*, **62**, 021501(R), 2000). *Phys. Rev. D*, **63**:029903(E), 1, 2000.

[74] T. Damour, P. Jaranowski, and G. Schäfer. Dimensional regularization of the gravitational interaction of point masses. *Phys. Lett. B*, **513**, 147–155, 2001.

[75] L. Blanchet and G. Faye. On the equations of motion of point-particle binaries at the third post-Newtonian order. *Phys. Lett. A*, **271**, 58–64, 2000.

[76] L. Blanchet and G. Faye. General relativistic dynamics of compact binaries at the third post-Newtonian order. *Phys. Rev. D*, **63**:062005, 1–43, 2001.

[77] L. Blanchet, T. Damour, and G. Esposito-Farése. Dimensional regularization of the third post-Newtonian dynamics of point particles in harmonic coordinates. *Phys. Rev. D*, **69**:124007, 1–51, 2004.

[78] Y. Itoh and T. Futamase. New derivation of a third post-Newtonian equation of motion for relativistic compact binaries without ambiguity. *Phys. Rev. D*, **68**:121501(R), 1–5, 2003.

[79] Y. Itoh. Equation of motion for relativistic compact binaries with the strong field point particle limit: Third post-Newtonian order. *Phys. Rev. D*, **69**:064018, 1–43, 2004.

[80] P. Jaranowski and G. Schäfer. Radiative 3.5 post-Newtonian ADM Hamiltonian for many-body point-mass systems. *Phys. Rev. D*, **55**, 4712–4722, 1997.

[81] Ch. Königsdörffer, G. Faye, and G. Schäfer. Binary black-hole dynamics at the third-and-a-half post-Newtonian order in the ADM formalism. *Phys. Rev. D*, **68**:044004, 1–19, 2003.

[82] L. Blanchet, T. Damour, B. R. Iyer, C. M. Will, and A. G. Wiseman. Gravitational-radiation damping of compact binary systems to second post-Newtonian order. *Phys. Rev. Lett.*, **74**, 3515–3518, 1995.

[83] L. Blanchet. Gravitational-wave tails of tails. *Class. Quantum Grav.*, **15**, 113–141, 1998.

[84] L. Blanchet. Corrigendum: Gravitational-wave tails of tails (*Class. Quantum Grav.* **15**, 113, 1998). *Class. Quantum Grav.*, **22**, 3381, 2005.

[85] L. Blanchet, B. R. Iyer, and B. Joguet. Gravitational waves from inspiralling compact binaries: Energy flux to third post-Newtonian order. *Phys. Rev. D*, **65**:064005, 1–41, 2002.

[86] L. Blanchet, G. Faye, B. R. Iyer, and B. Joguet. Gravitational-wave inspiral of compact binary systems to 7/2 post-Newtonian order. *Phys. Rev. D*, **65**:061501(R), 1–5, 2002.

[87] L. Blanchet, G. Faye, B. R. Iyer, and B. Joguet. Erratum: Gravitational-wave inspiral of compact binary systems to 7/2 post-Newtonian order (*Phys. Rev. D*, **65**, 061501(R), 2002). *Phys. Rev. D*, **71**:129902(E), 1–2, 2005.

[88] L. Blanchet, T. Damour, G. Esposito-Farése, and B. R. Iyer. Gravitational radiation from inspiralling compact binaries completed at the third post-Newtonian order. *Phys. Rev. Lett.*, **93**:091101, 1–4, 2004.

[89] K. G. Arun, L. Blanchet, B. R. Iyer, and M. S. S. Qusailah. Inspiralling compact binaries in quasi-elliptical orbits: the complete third post-Newtonian energy flux. *Phys. Rev. D*, **77**:064035, 1–24, 2008.

[90] H. Tagoshi, A. Ohashi, and B. Owen. Gravitational field and equations of motion of spinning compact binaries to 2.5 post-Newtonian order. *Phys. Rev. D*, **63**:044006, 1–14, 2001.

[91] G. Faye, L. Blanchet, and A. Buonanno. Higher-order spin effects in the dynamics of compact binaries. I. Equations of motion. *Phys. Rev. D*, **74**:104033, 1–19, 2006.

[92] T. Damour, P. Jaranowski, and G. Schäfer. Hamiltonian of two spinning compact bodies with next-to-leading order gravitational spin-orbit coupling. *Phys. Rev. D*, **77**:064032, 1–11, 2008.

[93] L. Blanchet, A. Buonanno, and G. Faye. Higher-order spin effects in the dynamics of compact binaries. II. Radiation field. *Phys. Rev. D*, **74**:104034, 1–17, 2006.

[94] G. Faye, L. Blanchet, and A. Buonanno. Erratum: Higher-order spin effects in the dynamics of compact binaries. II. Radiation field (*Phys. Rev. D*, **74**:104034, 2006). *Phys. Rev. D*, **75**:049903(E), 1–2, 2007.

[95] J. Steinhoff, S. Hergt, and G. Schäfer. Next-to-leading order gravitational spin(1)-spin(2) dynamics in Hamiltonian form. *Phys. Rev. D*, **77**:081501(R), 1–4, 2008.

[96] S. Hergt and G. Schäfer. Higher-order-in-spin interaction Hamiltonians for binary black holes from source terms of Kerr geometry in approximate ADM coordinates. *Phys. Rev. D*, **77**:104001, 1–15, 2008.

[97] J. Steinhoff, G. Schäfer, and S. Hergt. ADM canonical formalism for gravitating spinning objects. *Phys. Rev. D*, **77**:104018, 1–16, 2008.

[98] R. A. Porto and I. Z. Rothstein. Spin(1)spin(2) effects in the motion of inspiralling compact binaries at third order in the post-Newtonian expansion. *Phys. Rev. D*, **78**:044012, 1–14, 2008.

[99] R. A. Porto and I. Z. Rothstein. Next to leading order spin(1)spin(1) effects in the motion of inspiralling compact binaries. *Phys. Rev. D*, **78**:044013, 1–11, 2008.

[100] L. Blanchet, G. Faye, B. R. Iyer, and S. Sinha. The third post-Newtonian gravitational wave polarizations and associated spherical harmonic modes for inspiralling compact binaries in quasi-circular orbits. *Class. Quantum Grav.*, **25**:165003, 1–44, 2008.

[101] A. Gopakumar and B. R. Iyer. Second post-Newtonian gravitational wave polarizations for compact binaries in elliptical orbits. *Phys. Rev. D*, **65**:084011, 1–25, 2002.

[102] T. Damour, A. Gopakumar, and B. R. Iyer. Phasing of gravitational waves from inspiralling eccentric binaries. *Phys. Rev. D*, **70**:064028, 1–23, 2004.

[103] Ch. Königsdörffer and A. Gopakumar. Phasing of gravitational waves from inspiralling eccentric binaries at the third-and-a-half post-Newtonian order. *Phys. Rev. D*, **73**:124012, 1–24, 2006.

[104] T. Damour and A. Buonanno. Effective one-body approach to general relativistic two-body dynamics. *Phys. Rev. D*, **59**:084006, 1–24, 1999.

[105] T. Damour and A. Buonanno. Transition from inspiral to plunge in binary black hole coalescences. *Phys. Rev. D*, **62**:064015, 1–24, 2000.

[106] T. Damour, P. Jaranowski, and G. Schäfer. Determination of the last stable orbit for circular general relativistic binaries at the third post-Newtonian approximation. *Phys. Rev. D*, **62**:084011, 1–21, 2000.

[107] T. Damour. Coalescence of two spinning black holes: an effective one-body approach. *Phys. Rev. D*, **64**:124013, 1–22, 2001.

[108] T. Damour, P. Jaranowski, and G. Schäfer. Effective one body approach to the dynamics of two spinning black holes with next-to-leading order spin–orbit coupling. *Phys. Rev. D*, **78**:024009, 1–18, 2008.

[109] T. Damour. Introductory lectures on the effective one body formalism. *Int. J. Mod. Phys. A*, **23**, 1130–1148, 2008.

[110] H. Wahlquist. The Doppler response to gravitational waves from a binary star source. *Gen. Rel. Grav.*, **19**, 1101–1113, 1987.

[111] P. C. Peters and J. Mathews. Gravitational radiation from point masses in a Keplerian orbit. *Phys. Rev.*, **131**, 435–440, 1963.

[112] M. Zimmermann and M. Szedenits. Gravitational waves from rotating and precessing rigid bodies: Simple models and applications to pulsars. *Phys. Rev. D*, **20**, 351–355, 1979.

[113] M. Zimmermann. Gravitational waves from rotating and precessing rigid bodies. II. General solutions and computationally useful formulas. *Phys. Rev. D*, **21**, 891–898, 1980.

[114] B. Allen. The stochastic gravity-wave background: sources and detection. In J. A. Marck and J. P. Lasota, eds., *Relativistic Gravitation and Gravitational Radiation: Proceedings of the Les Houches School on Astrophysical Sources of Gravitational Radiation*. Cambridge: Cambridge University Press, 1997.

[115] B. Allen and J. D. Romano. Detecting a stochastic background of gravitational radiation: signal processing strategies and sensitivities. *Phys. Rev. D*, **59**:102001, 1–41, 1999.

[116] M. Fisz. *Probability Theory and Mathematical Statistics*. New York: Wiley, 1963.

[117] M. G. Kendall and A. Stuart. *The Advanced Theory of Statistics, Vol. II: Inference and Relationship*. London: Griffin, 1979.

[118] A. M. Yaglom. *An Introduction to the Theory of Stationary Random Functions*. Englewood Cliffs, NJ: Prentice Hall, 1962.

[119] L. H. Koopmans. *The Spectral Analysis of Time Series*. New York: Academic Press, 1974.

[120] M. B. Priestley. *Spectral Analysis and Time Series*. London: Academic Press, 1996.

[121] E. Wong. *Introduction to Random Processes*. New York: Springer-Verlag, 1983.

[122] R. S. Liptser and A. N. Shiryaev. *Statistics of Random Processes, Vols. I and II*. New York: Springer-Verlag, 1977.

[123] E. Wong and B. Hajek. *Stochastic Processes in Engineering Systems*. New York: Springer-Verlag, 1985.

[124] E. L. Lehmann. *Testing Statistical Hypothesis*. New York: Wiley, 1959.

[125] P. M. Woodward. *Probability and Information Theory with Applications to Radar*. Oxford: Pergamon Press, 1953.

[126] V. A. Kotelnikov. *The Theory of Optimum Noise Immunity*. New York: McGraw-Hill, 1959.

[127] L. A. Wainstein and V. D. Zubakov. *Extraction of Signals from Noise.* Englewood Cliffs, NJ: Prentice Hall, 1962.

[128] H. L. Van Trees. *Detection, Estimation and Modulation Theory. Part I.* New York: Wiley, 1968.

[129] R. N. McDonough and A. D. Whalen. *Detection of Signals in Noise*, 2nd edn. San Diego, CA: Academic Press, 1995.

[130] C. W. Helström. *Statistical Theory of Signal*, Detection, 2nd edn. Oxford: Pergamon Press, 1968.

[131] H. V. Poor. *An Introduction to Signal Detection and Estimation.* New York: Springer-Verlag, 1995.

[132] E. L. Lehmann. *Theory of Point Estimation.* New York: Wiley, 1983.

[133] R. Zieliński. Theory of parameter estimation. In A. Królak, ed., *Mathematics of Gravitation. Part II. Gravitational Wave Detection.* Institute of Mathematics, Polish Academy of Sciences. Warsaw: Banach Center Publications, pp. 209–220, 1997.

[134] J. Livas. Broadband search techniques for periodic sources of gravitational radiation. In B. F. Schutz, ed., *Gravitational Wave Data Analysis. Proceedings of the NATO Advanced Research Workshop held at St. Nicholas, Cardiff, Wales, July 6–9, 1987, NATO ASI Series, vol. C253.* Dordrecht: Kluwer, p. 217, 1989.

[135] A. Królak and M. Tinto. Resampled random processes in gravitational-wave data analysis. *Phys. Rev. D*, **63**:107101, 1–4, 2001.

[136] J. Edlund, M. Tinto, A. Królak, and G. Nelemans. White-dwarf–white-dwarf galactic background in the LISA data. *Phys. Rev. D*, **71**:122003, 1–16, 2005.

[137] T. Bayes. An essay towards solving a problem in doctrine of chances. *Phil. Trans. Roy. Soc.*, **53**, 293–315, 1763.

[138] J. Neyman and E. Pearson. On the problem of the most efficient tests of statistical hypothesis. *Phil. Trans. Roy. Soc. Ser. A*, **231**, 289–337, 1933.

[139] A. Wald. Contribution to the theory of statistical estimation and testing of hypotheses. *Ann. Math. Stat.*, **10**, 299–326, 1939.

[140] J. von Neumann. Zur Theorie der Gesellschaftsspiele. *Math. Ann.*, **100**, 295–320, 1928.

[141] O. Morgenstern and J. von Neumann. *The Theory of Games and Economic Behavior.* Princeton, NJ: Princeton University Press, 1947.

[142] M. H. A. Davis. A review of statistical theory of signal detection. In B. F. Schutz, ed., *Gravitational Wave Data Analysis. Proceedings of the NATO Advanced Research Workshop held at St. Nicholas, Cardiff, Wales, July 6–9, 1987, NATO ASI Series, vol. C253.* Dordrecht: Kluwer, pp. 73–94, 1989.

[143] S. Kullback and R. A. Leibler. On information and sufficiency. *Ann. Math. Stat.*, **22**, 79–86, 1951.

[144] J. B. Budzyński, W. Kondracki, and A. Królak. Applications of distance between probability distributions to gravitational wave data analysis. *Class. Quantum Grav.*, **25**:015005, 1–18, 2008.

[145] C. E. Shannon. A mathematical theory of communication. *Bell Sys. Tech. J.*, **27**, 623–656, 1948.

[146] E. T. Jaynes. Information theory and statistical mechanics. *Phys. Rev.*, **106**, 620–630, 1957.

[147] C. Rao. Information and the accuracy attainable in the estimation of statistical parameters. *Bull. Calcutta Math. Soc.*, **37**, 81–89, 1945.

[148] H. Cramèr. *Mathematical Methods of Statistic*. Princeton, NJ: Princeton University Press, 1946.

[149] A. Bhattacharyya. On some analogues of the amount of information and their use in statistical estimation. *Sankhya*, **8**, 1–32, 1946.

[150] D. Nicholson and A. Vecchio. Bayesian bounds on parameter estimation accuracy for compact coalescing binary gravitational wave signals. *Phys. Rev. D*, **57**, 4588–4599, 1998.

[151] M. B. Priestley. *Non-linear and Non-stationary Time Series Analysis*. London: Academic Press, 1988.

[152] H. L. Hurd. Nonparametric time series analysis for periodically correlated processes. *IEEE Trans. Inform. Theory*, **35**, 350–359, 1989.

[153] D. J. Thomson. Spectrum estimation and harmonic analysis. *Proc. IEEE*, **70**, 1055–1096, 1982.

[154] S. M. Kay. *Modern Spectral Estimation: Theory and Application*. Englewood Cliffs, NJ: Prentice-Hall, 1988.

[155] W. A. Gardner. *Statistical Spectral Analysis. A Nonprobabilistic Theory*. Englewood Cliffs, NJ: Prentice-Hall, 1988.

[156] D. B. Percival and A. T. Walden. *Spectral Analysis for Physical Applications: Multitaper and Conventional Univariate Techniques*. Cambridge: Cambridge University Press, 1993.

[157] D. R. Brillinger. *Time Series: Data Analysis and Theory (Expanded Edition)*. San Francisco, CA: Holden-Day, 1981.

[158] M. S. Bartlett. Periodogram analysis and continuous spectra. *Biometrika*, **37**, 1–16, 1950.

[159] R. B. Blackman and J. W. Tukey. *The Measurement of Power Spectra from the Point of View of Communication Engineering*. New York: Dover Publications, 1958.

[160] P. J. Daniell. Discussion on the symposium on autocorrelation in time series. *J. Roy. Statist. Soc. (Suppl.)*, **8**, 88–90, 1946.

[161] P. D. Welch. The use of fast Fourier transform for estimation of power spectra. *IEEE Trans. Audio Electroacustics*, **15**, 70–73, 1970.

[162] B. Allen and A. Ottewill. Multi-taper spectral analysis in gravitational wave data analysis. *Gen. Rel. Grav.*, **32**, 385–398, 2000.

[163] D. Slepian. Prolate spheroidal wave functions, Fourier analysis, and uncertainty – V: The discrete case. *Bell Syst. Tech. J.*, **57**, 1371–1430, 1978.

[164] A. Schuster. On the investigation of hidden periodicities with application to a supposed 26 day period of meteorological phenomena. *Terrest. Magn.*, **3**, 13–41, 1898.

[165] R. A. Fisher. Test of significance in harmonic analysis. *Proc. Roy. Soc. London A*, **125**, 54–59, 1929.

[166] T. W. Anderson. *The Statistical Analysis of Time Series*. New York: John Wiley & Sons, 1971.

[167] A. F. Siegel. Testing for periodicity in a time series. *J. Amer. Statist. Assoc.*, **75**, 345–348, 1980.

[168] M. J. Hinich. Testing for Gaussianity and linearity of a stationary time series. *J. Time Series Anal.*, **3**, 169–176, 1982.

[169] P. R. Saulson. *Fundamentals of Interferometric Gravitational Wave Detectors.* Singapore: World Scientific, 1994.

[170] LISA: Pre-phase A report, December 1998. Technical Report MPQ 223, Max-Planck-Institut für Quantenoptik, Garching, Germany, 1998.

[171] S. Rowan and J. Hough. Gravitational wave detection by interferometry (ground and space). *Liv. Rev. Relat.*, **3**, 2000.

[172] J. W. Armstrong. Low-frequency gravitational wave searches using spacecraft Doppler tracking. *Liv. Rev. Relat.*, **9**, 2006.

[173] E. Amaldi, O. Aguiar, M. Bassan, *et al.* First gravity wave coincident experiment between resonant cryogenic detectors: Louisiana–Rome–Stanford. *Astron. Astrophys.*, **216**, 325–332, 1989.

[174] Z. A. Allen, P. Astone, L. Baggio, *et al.* First search for gravitational wave bursts with a network of detectors. *Phys. Rev. Lett.*, **85**, 5046–5050, 2000.

[175] P. Astone, M. Bassan, P. Bonifazi, *et al.* Long-term operation of the Rome "Explorer" cryogenic gravitational wave detector. *Phys. Rev. D*, **47**, 362–375, 1993.

[176] M. Tinto and S. V. Dhurandhar. Time-delay interferometry. *Liv. Rev. Relat.*, **8**, 2005.

[177] D. A. Shaddock, M. Tinto, F. B. Estabrook, and J. W. Armstrong. Data combinations accounting for LISA spacecraft motion. *Phys. Rev. D*, **68**:061303, 1–4, 2003.

[178] M. Tinto, F. B. Estabrook, and J. W. Armstrong. Time delay interferometry with moving spacecraft arrays. *Phys. Rev. D*, **69**:082001, 1–10, 2004.

[179] C. Meyer. *Matrix Analysis and Applied Linear Algebra.* Philadelphia, PA: SIAM, 2000.

[180] J. I. Marcum. A statistical theory of target detection by pulsed radar: Mathematical appendix. Technical Report Research Memorandum RM-753-PR, Rand Corporation, Santa Monica, CA, 1948.

[181] J. I. Marcum. Table of Q functions. Technical Report Research Memorandum RM-339-PR, Rand Corporation, Santa Monica, CA, 1950.

[182] R. Adler. *The Geometry of Random Fields.* New York: Wiley, 1981.

[183] R. J. Adler and J. E. Taylor. *Random Fileds and Geometry.* New York: Springer-Verlag, 2007.

[184] S. V. Dhurandhar and B. F. Schutz. Filtering coalescing binary signals: issues concerning narrow banding, thresholds, and optimal sampling. *Phys. Rev. D*, **50**, 2390–2405, 1994.

[185] H. Cramér and M. R. Leadbetter. *Stationary and Related Stochastic Processes (Sample Function Properties and Their Applications).* New York: John Wiley & Sons, 1967.

[186] K. J. Worsley. Local maxima and the expected Euler characteristic of excursion sets of χ^2, F and t fields. *Adv. App. Prob.*, **26**, 13–42, 1994.

[187] A. Buonanno, Y. Chen, and M. Vallisneri. Detection template families for gravitational waves from the final stages of binary–black-hole inspirals: Nonspinning case. *Phys. Rev. D*, **67**:024016, 1–50, 2003.

[188] T. Tanaka and H. Tagoshi. Use of new coordinates for the template space in a hierarchical search for gravitational waves from inspiraling binaries. *Phys. Rev. D*, **62**:082001, 1–8, 2000.

[189] P. Jaranowski and A. Królak. Data analysis of gravitational-wave signals from spinning neutron stars. III. Detection statistics and computational requirements. *Phys. Rev. D*, **61**:062001, 1–32, 2000.

[190] P. Astone, K. M. Borkowski, P. Jaranowski, and A. Królak. Data analysis of gravitational-wave signals from spinning neutron stars. IV. An all-sky search. *Phys. Rev. D*, **65**:042003, 1–18, 2002.

[191] R. Prix. Search for continuous gravitational waves: Metric of the multi-detector $\mathcal{F}$-statistic. *Phys. Rev. D*, **75**:023004, 1–20, 2007.

[192] P. Astone, J. A. Lobo, and B. F. Schutz. Coincidence experiments between interferometric and resonant bar detectors of gravitational waves. *Class. Quantum Grav.*, **11**, 2093–2112, 1994.

[193] T. A. Prince, M. Tinto, S. L. Larson, and J. W. Armstrong. LISA optimal sensitivity. *Phys. Rev. D*, **66**:122002, 1–7, 2002.

[194] K. R. Nayak, A. Pai, S. V. Dhurandhar, and J.-Y. Vinet. Improving the sensitivity of LISA. *Class. Quantum Grav.*, **20**, 1217–1232, 2003.

[195] L. S. Finn. Aperture synthesis for gravitational-wave data analysis: deterministic sources. *Phys. Rev. D*, **63**:102001, 1–18, 2001.

[196] P. Jaranowski and A. Królak. Optimal solution to the inverse problem for the gravitational wave signal of a coalescing binary. *Phys. Rev. D*, **49**, 1723–1739, 1994.

[197] A. Pai, S. Bose, and S. V. Dhurandhar. Data-analysis strategy for detecting gravitational-wave signals from inspiraling compact binaries with a network of laser-interferometric detectors. *Phys. Rev. D*, **64**:042004, 1–30, 2001.

[198] Y. Gürsel and M. Tinto. Nearly optimal solution to the inverse problem for gravitational-wave bursts. *Phys. Rev. D*, **40**, 3884–3938, 1989.

[199] P. Astone, D. Babusci, L. Baggio, *et al.* Methods and results of the IGEC search for burst gravitational waves in the years 1997–2000. *Phys. Rev. D*, **68**:022001, 1–33, 2003.

[200] B. F. Schutz and M. Tinto. Antenna patterns of interferometric detectors of gravitational waves – I. Linearly polarized waves. *Mon. Not. R. Astron. Soc.*, **224**, 131–154, 1987.

[201] N. L. Christensen. Measuring the stochastic gravitational-radiation background with laser-interferometric antennas. *Phys. Rev. D*, **46**, 5250–5266, 1992.

[202] É. É. Flanagan. Sensitivity of the laser interferometric gravitational wave observatory to a stochastic background, and its dependence on the detector orientations. *Phys. Rev. D*, **48**, 2389–2408, 1993.

[203] J. T. Whelan. Stochastic gravitational wave measurements with bar detectors: dependence of response on detector orientation. *Class. Quantum Grav.*, **23**, 1181–1192, 2006.

[204] B. Allen and A. C. Ottewill. Detection of anisotropies in the gravitational-wave stochastic background. *Phys. Rev. D*, **56**, 545–563, 1997.

[205] P. Astone, D. Babusci, M. Bassan, *et al.* All-sky upper limit for gravitational radiation from spinning neutron stars. *Class. Quantum Grav.*, **20**:S665–S676, 2003.

[206] P. Jaranowski and A. Królak. Data analysis of gravitational-wave signals from spinning neutron stars. II. Accuracy of estimation of parameters. *Phys. Rev. D*, **59**:063003, 1–29, 1999.

[207] J. H. Conway and N. J. A. Slane. *Sphere Packings, Lattices and Groups.* New York: Springer-Verlag, 1999 3rd edn.

[208] J. W. Cooley and J. W. Tukey. An algorithm for the machine calculation of complex Fourier series. *Math. Comput.*, **19**, 297–301, 1965.

[209] E. O. Brigham. *Fast Fourier Transform and Its Applications.* Englewood Cliffs, NJ: Prentice-Hall, 1988.

[210] C. de Boor. *A Practical Guide to Splines.* New York: Springer-Verlag, 1978.

[211] S. Mohanty and S. V. Dhurandhar. Hierarchical search strategy for the detection of gravitational waves from coalescing binaries. *Phys. Rev. D*, **54**, 7108–7128, 1996.

[212] S. D. Mohanty. Hierarchical search strategy for the detection of gravitational waves from coalescing binaries: extension to post-Newtonian wave forms. *Phys. Rev. D*, **57**, 630–658, 1998.

[213] S. A. Sengupta, S. V. Dhurandhar, and A. Lazzarini. Faster implementation of the hierarchical search algorithm for detection of gravitational waves from inspiraling compact binaries. *Phys. Rev. D*, **67**:082004, 1–14, 2003.

[214] R. P. Brent. *Algorithms for Minimization Without Derivatives.* Englewood Cliffs, NJ: Prentice-Hall, 1973.

[215] J. A. Nelder and R. Mead. A simplex method for function minimization. *Computer J.*, **7**, 308–313, 1965.

[216] J. C. Lagarias, J. A. Reeds, M. H. Wright, and P. E. Wright. Convergence properties of the Nelder–Mead simplex method in low dimensions. *SIAM J. Optimiz.*, **9**, 112–147, 1998.

[217] N. Christensen, R. J. Dupuis, G. Woan, and R. Meyer. Metropolis–Hastings algorithm for extracting periodic gravitational wave signals from laser interferometric detector data. *Phys. Rev. D*, **70**:022001, 1–7, 2004.

[218] R. Umstätter, N. Christensen, M. Hendry, *et al.* LISA source confusion: identification and characterization of signals. *Class. Quantum Grav.*, **22**:S901–S902, 2005.

[219] C. Röver, R. Meyer, and N. Christensen. Coherent Bayesian inference on compact binary inspirals using a network of interferometric gravitational wave detectors. *Phys. Rev. D*, **75**:062004, 1–11, 2007.

[220] K. D. Kokkotas, A. Królak, and G. Tsegas. Statistical analysis of the estimators of the parameters of the gravitational-wave signal from a coalescing binary. *Class. Quantum Grav.*, **11**, 1901–1918, 1994.

[221] R. Balasubramanian, B. S. Sathyaprakash, and S. V. Dhurandhar. Gravitational waves from coalescing binaries: detection strategies and Monte Carlo estimation of parameters. *Phys. Rev. D*, **53**, 3033–3055, 1996.

[222] D. C. Rife and R. R. Boorstyn. Single tone parameter estimation from discrete-time observations. *IEEE Trans. Inform. Theory*, **20**, 591–598, 1974.

[223] B. Abbott, R. Abbott, R. Adhikari, *et al.* Analysis of LIGO data for gravitational waves from binary neutron stars. *Phys. Rev. D*, **69**:122001, 1–16, 2004.

[224] B. Abbott, R. Abbott, R. Adhikari, *et al.* Setting upper limits on the strength of periodic gravitational waves from PSR J1939+2134 using the first science data from the GEO 600 and LIGO detectors. *Phys. Rev. D*, **69**:082004, 1–16, 2004.

[225] R. J. Dupuis and G. Woan. Bayesian estimation of pulsar parameters from gravitational wave data. *Phys. Rev. D*, **72**:102002, 1–9, 2005.

[226] J. W. Armstrong, F. B. Estabrook, and M. Tinto. Time-delay interferometry for space-based gravitational wave searches. *Ap. J.*, **527**, 814–826, 1999.

[227] M. Vallisneri. Synthetic LISA: Simulating time delay interferometry in a model LISA. *Phys. Rev. D*, **71**:022001, 1–18, 2005.

Index

χ^2 distribution, 104, 107, 108, 110, 113, 138, 154, 210
  noncentral, 138, 155
χ^2 test, 109–110
σ-algebra, 52
  Borel, 52

A_k^* lattice, 198
  thickness of, 198, 199
a priori probability, *see* prior
amplitude parameters, *see* extrinsic parameters
autocorrelation function, *see* stochastic process, autocorrelation of
autocorrelation of time series, 61, 62
autocovariance function, *see* stochastic process, autocovariance of

balance equation, 25, 41
Bartlett procedure, 104
barycentric time, 202
Bayes risk, *see* risk, Bayes
Bayes rule, *see* decision rule, Bayes
Bayes theorem, 87
beam-pattern functions
  of Earth-based bar, 130, 150, 151, 225, 227–228

of Earth-based interferometer, 129, 150, 151, 158, 162, 170, 225–227
  of LISA, 127, 222–224
Bessel functions, 138
Bhattacharyya bound, 89
bicoherence, 112, 113
big-bang nucleosynthesis, 29
binomial distribution, 55, 83
bispectrum, 111–113
Blackman–Tukey procedure, 105–106
Bochner theorem, 60
body-centered-cubic (bcc) covering, 198
Borel σ-algebra, 52
Brent's algorithm, 207
Brillouin zone, 200
Brown process, 91
Brownian bridge, 92, 110
  auotocorrelation of, 92

Cameron–Martin formula, 72, 93
cardinality of a set, 53
Cauchy–Schwarz inequality, 76, 89, 154
central limit theorems, 56
chirp, 27
  Fourier transform of, 215–217
  polarization waveforms, 35
  with radiation-reaction effects, 36–39

Printed in the United States
by Baker & Taylor Publisher Services